Renewable Energy Systems

The way forward

Online at: https://doi.org/10.1088/978-0-7503-6179-8

IOP Series in Renewable and Sustainable Power

The IOP Series in Renewable and Sustainable Power aims to bring together topics relating to renewable energy, from generation to transmission, storage, integration, and use patterns, with a particular focus on systems-level and interdisciplinary discussions. It is intended to provide a state-of-the-art resource for all researchers involved in the power conversation.

Series Editor
Professor David Elliott
Open University, UK

About the Editor
David Elliott is Emeritus Professor of Technology Policy at the Open University, where he developed courses and research on technological innovation, focusing on renewable energy policy. Since retirement, he has continued to write extensively on that topic, including a series of books for IOP Publishing and a weekly blog post for *Physics World* (physicsworld.com/author/david-elliott)

About the Series
Renewable and sustainable energy systems offer the potential for long-term solutions to the world's growing energy needs, operating at a broad array of scales and technology levels. The IOP Series in Renewable and Sustainable Power aims to bring together topics relating to renewable energy, from generation to transmission, storage, integration, and use patterns, with a particular focus on systems-level and interdisciplinary discussions. It is intended to provide a state-of-the-art resource for all researchers involved in the power conversation.

We welcome proposals in all areas of renewable energy including (but not limited to) wind power, wave power, tidal power, hydroelectric power, PV/solar power, geothermal power, bioenergy, heating, grid balancing and integration, energy storage, energy efficiency, carbon capture, fuel cells, power to gas, electric/green transport, and energy saving and efficiency.

Authors are encouraged to take advantage of electronic publication through the use of colour, animations, video, data files, and interactive elements, all of which provide opportunities to enhance the reader experience.

A list of recently published and forthcoming titles published in this series can be found here: https://iopscience.iop.org/bookListInfo/iop-series-in-renewable-and-sustainable-power#series.

Renewable Energy Systems

The way forward

Edited by
David S-K Ting
MAME, University of Windsor, Windsor, Ontario, Canada

Jacqueline A Stagner
Faculty of Engineering, University of Windsor, Windsor, Ontario, Canada

IOP Publishing, Bristol, UK

ISBN 978-0-7503-6179-8 (ebook)
ISBN 978-0-7503-6177-4 (print)
ISBN 978-0-7503-6180-4 (myPrint)
ISBN 978-0-7503-6178-1 (mobi)

DOI 10.1088/978-0-7503-6179-8

Version: 20250401

IOP ebooks

British Library Cataloguing-in-Publication Data: A catalogue record for this book is available from the British Library.

Published by IOP Publishing, wholly owned by The Institute of Physics, London

IOP Publishing, No.2 The Distillery, Glassfields, Avon Street, Bristol, BS2 0GR, UK

US Office: IOP Publishing, Inc., 190 North Independence Mall West, Suite 601, Philadelphia, PA 19106, USA

To everyone who strives forward, endeavouring to brighten tomorrow.

Contents

Preface

The way forward is to brighten tomorrow. To do so, further progress in renewable energy systems is a must. With myriad efforts rightly invested into greening energy in recent years, it is timely to scrutinize the state-of-the-art to better shape the way forward. G T Reader breaks the ice with an overarching chapter, 'The future of renewable energy systems—a long and winding transition pathway?' He puts the often-taken-for-granted household-term, 'renewable energy' into question. The United Nations asserts that the adoption of renewables could create new jobs and boost the global economy, in addition to numerous other benefits including solving the problems associated with anthropogenic climate change. Reader concludes that conversion efficiencies, energy storage methods and capabilities, capacity factors, affordability, access, dispatchability, land-use, the provision of clean cooking facilities, and improved air quality for all remain as obstacles to overcome.

The Sun capacitates photosynthesis, sustaining life on Earth. In colder regions, plants must stay unfrozen to photosynthesize and grow. The conventional approach of burning natural gas to keep horticulture greenhouses warm is unsustainable. We must design greenhouses that tap more fully into solar thermal during the cold seasons. A prerequisite for designing such greenhouses is to have a systematic set of guidelines to assess standard building energy models. Accordingly, G Khanuja, R Ruparathna and D S-K Ting report 'A methodological framework for modeling passive solar greenhouses' in chapter 2. This framework can be employed to methodically check a model's accuracy in predicting energy consumption by comparing predicted results with measured values under dynamic solar loading.

Hastening net-zero buildings on a global scale should be a priority because buildings utilize approximately 40% of global energy. Proper management of the readily available solar energy is the key. In chapter 3, E Koçak and E Aylı delineate the latest status in reducing building energy consumption via passive architecture, solar chimneys, Trombe walls, photovoltaics, and hybrid methods. Active methods such as building-integrated photovoltaic systems and building-integrated photovoltaic and thermal systems are most promising. To facilitate widespread application of these efficacious active methods, efforts should be invested to lower the capital cost.

Photovoltaic systems are the most direct way to harness the clean and abundant solar energy and convert it into electricity. Other than issues in the manufacturing process, one operational challenge is the creation of an undesirable heat-island effect. The degree of the heat-island effect is a direct function of the panel's surface temperature and, thus, mitigation relies on accurate monitoring of this temperature. In chapter 4, D T Cannon and A Vasel-Be-Hagh discuss the challenge of *in situ* measurements of the surface temperature of photovoltaic panels. A complication can arise from the coupling of solar irradiance and wind-induced convection. They evaluate the value gauged by a single direct-contact surface heat flux sensor and find this cost-effective, low-maintenance approach to be reliable and superior to more sophisticated methods using pyranometers and pyrgeometers, which are sensitive to clouds and edge effects.

Chapter 5 is about the 'Production of clean fuels by thermochemical conversion and photocatalytic water splitting'. In this chapter, L Puri, N Mia, Y Hu and G F Naterer enlighten us with a promising solution to conventional carbon-based fossil fuels. They focus on bio-oil produced from pyrolysis and hydrothermal liquefaction, and hydrogen production by solar based water splitting over a photocatalyst. They highlight, among other items, the significance of synthesizing a semiconductor-based photocatalyst with a suitable bandgap, crystallinity, crystallite size, Fermi level, dimensionality, and apparent quantum efficiency.

Imagine wind turbine towers as tall buildings such as skyscrapers. This would mean more than furthering renewable penetration but the beautification of both renewable energy systems and urban architectures. H E Ilgin and M H Gunel reveal this promising future in chapter 6, 'A proposal for the typological categorization of architectural integration approaches for wind energy in tall towers'. They delve into the prospects of wind energy integration with urban settings, particularly from an architectural perspective. An all-encompassing and distinctive typological framework is presented.

What about meeting the energy demand in the Congo Basin? N P Awazi communicates this in chapter 7, 'Meeting the energy needs of the urban and rural poor in the Congo Basin through renewables: challenges and prospects.' The Congo Basin is blessed with abundant renewable energy resources, including solar, hydro, biomass, wind, and geothermal. Nevertheless, most of the energy needs of the urban and rural poor in the Congo Basin remain unmet. The main hurdles include poor governance, weak institutional and policy frameworks, high transition costs, inadequate infrastructure, and limited technology transfer.

The price of energy is much more complex than the intercept between the linear lines of supply and demand. As such, energy contracting has been a specialized field set aside for trained experts. GENIUS is a negotiation simulator platform that can ease the task significantly. A Bertuccio, J Stagner and R Carriveau expound on this in chapter 8, 'Autonomous agent power contracting'. They argue that artificial intelligence can be invoked to help achieve less biased, more ethical, and preferred solutions. Such automated negotiating systems can enhance market efficiency and allow for dynamic price adjustments based on real-time supply and demand changes.

Air transport consumes a significant amount of fossil fuels, accounting for a large share of carbon dioxide production. Replacing carbon-based fuels with hydrogen is one means to resolve the issue. This hydrogen solution, however, may result in a net warming effect on the planet. Much work is needed to get to the bottom of this, and computational fluid dynamics is a viable approach to scrutinize this issue. This being the case, a timely chapter, 'A prerequisite to computational fluid dynamics of airplane condensation trials', is furnished by D Roland and A Vasel-Be-Hagh as chapter 9.

Around the globe, people are moving to cities. With rapid urbanization, how do we ensure sustainability in both urban and rural areas? This is the topic of the final chapter of this volume, chapter 10, 'Sustainability transitions: comparing urban sophistication and rural simplicity'. A S Permana and C Potipituk contend that resource management and waste reduction are challenges facing urban areas, whereas renewable energy and green buildings present innovative solutions. They suggest sustainable agriculture for rural areas but acknowledge that the lack of resources is a challenge. The solution is to bridge the gap between urban and rural sustainability via knowledge sharing, collaboration, policy coordination, and community engagement. The way forward is to foster innovation and community empowerment for a resilient and sustainable world.

The ten timely chapters constituting this book provide a glimmer of hope on the way forward with renewable energy systems. There remains much room for hastened work at this front. Together, we can brighten tomorrow. We forge forward auspiciously.

Acknowledgements

The dream of realizing this book would have vaporized if not for Providence from above and the resilient experts who compiled the ten chapters. A big round of applause goes to the anonymous reviewers who enhanced the quality of these chapters.

It would be amiss if we did not promulgate the superb IOP team—Caroline Mitchell was the first to take the risk of inviting us to edit a volume. We hope that you share the joy of the journey from the onset to many years beyond crossing the finish line. We look forward to the opportunity to collaborate on another timely volume on bettering tomorrow.

David S-K Ting and Jacqueline A Stagner
Turbulence and Energy Laboratory
University of Windsor, ON, Canada

Editor biographies

David S-K Ting

David S-K Ting studied combustion and turbulence, followed by convection heat transfer and fluid–structure interactions, prior to joining the University of Windsor. Dr Ting is the founder of the Turbulence and Energy Laboratory. Professor Ting has been supervising research students, primarily on turbulence, energy, and sustainability, for 27 years. To date, he has co-supervised over 90 MASc and PhD students, co-authored more than 180 scientific journal papers, co-edited over 30 volumes, and authored five textbooks.

Jacqueline A Stagner

Jacqueline A Stagner is the Undergraduate Programs Coordinator in the Faculty of Engineering at the University of Windsor and an adjunct faculty member in the Department of Mechanical, Automotive, and Materials Engineering. Dr Stagner co-advises students in the Turbulence and Energy Lab. She is also an instructor within the Faculty of Engineering and works on various curriculum-development projects.

List of contributors

Nyong Princely Awazi
Department of Forestry and Wildlife Technology, the University of Bamenda, Cameroon

Ece Ayli
Department of Mechanical Engineering, Cankaya University, Ankara, Turkey

Antonio Bertuccio
Mechanical, Automotive and Materials Engineering, University of Windsor, Windsor, ON, Canada

Daniel Trevor Cannon
Mechanical Engineering, Tennessee Tech University, Cookeville, TN, USA

Rupp Carriveau
Civil and Environmental Engineering, University of Windsor, Windsor, ON, Canada

Mehmet Halis Gunel
Department of Architecture, Faculty of Architecture, Middle East Technical University, Ankara, Turkey

Yulin Hu
Faculty of Sustainable Design Engineering, University of Prince Edward Island, Charlottetown, PE, Canada

Hüseyin Emre Ilgın
School of Architecture, Faculty of Built Environment, Tampere University, Tampere, Finland

Gurpreet Khanuja
University of Windsor, Windsor, ON, Canada

Eyup Koçak
Department of Mechanical Engineering, Cankaya University, Ankara, Turkey

Nasim Mia
Faculty of Sustainable Design Engineering, University of Prince Edward Island, Charlottetown, PE, Canada

Greg Naterer
Faculty of Sustainable Design Engineering, University of Prince Edward Island, Charlottetown, PE, Canada

Ahmadreza Permana
Department of Civil Engineering, Faculty of Engineering, King Mongkut Institute of Technology Ladkrabang, Bangkok, Thailand

Chantamon Potipituk
Faculty of Architecture and Design, Rajamangala University of Technology Rattanakosin, Bangkok, Thailand

Lokeshwar Puri
Faculty of Sustainable Design Engineering, University of Prince Edward Island, Charlottetown, PE, Canada

Graham T Reader
University of Windsor, Windsor, ON, Canada

Devin Roland
Mechanical Engineering, University of South Florida, Tampa, FL, USA

Rajeev Ruparathna
University of Windsor, Windsor, ON, Canada

Jacqueline A Stagner
Faculty of Engineering, University of Windsor, Windsor, ON, Canada

David S-K Ting
University of Windsor, Windsor, ON, Canada

Ahmadreza Vasel-Be-Hagh
Mechanical Engineering, Tennessee Tech University, Cookeville, TN, USA
and
Mechanical Engineering, University of South Florida, Tampa, FL, USA

Contributor biographies

Dr Nyong Princely Awazi is a Senior Lecturer at the University of Bamenda, Cameroon, with a PhD in Agroforestry and Valuation of Ecosystem Services. He has expertise in agroforestry, forestry, ecotourism, climate change, and biodiversity conservation. Since 2014, he has worked as a researcher and consultant for different national, regional, and international institutions. Dr Awazi has (co) authored over 90 publications and reviewed more than 650 manuscripts for prestigious journals. He serves on several editorial boards and is a member of different international associations. He is also a climate change and biodiversity conservation consultant.

Associate Professor Dr Ece Aylı, born in Ankara in 1989, completed her secondary education at TED Ankara College in 2006. She pursued higher education at TOBB University of Economics and Technology, earning her BSc in 2010, MSc in 2012, and PhD in 2016 from the Department of Mechanical Engineering. In 2017 she joined the faculty of Çankaya University, Department of Mechanical Engineering, where she currently serves as an associate professor. Dr Aylı specializes in the fields of heat transfer, renewable energy systems, and soft computing methodologies, with a focus on applying computational techniques to solve complex engineering problems. Her research contributions span numerical heat transfer, optimization of energy systems, and the integration of artificial intelligence into thermal-fluid applications.

Antonio Bertuccio is an undergraduate mechanical engineering student and researcher enrolled at the University of Windsor. Antonio's interest in energy extends to a variety of fields, such as power contracting and battery thermal management systems for electric vehicles. He is also involved as a lead in the University of Windsor, Formula Electric (FSAE) team.

Rupp Carriveau is the Director of the Environmental Energy Institute and Co-Director of the Turbulence and Energy Lab at the University of Windsor. Professor Carriveau is a founder of the Offshore Energy and Storage Society and an IEEE Ocean Energy Technology chair. Dr Carriveau works with students and industry to create sustainable pathways to energy and food solutions to empower people and the planet.

Mehmet Halis Gunel pursued his studies in civil engineering at Middle East Technical University (METU), graduating in 1982. He continued his graduate education at the same institution, earning an MSc in 1984 and a PhD in Structures in 1995. Between 1982 and 1984, as well as 1986–1989, he served as an assistant in the Structural Mechanics Division of the Civil Engineering Department, METU. From 1989 to 1991, he worked as a project manager at Prokon Consultants Company. Subsequently, he joined the Architecture Department, METU, where he has been teaching and is now an emeritus professor. His expertise includes reinforced concrete, tall buildings, prefabrication, and structural design in architecture.

Dr Yulin Hu is an assistant professor in the Faculty of Sustainable Design Engineering at the University of Prince Edward Island. Dr Hu received a PhD degree in Chemical and Biochemical Engineering from Western University in 2018, and an MSc degree in Bioresource Engineering from McGill University in 2015. Dr Hu has made major contributions to the research and development of biomass valorization, biomass thermochemical conversion, and value-added bioproducts, resulting in 56 peer-reviewed articles, nine book chapters, and seven invited international and national talks.

Hüseyin Emre Ilgın received his PhD in Building Sciences in Architecture from Middle East Technical University (METU) in 2018. Since 2019, he has been conducting post-doctoral research on wood construction at Tampere University. Dr Ilgın worked as a Marie Skłodowska-Curie post-doctoral research fellow on dovetailed massive wood board elements between 2021 and 2023. He has over 120 scientific publications, mostly on tall building design and wood construction.

Gurpreet Khanuja earned an MASc from the University of Windsor, focusing on optimizing energy performance and life-cycle costs for controlled environment agriculture facilities. Her research advanced sustainable, energy-efficient indoor farming solutions by integrating engineering principles with environmental stewardship. During her studies, she published two book chapters and authored a thesis, demonstrating a strong commitment to advancing knowledge in her field.

Dr Eyup Koçak completed his BSc in Mechanical Engineering at Gazi University in 2014 and his MSc in the same field at Gazi University in 2017. He furthered his education by conducting research in mechanical engineering at the University of Aveiro between 2016 and 2017. Dr Koçak earned his PhD in Mechanical Engineering from Çankaya University in 2023. Dr Koçak's research interests encompass heat transfer, renewable energy systems, and fluid dynamics. He has contributed to various projects, including the design and analysis of a horizontal type Francis turbine and the investigation of the aeroacoustic and aerodynamic performance of biomimetic wing structures.

Mr Nasim Mia is an MSc student in the Faculty of Sustainable Design Engineering at the University of Prince Edward Island, Charlottetown, PE, Canada. His master's research focuses on solar energy-driven hydrogen generation through the photocatalytic water-splitting method. Prior to this, Nasim earned a bachelor's degree in electrical and electronic engineering and worked on developing sustainable energy devices.

Dr Greg Naterer is the Vice-President, Academic and Research, at the University of Prince Edward Island. His research interests lie in energy systems, thermodynamics, and heat transfer, including three books and over 300 refereed articles in these fields. He received the Julian C Smith Medal of the Engineering Institute of Canada, among other awards. Dr Naterer is the Editor-in-Chief of the AIAA *Journal of Thermophysics and Heat Transfer*. Previously, as a Tier 1 Canada Research Chair in Advanced Energy Systems, he led an international team on the development of thermochemical copper–chlorine cycle of hydrogen production.

Dr Ariva Sugandi Permana, an academic staff member at the Department of Civil Engineering, School of Engineering, King Mongkut's Institute of Technology Ladkrabang in Bangkok, Thailand, holds a PhD in urban planning and environmental management from the Asian Institute of Technology. His research interests are diverse and encompass sustainable water resources management, sustainable building design, urban planning, and environmental sustainability. Having completed his PhD in Thailand, he considers it his second home, further enriching his commitment to sustainable practices in the region.

Dr Chantamon Potipituk, an Assistant Professor in Architecture at the Rajamangala University of Technology Rattanakosin, holds a PhD in urban environmental management from the Asian Institute of Technology. Her research interests span the urban environment, sustainable development, environmental impact assessment, and energy saving, with a particular focus on low carbon communities. She is deeply passionate about enhancing research methodology in environmental management and making significant academic contributions in this field.

Mr Lokeshwar Puri is an MSc student in the Faculty of Sustainable Design Engineering at the University of Prince Edward Island (UPEI), Charlottetown, PE, Canada. His master's research focuses on biomass pyrolysis for biochar and bio-oil production. Prior to joining UPEI, Lokeshwar received a BSc in Chemical Engineering.

Dr Graham Reader is Professor in the Department of Mechanical, Materials and Automotive Engineering at the University of Windsor, ON, Canada. He served as the Dean of the Faculty of Engineering from 1999 to 2010 and as Special Advisor (CEI) to the President, 2010–2011. Dr Reader has been involved in powertrain research since 1970 and his particular areas of interest are underwater technology, diesel engines, Stirling engines, and alternative fuels. His earlier research work was concerned with power systems for underwater vehicles. His research has been sponsored by various organizations including NSERC, Imperial Oil, The Ford Motor Company, NATO, the United Nations, the UK Ministry of Defence, the US Office of Naval Research, The International Engine and Truck Corporation, AUTO21, the Ontario Research Fund, the UK Science and Engineering Research Council (EPSRC), and the Universities of Windsor and Calgary. Dr Reader has published a number of books in English and Russian, and over 350 articles and papers. He holds five patents and is an award-winning military historian, poet, and songwriter.

Devin Roland was born and raised in Tennessee, USA. He attended Tennessee Tech University for his undergraduate and master's degrees in mechanical engineering before moving to the University of South Florida to pursue a PhD in the same field. At the time of this publication, his PhD research is in progress, focused on contrail simulation through the application of computational fluid dynamics.

Rajeev Ruparathna is an Associate Professor in the Department of Civil and Environmental Engineering at the University of Windsor. His research focuses on life-cycle thinking in civil infrastructure planning, the circular economy, and building information modeling (BIM). He has authored over 75 peer-reviewed journal and conference papers and has successfully supervised more than 15 MASc and 2 PhD students. Dr Ruparathna actively collaborates with numerous public and private sector organizations and holds multiple research grants, including funding from NSERC and MITACS.

Dr Ahmad Vasel-Be-Hagh is an Associate Professor of Mechanical Engineering and the Founder of the Thermofluids Discovery Lab at the University of South Florida. He develops analytical models and numerical simulations and conducts laboratory and field experiments to enhance our understanding of fluid mechanics, leading to more accurate predictions and control of fluid flows. His research, funded by NASA, the US National Science Foundation, and the Tennessee Valley Authority, focuses on applications in energy and aerospace. He has published over 50 articles and has edited nine books, along with several special issues of academic journals.

IOP Publishing

Renewable Energy Systems
The way forward
David S-K Ting and Jacqueline A Stagner

Chapter 1

The future of renewable energy systems—a long and winding transition pathway?

Graham T Reader

Many sources of renewable energy have been identified as replacements for fossil fuels. But what exactly is a renewable energy source? Of the numerous attempts to provide a definitive answer to that question none are wholly satisfactory but, generally, it appears that a renewable source is any that is not carbon based, except for biomass, and its supply can be sustained naturally or by human action, e.g. tree-planting. In the utopian, or unconstrained, vision it is claimed that the use of renewables can solve, not just manage, all the problems associated with anthropogenic climate change. Indeed, the United Nations asserts that renewables are safer, cheaper, and healthier, and their adoption could create millions of new jobs and, by 2050, have a huge positive financial impact on the global economy. With such appealing virtues it is surprising that renewables provide less than a third of the world's generated electricity and far less of the overall global primary energy consumption, despite significant increases in recent years. Is this because the claims are overstated or is the impatience, and sometimes intolerance, of the protagonists of renewable energy transitions detrimental to more general acceptance? If the laudable aim of the eradication of global poverty, largely through sustainable development, is to be achieved, and if renewable energy sources can play indispensable roles in its realization, what practical steps need to be taken towards universal credibility and access? What tools do policy-makers have at their disposal when planning the way ahead for renewables? Would more pragmatic, albeit imperfect, approaches to determining the nature of future energy mixes based on national usage patterns be valuable, rather than endlessly debating the efficacy of climate models? These matters are discussed in this chapter.

doi:10.1088/978-0-7503-6179-8ch1

1.1 Introductory remarks

By 2022 CE[1] the world's human population had increased to almost 8 billion, compared with just over 1 billion in 1800. During this same period, the consumption of primary energy also increased, but over 30-fold, i.e. four times faster than population growth. The main global source of this energy, since the early twentieth century, was traditional fossil fuels—coal, oil, and gas—an energy-mix dominance that persists. Nevertheless, efforts have been and are being made to transition away from fossil energy to affordable alternative sources, the most recent undertakings being a complete transition to renewable energy sources. Why all these attempts? Over the past 50 years three main motivations have emerged for the replacement of traditional fossil fuel energy sources: the fear of rapidly diminishing reserves, the need to combat air pollution, and the protection of the Earth's environment by the mitigation of anthropogenic climate change[2].

Fears about rapidly diminishing reserves is not recent. Concerns about the supply of oil and gas not meeting demand were initially articulated in the early 1920s, especially in the United States of America[3] (Nordhauser 1973). Yet, ironically, in the search for alternative fuels, improvements in crude oil refining technologies coupled with advances in the development and manufacture of internal combustion engines (ICE) and other combustion devices resulted in gasoline[4], diesel and kerosene[5] fuels becoming the affordable and preferred *alternatives* to peanut oil, coal dust, wood and whale oil for on-road vehicles, home heating and lighting (Smil 2010, 2017). Moreover, the phenomenal expansion of railways in many countries and the advent of grid electricity production by thermal power stations in the late nineteenth and early twentieth century also increased the demand and consumption of another type of fossil fuel, coal. Furthermore, by the 1970s, the use of natural gas started to become increasing popular both for home heating, as well as cooking, and as a replacement fuel for coal in utility scale electrical power generation, so that in 2022 the global consumption of natural gas rivalled that of coal (Ritchie *et al* 2024a). However, there have almost always been concerns that since fossil fuels are a finite resource they would be exhausted sooner rather than later, especially oil, a situation compounded by a rapidly growing global population, increasing ownership of personal vehicular transportation, and the overall accompanying demands for more energy. But, as will be discussed later in this chapter, the exhaustion of the Earth's store of fossil fuels could be later rather than sooner.

The second set of concerns, about pollution as a result of the widespread use of fossil fuels, became more evident from the 1950s onwards as the associated adverse

[1] Dates given in this chapter refer to the Common Era (CE) of the Gregorian calendar system formerly known in many countries as the Anno Domini Era (AD).

[2] In the remainder of the chapter the term anthropogenic climate change will be foreshortened to climate change as prescribed by the United Nations Framework Convention on Climate Change (UNFCCC).

[3] Referred to as the United States or, if an official document or in a title, US in this chapter.

[4] Confusingly called 'gas' especially by North American motorists but 'petrol' in other regions, e.g. the United Kingdom.

[5] Also know as paraffin.

environmental and health effects could be ignored no longer by the public and politicians alike. Albeit, the problems with the emissions from burning coal, for example, were known as long ago as the thirteenth century in parts of England and Scotland and resulted in a Royal decree banning its use (Brake 1975). Nevertheless, the generation of smoke and smog were considered more of a nuisance than a health hazard until the infamous four day London 'pea-souper' of December 1952 that resulted in the deaths of thousands with associated illnesses in the tens of thousands (Martinez 2024). Legally enforceable 'Clean Air' parliamentary acts followed in the United Kingdom with similar versions being developed in other global regions and countries, such as by the State of California and the Federal Government of the United States (Rogers 2016). Burning high sulphur coal was identified as the major human-activity culprit of the London smog. As urban areas globally became larger and more densely populated, coupled with concomitant increases in the number on-road vehicles where the interaction of tailpipe emissions with sunlight generated significant amounts of another form of smog, i.e. photochemical smog, the combination prompted further clean air regulations (Lattanzio 2022), and even more air pollution regulations involving fossil fuel usage were to follow.

A couple of decades after the pea-souper, and as interest in environmental matters became more commonplace, there was growing public awareness of the soon to be called 'lead scandal' involving leaded gasoline. The issue involved the addition of an engine knock-inhibitor, tetra-ethyl lead (TEL) ($Pb(C_2H_5)_4$), to gasoline and the association of the subsequent vehicular tailpipe emissions with adverse health effects experienced by those living close to roads and highways, particularly on young children (Lanphear *et al* 2018). The use of the lead additive began in 1923 as a method of reducing knock in gasoline-fuelled internal combustion engines to improve their performance (Kovarik 2005). However, just before and almost immediately after its introduction a number of scientists raised concerns about the potential of lead poisoning from the subsequent emissions (Needleman 2000, Rosner and Markowitz 1985).

In the United States these health fears were dismissed by many others in the scientific community, as well as by industry and the government. Even so, the United States Environmental Protection Agency (USEPA) began, in the early 1970s, to advocate the phasing out of leaded gasoline and, with government approval, introduced new tailpipe emission standards. Even so, it would be another 23 years before leaded-gasoline was banned for use in on-road vehicles in the United States (Bryson 2003). Other countries also banned its use for new on-road vehicles, many doing so prior to the United States. For example, in 1970, Japan became the first country to legally ban leaded-gasoline. By July 2021 Algeria became the last country to enact such a ban (Ritchie 2022a). However, leaded-gasoline is still available for use in off-road vehicles, aviation, and motor sports, albeit the percentage of TEL in the fuel mix is often much lower than pre-1996 leaded gasoline (Wikipedia Contributors 2024d).

In general, the use of liquid, gaseous, and solid carbonaceous fuels, such as traditional biomass and coal, and their emissions have become linked with anthropogenic air pollution and diminished air quality, and are a primary cause of premature deaths and illnesses involving millions of people annually (Roser 2021a). Electricity generation and the dominance of petroleum products in land, sea, and air transportation applications make significant contributions to the air pollution encountered in the outdoor environment, but the use of solid fuels in heating and cooking activities produces indoor air pollution levels that overshadow the adverse health effects of outdoor pollutants. Globally, in many instances, the fuels used in indoor household activities by billions of people are considered, somewhat contentiously, to be renewable energy sources, e.g. firewood (Johnson 2009, Mehetre *et al* 2017). These are further discussed in section 1.4, dealing with possible obstacles to enhancing the renewable share of future energy mixes. However, in this chapter the discussion starts with perhaps the most important driver of a non-fossil-fuel energy future—climate change.

1.1.1 Climate change mitigation—the third driver

The third driver of the thrust to transition away from fossil fuels is the strongly held conviction that this is the only way to mitigate climate change. The rationale is that the generation of so-called greenhouse gases (GHGs) from human activities since the start of the industrial age and the gradual transition to fossil fuels for energy generation has reached such concentrations in the twenty-first century that the average terrestrial surface temperature and ocean temperatures are rising to damaging levels, as evidenced by rising mean sea levels, shrinking glaciers, polar ice caps and snow cover, ocean acidification, and an increasing frequency of destructive weather events, including intense rainfall events and wildfires (Feigin *et al* 2023, Lindwall 2022). Indeed, the USEPA have identified 50 indicators of climate change covering the aforementioned adverse effects, together with their potential impact on human and societal well-being (USEPA 2021, 2023a). Two of the many identified GHGs, carbon dioxide (CO_2) and methane (CH_4), are the ones that attract the most interest for their roles in increasing the greenhouse-effect, albeit water vapour is the most abundant GHG. The sources of the emissions of CO_2 and CH_4 include those generated by natural phenomena and by human activities. In terms of the latter about 75% of CO_2 emissions are associated with the combustion of fossil fuels (United Nations 2024).

Thus, as the increase in atmospheric CO_2 concentrations is judged by many to be the main cause of global warming, if the use of fossil fuels is eliminated, then, according to climate models, it should follow that global warming can be limited to no more than a 1.5 °C increase or at least less than 2.0 °C over pre-industrial levels by the end of this century (IPCC 2018). The expectation is then that any further environmentally harmful impacts of climate change will be avoided and maybe in future centuries the existing impacts can be reversed (Hawken 2018, Shu 2019). Many political and scientific leaders, and national and international energy agencies have advocated several key strategies and solutions to achieving pathways for combating climate change including the eventual abolition of the exploitation of fossil fuel energy sources, in essence a 'leave them in the ground' approach (Carrington 2021, Welsby *et al* 2021).

Furthermore, there appears to be a pervasive global acceptance, mainly evidenced by frequent rhetoric and political declarations rather than by vigorous activity, that the likelihood of achieving the somewhat optimistic scenario of climate change abatement and future reversal can only achieved by an almost complete transition from fossil fuels to renewable energy sources by 2050, especially for electricity generation (United Nations 2023). To achieve this 'net-zero' situation, in which any residual carbon emissions that are produced are balanced by an equivalent amount being taken out of the atmosphere, the UN's Climate Champions Team, under the auspices of the United Nations Framework Convention on Climate Change (UNFCCC), have determined a target of at least 63% of global electricity being generated from renewable sources by 2030 (Climate Champions 2023, United Nations Climate Action 2023a). However, countries such as India and China clearly consider these target dates to be premature (Divadkar *et al* 2021, L 2023). Notwithstanding the projections of the 'accepted' climate models, the link between GHG emissions, especially human generated CO_2, and global warming are still statistics-based correlations and the result of modellers 'forcing factors' rather than scientifically proven causation—albeit measured atmospheric concentrations of CO_2 and surface temperatures are increasing. Nevertheless, it is the concerns surrounding the possible negative impacts of future climate change scenarios that are the main drivers of the pursuit to decarbonize the global energy supply (Quaschning 2019).

Arguably, the present thrust in pursuing a net-zero global environment is, for the most part, because the 2030 targets embedded in the 2015 trio of UN climate resolutions, i.e. the Sendai Framework for Disaster Risk Reduction, Transforming Our World: the 2030 Agenda for Sustainable Development, and the Paris Agreement [on climate change], have not been met (United Nations 2015a, United Nations Climate Action 2023b, United Nations General Assembly 2015a, United Nations–UNDRR 2015). All three resolutions, and their associated goals and targets, were to be achieved by 2030, so in a sense the goals have been given an additional 20 years for most, but not all, countries to achieve them. This delay in UN action deadlines is by no means an unusual occurrence, particularly with regard to environmental matters. For example, comments on renewable and non-renewable energy sources were contained in the 1972 UN Stockholm Conference on the human environment together with elements that would be manifest in the 2015 treaties, e.g. disaster reduction, a 'price on pollution' from non-renewable sources, human development and climate (United Nations 1973). Now considered the landmark global environmental conference, perhaps the more important outcomes of the Stockholm conference were the establishment of the United Nations Environmental Programme (UNEP) and the beginnings of the pathways to international environmental laws (UNEP 2024). The Stockholm conference publications of its summary declaration and final report overtly marked a change in the UN's perspective on how environmental challenges from unrestrained human activities were impacting some of the core principles of the original UN charter. It had taken almost a quarter of century after the UN's formation for its members to focus on global environmental issues and it would be a similar timeline before any action was taken, as described in the following paragraphs.

The world emerged from the traumas of the Second World War with a widely held desire to avoid future wars and armed conflicts, resulting in the formation of the United Nations (UN) including a universal charter of principles outlined in the 111 articles of the agreed declaration (United Nations 1945, 2015b). At the time there were no specific mentions of the global environment or climate change. Indeed, for the first two decades after the end of the 1939–1945 war the intonation was for economic growth. However, concerns about the environmental impacts of economic growth began to surface in the scientific literature and UN discussions (Dell *et al* 2008, Thant 1968, United Nations 1968). By the 1970s, some scientists were expressing concerns about a cooling climate, albeit that the media of the day embellished these concerns announcing the possibility of the Earth entering another ice age (Peterson *et al* 2008). Even so, it led to numerous proposals and suggestions for delaying this onset including the idea of reducing the reflective properties of polar ice with some form of black coating. The concerns were not taken that seriously by policymakers but when, a decade later, scientists raised climate alarms about global warming, many governments paid far more attention.

In 1987, the General Assembly of the UN discussed a resolution aimed at increasing '[i]nternational co-operation in the field of the environment' (United Nations 1987). A year later, following a motion from the Government of Malta, UN resolution 43/53 calling for the '[p]rotection of global climate for present and future generations of mankind', was passed (Government of Malta/UN 1988, Magalhães 2020, United Nations 1988). The Maltese resolution identified increasing concentrations of 'carbon dioxide and other greenhouse gases', which could cause sea levels to rise and global warming, whereas the final UN resolution, while endorsing these issues, only mentioned greenhouse gases. There was no explicit mention of fossil fuels or renewable energy sources. Moreover, the terms 'climate heritage' and 'sustainable development' were amended in the UN resolution to 'climate concern' and 'development'. The changes, seemingly innocuous, had significant legal ramifications, apparently, in the enforcement of global environmental laws (Magalhães 2020). In many instances the legality of resolutions passed by the UN's General Assembly becomes a mute point if the resolution is a 'recommendation', as legally this is not binding but is a hortatory declaration, i.e. a way of encouraging or persuading nations to adopt a particular course of action or actions (Acosta 2015). Whatever the legal status of the early climate resolutions they were an indication that the UN was starting to focus on global environmental issues[6].

Following the climate concerns raised in the 1970s and 1980s there was general agreement that if the global climate was to be protected there was an obvious need to investigate and assess the global climate and find out whether it was changing and, if so, what was the cause or causes. This resulted in 1988 in the endorsement and establishment of the Intergovernmental Panel on Climate Change (IPCC) by the combined effort of the UNEP and the World Meteorological Organization (WMO). The IPCC's stated remit was '[t]o provide international scientific co-ordinated

[6] In terms of 'climate laws' some are claimed by the UN to be legally binding, such at the Kyoto Protocol's emission reduction targets.

assessments of the magnitude, timing and potential environmental and socio-economic impact of climate change and realistic response strategies' (United Nations 1988). There then followed a series of UN resolutions dealing with climate matters culminating in the establishment of the UNFCCC at the 1992 United Nations Conference on Environment and Development held in the Brazilian city of Rio de Janeiro. The intention of this framework was to establish a structure for intergovernmental efforts to engage in the challenges posed by climate change. The convention came into force in 1994 with the ultimate objective 'to stabilize greenhouse gas concentrations … at a level that would prevent dangerous anthropogenic (human induced) interference with the climate system' (United Nations 1992). To assess and scrutinize the efforts to achieve this objective by the signatories of the convention, a Conference of the Parties, frequently referred to as COP, was to be created and hold annual meetings.

A key outcome of the convention was that henceforth 'climate change' would be a phrase only attributable to human activity that 'alters the composition of the global atmosphere', any other effects on climate, such as plate tectonics, orbital cycles which drive long-term changes in the Earth's climate, and volcanic emissions that can impact short-term weather patterns, would now be referred to as 'climate variabilities'. It would not be unreasonable to ask, as there has only been any human activity for, at the most, the last 0.1% of the Earth's existence, does this mean that climate change is a relatively new phenomenon? The answer is both yes and no. Changes in climate have taken place, constantly driven by natural causes including collisions with external planetary bodies, but with the increasing use of carbonaceous fuels and the associated rise in CO_2 emissions since the 1750s, and especially the 1950s, surface temperatures have risen at a rate over a timeline that is proving insufficient for the Earth's normal climate system to adjust (National Academy of Sciences 2020, Tso 2021). It would seem that to differentiate and highlight between natural changes in climate and those caused by human activity the UNFCCC chose the terms 'variability' and 'change'. These particular terminology distinctions are likely not wholly appreciated by the general public even today and they have tended and continue to cause some unfortunate confusion in any dialogues involving climate discussions. Nevertheless, having defined climate change as due solely to human activity the next step was to call for the replacement of fossil-fuels by energy sources identified as 'renewable'.

1.1.2 The advent of the renewable energy transition

Apart from climate change, the UNFCCC, in the 26 articles of its Convention, also established other definitions for items such as 'emissions' and 'greenhouse gases' together with statements outlining, in non-specific language, the commitments that the signatories would be making 'to protect the climate system for present and future generations', including the promotion of sustainable development (United Nations 1992). The Convention also states that precautionary measures should be taken to minimize or prevent the adverse effects of climate change and that when the threats of these effects are considered as serious or irreversible 'lack of full scientific certainty should not be used as a reason for postponing such measures'. Moreover, with regard to the acceptance of the need for an eventual transition to renewable energy sources, the most relevant article of the convention is the principles article 3.3, which states, 'policies and measures to deal with climate change should be cost-effective so as to ensure global benefits at the lowest possible cost'. The cost factor is one of the two key criteria that Smil has espoused for a successful energy transition, the other been availability (Smil 2010). A codicil to source 'availability' is the reliability of supply, especially in terms of electricity generation. All these factors are evident in the text of the UNFCCC's Convention and, unmistakably, they provided the seeds of the 2015 treaties, as mentioned earlier.

The UNFCCC Convention was a relatively short 23 page publication which concisely described an environmental 'framework' involving agreed definitions, principles, and commitments. However, several other documents were eventually published from the deliberations of the 12 day Rio Conference (United Nations 1993). The most important of these other documents was Agenda 21, described by the UN as a comprehensive action plan to address the concerns about the human impact on the environment, with the key objective of achieving global sustainable development in the twenty-first century through the integration of global environmental and development challenges, present at the time, and for the future (United Nations: Sustainable Development 1992). In this Agenda, the terms 'renewable' and 'sustainable' are mentioned on numerous occasions throughout the 351 page document. Consequently, the emphasis on the global promotion of 'new and renewable' energy sources is evident and these sources were identified as 'solar thermal, solar photovoltaic, wind, hydro, biomass, geothermal, ocean, animal and human power', all of which were considered, at the time, to be 'environmentally sound'.

Despite the promotion of renewable energy sources to address the perceived global environmental and development concerns, there was very little noticeable change in the national shares of primary energy consumption from such sources until the middle of the first decade of the new millennium, as illustrated by the exemplars shown in figure 1.1 (Ritchie *et al* 2024c). It would be three decades following the 1972 UN conference, and a decade after the Rio conventions, that the share of renewable energy sources in the global energy mix began to increase. This is not surprising, since the transition away from traditional sources to renewable sources could not happen overnight, given that any energy source requires appropriate energy conversion devices as well as the necessary infrastructure. The energy pathways from source to conversion are illustrated in figure 1.2. The scale of the challenge can be appreciated by the fact that less than 0.5% of the global energy consumption was provided from renewable sources, not including hydropower and traditional biomass, whereas over 77% came from conventional fossil fuels (Ritchie *et al* 2024c).

However, if hydropower and traditional biomass are included in the energy consumption shares as renewables, then the percentage share rises from 0.5% to 17%. The reason they are not included is that they do not wholly meet the 'environmentally sound' criterion of Agenda 21. The portion of hydropower energy

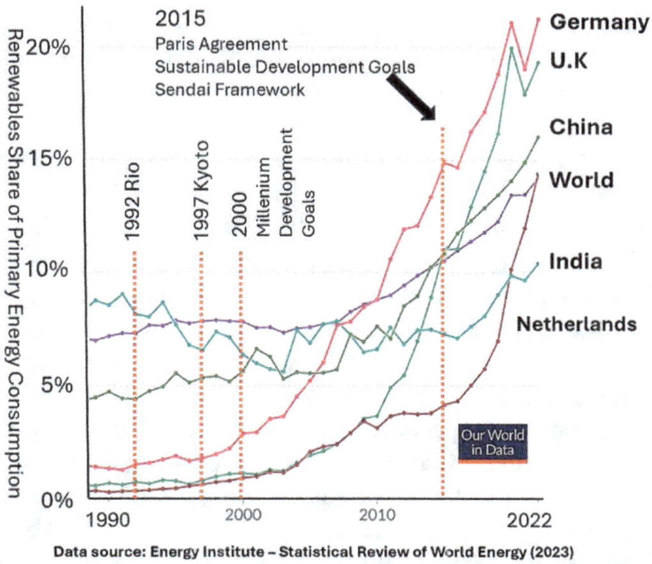

Figure 1.1. The global and exemplar rises of renewable primary energy share 1990–2022 modified. (Reproduced from Ritchie *et al* 2024a. CC BY 4.0.)

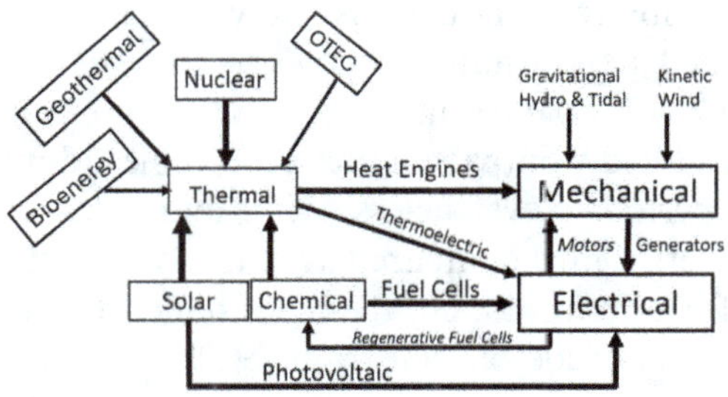

Figure 1.2. Energy sources, types and converters. (Source: The author).

that is considered to be environmentally sound depends upon national policies regarding the output of the associated power plant, and these are not consistent (California Energy Commission 2024, Kumar *et al* 2012). Traditional biomass, e.g. wood and animal dung, is specified by many countries, albeit contentiously, as being carbon-neutral, but is identified as an unclean cooking fuel if used directly for such purposes (Camia *et al* 2021, Harvey and Heikkinen 2018, IEA 2023a, The European Commission 2022). Ironically, if traditional biomass is used in the form of wood pellets as the fuel for the thermal generation of electrical power, then it is identified as a renewable source, especially in Europe (European Commission, Directorate-General for Energy 2024, Ireland 2022). These energy 'accounting' quirks are discussed in sections 1.2 and 1.3. Nevertheless, regardless of whether the share of renewables in the 1992 global consumption energy mix was 0.5% or 17%, to replace the 77% fossil fuel source share in a timely manner was a colossal task which would only be made even more difficult if nuclear fuel sources were also to be replaced.

Despite the obvious low share of renewables in the 1992 global energy mix, as illustrated in figure 1.1, some countries, notably Brazil, but also Canada, Sweden, and New Zealand, already had national energy mixes involving significant levels of renewable sources. These particular renewable shares were almost exclusively from hydropower sources, whose categorization as renewable was to become increasingly challenged. Following the Kyoto Protocol agreement in 1992, specifying the level of national emission reductions compared to their 1990 levels, which did not come in force until 2005 and the UN's 2000 Millennium Declaration and accompanying Development Goals that also called for a limit to greenhouse emissions by 2015, the percentage of renewables from non-contentious sources such as wind and solar in the global energy mixes started to increase. It was noticeable that, between 2008 and 2015, the political leaders and government policies in some European countries, such as the United Kingdom and especially Germany, began to enthusiastically embrace a transition to renewable energy sources, wind power being the favoured renewable, although Germany started to place increasing emphasis on solar power as well, as shown in table 1.1 (Ritchie *et al* 2024c, Ritchie and Rosado 2024b).

Indeed, for the period 2008–15, efforts to increase primary energy from wind sources were made in many countries, such as the United Kingdom and Chile, while Australia and Japan concentrated on solar power sources. However, the pace of renewable development between 2015 and 2022, in terms of wind and solar sources, has notably declined globally, although, as shown in table 1.1, exceptions include Chile, which accelerated its consumption of solar power. However, it has to be remembered that the COVID pandemic, because of the financial ramifications of the global response, undoubtedly impacted the renewable adoption rate during this second selected time period. Nevertheless, wind and solar in 2022 accounted for 5.3% of the global energy mix compared to slightly less than 0.5% in 2008, an impressive increase of over ten-fold. On the other hand, the percentage of fossil-fuel sources, i.e. coal, gas, and crude oil, fell from 79.5% to 76.7%, the majority of the

Table 1.1. Exemplars of countries adopting wind and solar sources 2008–2022 (Ritchie *et al* 2024c, Ritchie and Rosado 2024b).

Country/ region	2008 %		2015 %		2022 %		2015/2008		2022/2015	
	Wind	Solar	Wind	Solar	Wind	Solar	Wind	Solar	Wind	Solar
Germany	2.96	0.32	5.74	2.71	9.56	4.64	1.94	8.5	1.67	1.71
Spain	5.16	0.4	8.43	2.37	10.23	5.51	1.63	5.9	1.21	2.32
United Kingdom	0.78	<0.01	4.77	0.89	10.29	1.79	6.12	N/A	2.16	2.01
Chile	0.03	0	1.37	0.88	4.91	7.59	45.67	N/A	3.58	8.63
Australia	0.63	0.03	1.99	1.04	4.97	6.1	3.16	34.7	2.50	5.87
Japan	0.14	0.12	0.27	1.76	0.43	5.39	1.93	14.7	1.6	3.06
World	0.45	0.03	1.47	0.45	3.27	2.06	3.27	15	2.22	4.58

share decrease taking place in the 2015–2022 period, i.e. after the Paris Agreement and Agenda 2030 became almost universally accepted UN resolutions.

Has then the use of renewables such as wind and solar facilitated the reduction of the share of fossil-fuels in the global energy mix? The mathematics suggest that the answer is yes, but as there are many other forms of renewable sources, the increased use of wind and solar are not the only reasons for the fossil-fuel share diminishing. Nevertheless, in terms of actual consumption, measured in terawatt-hours (TWh), fossil-fuel sources provided over 15% more in 2022 than in 2008, generating more CO_2 emissions not less as called for in the Paris Agreement. In fact, in terms of primary energy consumption, the total amount since 2008 has increased by over 20%, a rate greater than the 17% growth in global population over the same period (Ritchie and Rosado 2024b, Worldometers 2024). To meet this demand all forms of energy source consumption have increased apart from nuclear power, which has decreased both in terms of total output and percentage share. So, arguably, the increased use of renewables has not only led to a reduction in the share of fossil-fuel energy consumption but also nuclear power which, while not considered to be a source of GHGs, is also not a renewable source. However, in the net-zero scenario, as mentioned previously, there is an acknowledgement that not all GHG emissions will be eradicated by 2050 and those that persist, i.e. the 'residuals', will need to be sequestrated before entering the atmosphere.

The implication of this scenario is that it will be beyond 2050 if renewables are ever to be the sole sources of energy. For some, the global utilization of renewables has been agonizingly slow despite the fact that primary energy consumption from renewable sources, not including any contributions from hydropower, has more than doubled since the passing of the three 2015 treaties (Ritchie *et al* 2024a). The concerns about the perceived lack of progress are driven by fears of increased global warming caused by anthropogenic GHG emissions which have been rising since 2015 (Levin *et al* 2023, Lo 2023, Subsidiary Body for Scientific and Technological Advice 2023). Indeed, a lengthy 2024 report, 'Climate risk assessment', carried out by the European Environmental Agency (EEA) has called for urgent action to be taken as even with the IPCC's low-emissions scenario, known as Shared Socioeconomic Pathway (SSP) 1–2.6, the global temperature anomaly will reach over 2 °C and with very high emissions, SSP 5–8.5, the temperature could be 7 °C or higher by the end of the century, causing catastrophic climate damage through a variety of impacts (European Environmental Agency 2024). Renewables feature in the many sections of the EEA's report as part of the menu of solutions and it is stressed that unabated climate change poses dangers to the effective employment of such sources, especially those which are 'water intensive'. Regardless of the motivation to adopt more renewables as soon as possible, there are numerous instances in the literature and media that claim the 2050 net-zero target is simply unachievable and unrealistic (Dyke *et al* 2021, Jaremko 2023). Obviously two of the largest polluters, China and India, agree that the net-zero timeline is too short.

Perhaps the transition to a green economy, based largely on renewable energy sources, could be accelerated if the insatiable consumption of energy by a growing population were lessened? Improving the efficiency of energy conversion would be instrumental in such an endeavour for all converter forms. Moreover, improvements in the so-called 'end-use efficiency', i.e. the percentage of energy used by an application compared with the energy supplied, can also be contemplated, particularly in the industrial and transportation sectors. In the latter case, using the United States 2022 data as an example, the end-use efficiency is only 21% whereas in their residential sector it is 65% (Lawrence Livermore National Laboratory 2023). Overall, only about 33% of the primary energy source inputs are converted into actual energy consumption, referred to as 'energy services' on the Sankey energy flow diagram produced by the US's Lawrence Livermore National Laboratory (LLNL), figure 1.3 (Lawrence Livermore National Laboratory 2023). On the face of it, there appear to be many opportunities to improve energy-use efficiencies but the laws of thermodynamics and physics, together with materials limitations, exacerbate attempts at energy efficiency improvements. This will be discussed in section 1.4. Additionally, for renewables such as wind and solar power, geographic location, prevailing weather, and climate all play roles in the actual energy outputs of the attendant conversion devices, circumstances which can also impact of the power outputs of *quasi*-renewables such as hydropower.

Notwithstanding the challenges of energy efficiency improvements, renewables such as wind and solar also suffer from intermittent power generation, i.e. the familiar 'the Sun don't shine, the wind don't blow', epithet frequently quoted by those likely less than enthusiastic about the renewable energy transition. However, the observation is factually accurate, especially for solar sources, in that when there is an absence of sunlight no power is produced. There will also be periods during the daytime availability of sunlight when the amount of electrical power produced is more than needed and, if fed into the supply system, could destabilize the existing grid of a utility scale supply system. In such circumstances the solar power generation

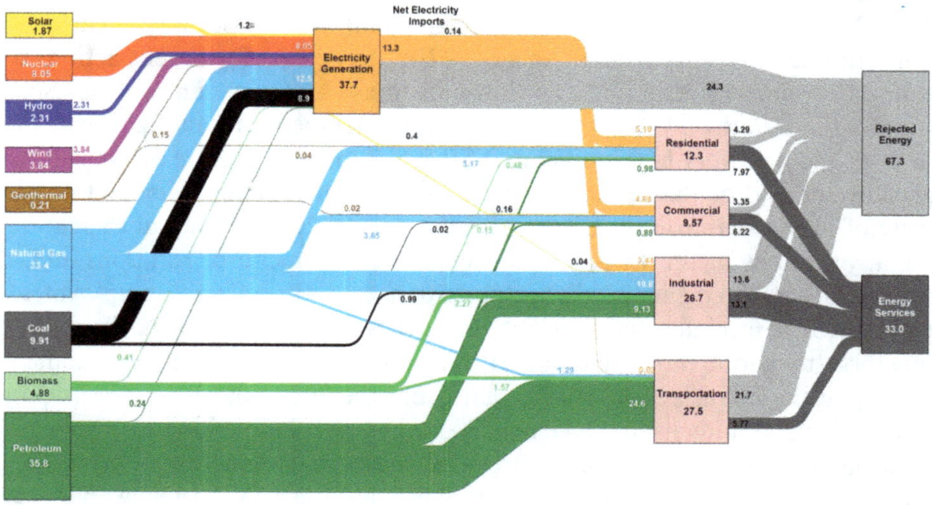

Figure 1.3. Sankey diagram of US energy flows in 2022. (Reproduced with permission from (Lawrence Livermore National Laboratory 2023).)

system, wholly or in part, has to be temporarily shut down or curtailed, meaning that the energy that could have been generated is lost or wasted. Curtailment could be avoided if (i) there was some means of storing the excess production, or (ii) an extended or smart grid was available so that the power produced in a region experiencing excess energy could be distributed to regions where the sunlight has disappeared and needs additional generation to met consumer demands. Similar situations can also occur with wind power systems. There is no doubt that improvements in energy storage capability and capacity, along with increased electrical distribution infrastructure, would have a marked impact on the better utilization of existing and future solar and wind power installations.

However, it is worth noting that to focus on solar and wind energy alone can be misleading. In the European Union (EU), for example, 59% of renewable energy consumption was biomass in 2021, lead by Germany (European Commission, Directorate-General for Energy 2024). The EU categorize biomass into three groups, solid, liquid, and gaseous biofuels, with solid biofuels, e.g. wood and wood products, providing almost 70% of renewable bioenergy consumption. Moreover, in 2018, liquid biofuels, such as biodiesel and biogasoline, provided a greater share of renewable energy consumption than solar sources. Thus, in discussing the future of renewables, it is crucial to consider sources other than solar and wind, including renewable bioenergy and hydropower as well as geothermal. Although not of significance presently, it also seems likely that some forms of ocean energy harvesting, such as tidal, wave and thermal sources, could provide a measurable share of renewable energy in the second half of this century. Consequently, in the discussions presented in section 1.2, a brief overview of other, and possibly additional, types of renewable sources, such as white hydrogen, are included.

1.2 Renewable energy sources

1.2.1 What is a renewable energy source?

As the literature on renewable energy continues to grow and become encyclopedic in scale, it is important that the terminology used coalesces to a universally accepted set of definitions, if only to avoid misunderstandings. While such an agreed lexicon has yet to be firmly established, a noticeable change in the literature is the tempering of the somewhat hyperbolic and staunchly pedantic language concerned with defining what resources constitute an acceptable renewable energy source (Gibon *et al* 2022). Maybe this language softening is the result of growing mental health concerns associated with 'climate anxiety' (Dodds 2021, Kurth and Pihkala 2023). Perhaps it is the enthusiastically advocated use of bioenergy in Europe that has been instrumental in this trend, for now the United Nations states that '[g]enerating renewable energy creates far lower emissions than burning fossil fuels', rather than the fervent zero emissions scenarios (United Nations Climate Action 2024). Another example of the change comes from the Inflation Reduction Act (IRA) signed into United States Law in 2022 by US President J R 'Joe' Biden as part of his administration's commitment to reducing the country's GHG emission by 50%–52% by 2030 compared to their level in 2005 (The White House 2023). The Act specified many financial incentives and inducements, especially with regard to renewable energies and alternative fuels, e.g. biodiesel, but does not claim by itself to enable the 50% target to be reached, but estimates that its implementation could achieve a 40% reduction which 'would slow global warming giving societies more time to adapt' (Johns Hopkins School of Advanced International Studies 2022, The White House 2023, US Department of Energy 2022).

Another trend has been the attempts to differentiate between renewable sources producing GHG emissions and those emitting zero emissions. Terms such as 'green energy' and 'clean energy' can be encountered. Exactly what is meant by the energy terms 'renewable', 'green', and 'clean' is to some extent reliant on the agency defining them. One large UK energy company agency defines green energy being produced by a method and from a source that 'causes no harm to the natural environment', while clean energy 'creates little or no GHG emissions', the latter being akin to how the United Nations defines renewable energy (National Grid 2023a, United Nations Climate Action 2024). Arguably, a more succinct representation, using electricity production as the example, is the most recent depiction of 'green power' by the USEPA (figure 1.4), indicating that green power sources are a subset of renewable energy sources and they offer, comparatively, the most environmental benefits of all energy sources (USEPA 2024a). However, the terms 'renewable', 'green', and 'clean' are frequently used interchangeability despite there being differences in definition, but the precise nature of these differences relies on how a specific agency uses them. In general, as shown in the examples in table 1.2, the definition for what constitutes renewable energy as used by authoritative sources has a common basis.

Nevertheless, interpretation nuisances can be encountered. For example, the United Nations Climate Action definition quoted in the previous paragraph also states that '[s]unlight and wind, for example, are such sources (renewable) that are constantly being replenished', but the Sun has been burning through its fuel source,

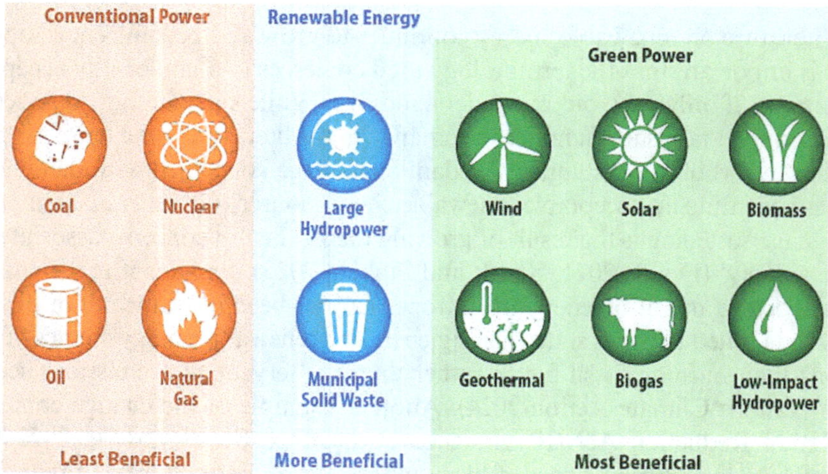

Figure 1.4. Green power. Courtesy of the US Environmental Agency's Green Power Partnership (USEPA 2024a).

Table 1.2. Example definitions of what renewable energy is.

Source	Definition
United Nations Economic and Social Council (2016)	'Renewable energy is energy that is derived from natural processes (e.g. sunlight and wind) that are replenished at a higher rate than they are consumed.'
The United States Energy Information Administration, USEIA (2023a)	'Renewable energy is energy from sources that are naturally replenishing but flow-limited; renewable resources are virtually inexhaustible in duration but limited in the amount of energy that is available per unit of time.'
National Geographic Society (2024a)	'Renewable energy comes from sources that will not be used up in our lifetimes, such as wind and solar.'
National Grid (2022a)	'Renewable energy is energy that comes from a source that won't run out. They are natural and self-replenishing, and usually have a low- or zero-carbon footprint.'

i.e. hydrogen, for billions of years and, although this process will continue for another 4–5 billion years, it will eventually be consumed, so it is not being constantly replenished. However, as the average human life-expectancy is just over 70 years, albeit it has doubled since 1900, and as there are three human generations in a century, there could be 5 million human generations before the Sun's hydrogen fuel source is exhausted. So, while it is not strictly accurate, it is realistic to assume that sunlight is a sustainable and renewable energy source.

Is there then a time limit on how long an energy source will last for it to be considered renewable and sustainable? For example, it has been estimated that natural gas and coal resources could last for centuries or longer. However, as 'proven reserves', basically what can be extracted economically, are different from resources, the time to depletion of reserves is far shorter, albeit the scale of reserves has remained largely unchanged for at least three decades, as further discussed in section 1.3.2.1 (Ritchie and Rosado 2024c, Society of Petroleum Engineers 2001). Despite the possible longevity of fossil fuels, they cannot be classified as renewables, partly because they are not replenished at a higher rate than they are consumed. Moreover, as previously discussed, the main driver of the transition to renewable from fossil sources is the global warming associated, potentially catastrophic, climate change. The global warming increases that have occurred since pre-industrial times have been declared as a direct result of increasing atmospheric GHG, especially CO_2, concentrations and the largest contributor to these rises has been fossil fuel combustion. While some renewables can produce GHG emissions, in general, their use drastically reduces the amount of such emissions. Thus, the extremely low rate of source replenishment, combined with the much larger scale of GHG emissions production, means that climate change mitigation cannot be achieved if the use of fossil fuels continues. That is, if, as asserted in the bulk of the scientific literature, anthropogenic GHG emissions are largely responsible for climate change (Crowley 2000, Hegeral *et al* 2019, IPCC 2018). As can be seen from figure 1.5, more people than may be imagined believe in climate change and that it is caused by humans (Ritchie 2024a).

Even so, many definitions of renewable energies do not mention emissions or climate change, as evidenced in the examples provided so far in this chapter, but in almost every instance, when not explicitly stated, a qualifying statement follows the definitions highlighting the benefits of the use of renewables in combating air pollution and global warming. The energy description 'sustainable' is almost invariably coupled with 'renewable' and regularly the terms are used interchangeably but, as with the 'clean', 'green', and 'renewable', descriptions, they are not identical. Moreover, analogous to the different definitions of 'renewable' that are available in the literature, there exist sundry explanations of what determines an energy source as being considered to be 'sustainable' (Halawa 2024, Mardani *et al* 2015). The lineage of all the definitions can be traced to the 1987 UN instigated 'Report of the World Commission on Environment and Development: Our Common Future', frequently called the Brundtland Report, which stated that '[h]umanity has the ability to make development sustainable *to ensure that it meets the needs of the present without compromising the ability of future generations to meet their own needs*' (Brundtland and Mansour 1987). The latter part of the statement, with the emphasis by the chapter author, is the basis for the majority of sustainable energy definitions.

So, a sustainable energy source is one that is never depleted and does not need to be replenished or renewed while meeting the energy needs and demands of all, indefinitely. Other common definitions also include that sustainable energy usage must do no, or little harm, to the environment. Applying these criteria can make it appear that sustainable energy sources and renewable energy sources are the same,

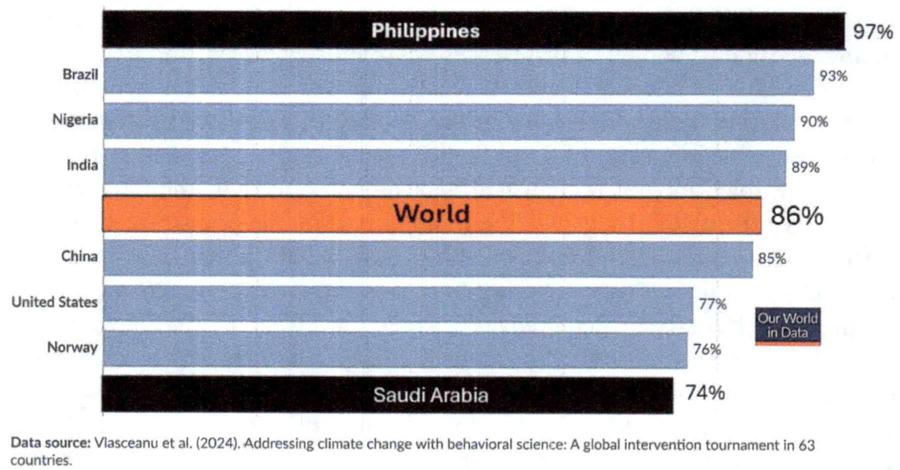

Figure 1.5. Percentage of people caring about climate change—exemplars. (Reproduced from (Ritchie 2024a). CC BY 4.0.)

but this is not the case. For instance, using the example definitions in table 1.2, if a renewable energy source is used at a greater rate than it is replenished then it will become unsustainable. The use of the appellation 'sustainable' when used to describe an energy source regularly has codicils reflecting political inferences regarding the achievement of a desired balance between economic growth, environmental stewardship, and social well-being. The unevenness of the use of the terms renewable and sustainable does not help in the collection of energy statistics, so citing the source of such data becomes essential. A good example of this need is the authoritative annual databases of global energy statistics published, until recently, by BP[7], which treated energy data associated with hydropower separately from other renewables' energy data. This is not unreasonable, since, as previously mentioned, some nations and data collection agencies consider *all* hydropower generated energy as renewable whilst others do not.

Thus, the quoted shares of primary energy supply and consumption attributed to renewable sources can differ significantly if hydropower is, or is not, included. The issues surrounding whether hydropower energy can be considered a renewable energy source have been focused on the generation of GHG emissions from large bodies of open water. As always, the situation is more complicated, and data analyses conducted by the IPCC and the International Hydropower Association (IHA) have found that many of the stated fears about the level of GHG emissions encountered in the hydropower renewable classification discussions are flawed, albeit there are some exceptions (European Union 2023, International Hydropower Association 2022a, Kilajian and Mercier-Blais 2019). Nevertheless, some nations still take a somewhat conservative view of what can and cannot be defined as renewable hydropower. In some jurisdictions it appears that the level of central and state government incentives for solar and wind investments could be adversely affected if too much hydropower is included in their 'renewable portfolio standards' (Stori 2020).

The renewable energy sources identified by the UN in 1992 remain the accepted types of such resources, albeit that not all hydropower and biomass energy is considered to be renewable. Nevertheless, in the increased utilization of renewable sources, solar, wind, biomass, and hydropower have dominated, attracting more financial investment than the other accepted forms. These significant investments have enabled many technological improvements to be made resulting in better system performances at increasingly competitive prices, especially for solar and wind systems. But what about the other identified renewable energy sources? There appears to be a general lack of public awareness about these other sources, particularly geothermal energy sources which, in 2023, generated over 95 TWh of electricity, sufficient in the United States to meet the average needs of almost 9 million households (IRENA 2023a, USEPA 2024c). Consequently, in discussing the 'other' four sources briefly in the next section, a more detailed outline of geothermal energy is presented.

[7] Formerly known as British Petroleum plc.

1.2.2 The 'other' renewable energy sources

Four other 'new and renewable' energy sources were identified in the 1992 Agenda 21, including human and animal power, geothermal, and ocean sources (United Nations: Sustainable Development 1992). Although human and animal energy have long been connected with the use of muscle power, in more recent times other ways of using these energy sources have or are being developed. In the case of energy from animals, their manure and other wastes can be converted to a gas, biomethane, by anaerobic decomposition and more recently a chemical process known as torrefaction has been advocated and used to produce solid fuel from animal manure (Adams 2021, Hong *et al* 2023, Nunes *et al* 2017). These animal energy sources are generally included in the biomass–biofuels energy category and are considered to be renewables, even if not necessarily green.

1.2.2.1 Human generated energy as renewable?

Human generated energy is commonplace as the mechanical power produced by muscles is encountered everyday especially in methods of transportation from walking and running to cycling to rowing a boat or propelling a wheel-chair. However, without a human's metabolic conversion of food the generation of energy would not be possible. Arguably then, it can appear that the human body is more of an energy converter than a primary energy source. This assertion is now being contested as blood circulation, breathing and body movements could be used to power tiny, nanoscale, devices such as pacemakers (Haeberlin *et al* 2020, Li *et al* 2018). Moreover, body heat can be captured to help warm buildings and public spaces, such as the the Mall of America in the state of Wisconsin, Stockholm Central railway station, and a Metro station in Paris (Lee 2020). As long as a sufficient number of people use these facilities then most certainly the energy source could be described as renewable and such innovative use of human body heat could become more common in the future, but the concept is not that new (Al-Habaibeh 2022).

1.2.2.2 Ocean and geothermal renewable energies

The two remaining renewable energies identified in Agenda 21 are geothermal energy and ocean energy. Geothermal, as its name implies, is heat energy emanating from the Earth itself. Ocean energy, frequently termed 'marine energy', can also be harvested in the form of heat, through a process known as ocean thermal energy conversion (OTEC) which make use of the temperature differences between the ocean surface and deeper water to power a dynamic heat converter, i.e. engine, but the kinetic and potential energy associated with tidal flows and waves can also be garnered using underwater 'current' or 'tidal' turbines and various wave energy devices (Chandrasekaran *et al* 2022, Lai 2022). Ocean energies, as renewable sources, have tremendous potential that is yet to be realized, while their technical feasibility and viability has been proven (Kilcher *et al* 2021, Uppal 2018). The intended outcome of ocean energy devices is the generation of electricity, an implicit indicator of achieving Sustainable Development Goal (SDG) 7 (United Nations

General Assembly 2015a). However, in 2021 ocean energy accounted for only 0.012% of electrical generation from renewable sources with a global output of 971 GWh, an amount which has been decreasing since 2017 and a situation likely indicative of the total global lack of annual financial commitments to marine energy development (International Renewable Energy Agency 2024, IRENA and CPI 2023). So, despite the known scientific potential of ocean energy, the prospects for its short-term and medium-term utilization are not encouraging.

Analogous to ocean energy's OTEC systems, geothermal energy stems from the temperature difference between the Earth's outer layer, the crust, and its central core, where temperatures are estimated to be several thousand degrees Celsius, about the same as that of the Sun's surface. Thus, as compared to the oceans, the temperature gradient is reversed with the temperature increasing with depth. Although the Earth has been cooling since its formation 4–5 billion years ago, the heat lost in the natural geological processes is replaced and replenished in part, about 50%, by the slow decay of naturally occurring radioactive particles such as potassium (^{4o}K), uranium (^{235}U and ^{238}U), and thorium (^{232}Th) present in the mantle and the core, figure 1.6 (Earle and Panchuk 2019, KamLAND Collaboration 2011, National Geographic Society 2024b). Starting at the solid inner core the heat is transferred to the mantle by conduction and convection, the liquid rocks in the mantle, i.e. magma, in turn transfers the heat to the crust, as illustrated in figure 1.7 (Energy Education 2024).

Some of this magma can escape directly to the surface via volcanic eruptions after which it becomes known as lava. When the magma intrudes into groundwater near the surface the heat boils the water and emits it through surface cracks, fissures, and vents, known as 'fumaroles', in the form of steam and various gases, such as sulphur gases and CO_2. If the body of water is at or near the surface, and the magma causes the temperature of the water to be much higher than the ambient air temperature, then hot springs are created together with a noticeable sulphur smell. A form of hot spring is a geyser, which spurts interment streams of steam and hot water through

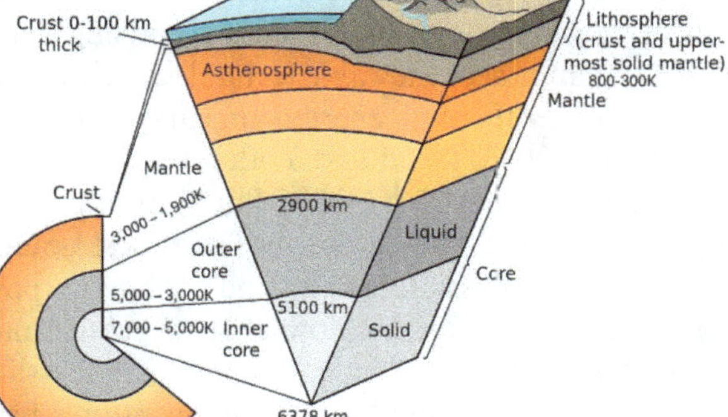

Figure 1.6. Earth's lithosphere, mantle and core with temperature ranges. (Image credit: USGS.)

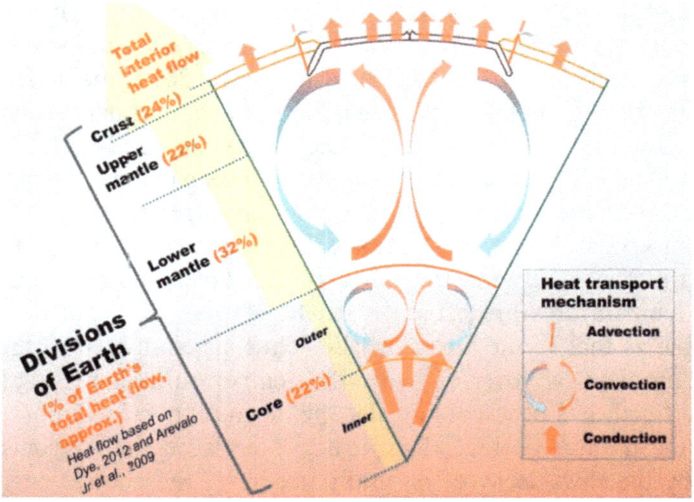

Figure 1.7. Heat transfer mechanisms between the core and crust (Energy Education 2024). (This image has been obtained by the author(s) from the Wikimedia website where it was made available by [Bkillil] under a CC BY-SA 3.0 licence. It is included within this article on that basis. It is attributed to [Bkillil].)

narrow surface vents, the most famous of the 500 or so geysers in the United States' Yellowstone National Park being Old Faithful, located in Wyoming (USDOI 2024). The Indigenous peoples of North America have known about these sources of geothermal energy for several millennia, using them for medicinal bathing, heating, and cooking.

Health and wellness aspects are also associated with the mineral content of certain hot-spring waters today with such spa facilities also known as 'onsens' in Japan. Records show that the Romans and Chinese also knew about and used hot springs over two thousand years ago and countries such as Japan and France have a long history of their use in some cases for what today would be called 'district heating' (National Geographic Society 2024a). There is a reasonable public awareness of hot springs, but rarely of their association with geothermal energy or as a source of renewable energy. Moreover, as most geothermal energy sources are found below the land surface where they are not visible, it is not surprising that the public appear to be less aware of them than other renewable technologies, albeit that the direct-use, i.e. non-electrical use, of geothermal energy appears to be attracting increased attention (Ediger and Akar 2023, Lund and Toth 2020).

The geothermal gradient is about 25 °C–30 °C km^{-1}, meaning the deeper the source the higher its temperature. The gathered energy from these sources can be used to generate electricity, or for low emission heating and cooling purposes, or both. It has become common practice to categorize below ground geothermal sources by their temperatures. Initially, there were only two commonly used classifications, 'high', i.e. source temperatures above 150 °C, and 'low' for temperatures below 150 °C. There can now be up to four categories, as shown in table 1.3 (IRENA and IGA 2023, Planete Energies 2021). The sources above 150 °C are favoured for the production of

Table 1.3. Geothermal energy categories by temperature.

Description	Temperature (°C)	References	Comments
High Medium Low	>150 90–150 <90	IRENA and IGA (2023)	• For flash steam plant >182 °C • For binary steam plant 107 °C–182 °C • Direct and indirect uses
Very low	<30	Planete Energies (2021)	▪ Heating and cooling, e.g. with residential heat pumps

electricity, which is invariably generated by the use of steam turbines. The quality of steam, as defined by its dryness fraction, i.e. a measure of how much water it contains, the lower the fraction the higher the water content. Wet steam used in a turbine lowers the machine's efficiency as well as having the potential to damage the turbine blades.

The first electricity generation geothermal power plants in the early twentieth century used a system whereby dry steam from a geothermal heat reservoir was used directly (Lund 2024). However, globally, geothermal reservoirs seldom contain only steam and more commonly are a mixture of hot water and steam or just hot water. To avoid the use of wet steam, two other types of geothermal energy system have been developed, referred to as flash steam and binary cycle systems (Lund 2024, NREL 2024). In the flash system any liquid water is extracted from the water vapour mix, whereas in the binary cycle system the extracted geothermal heated water is used to boil and vaporize another working fluid via a heat exchanger before the resulting dry vapour is then used to drive the turbine. In both cases any excess liquid water, including condensed water vapour, is returned to the original geothermal source where it can be reheated adding to the renewable and sustainable aspects of geothermal energy.

As illustrated in table 1.3, the condition that the heat contained in the porous rock groundwater aquifers must be above 150 °C to generate electricity is somewhat misleading. Moreover, geothermal energy exists in different forms other than heated water or steam reservoirs such as 'hot rocks' and what could be described as below surface shallow land as a result of the *in situ* geothermal gradient. Certain types of rocks or rock formations at depths from 3 to 10 km below the surface maybe at a sufficient temperature to produce geothermal energy, but may not be sufficiently porous or have a high enough permeability to naturally form hydrothermal reservoirs. However, using the type of drilling technology developed by the oil and gas industry, it is possible to construct human-created hydrothermal reservoirs by drilling into these rocks and injecting water to increase their permeability and enabling the water to be heated and pumped to the surface, as with natural hydrothermal energy. Such an arrangement is known as an 'enhanced geothermal system' (EGS) (Office of Energy Efficiency and Renewable Energy 2024). The heated water from an EGS installation can be used to produce electricity or for heating and cooling purposes in the same way as that from heated groundwater aquifers (USDOE 2023).

A constant challenge for exploiting geothermal sources is knowing where they are and what they can be used for, in keeping with the classifications given in table 1.2. In the first instance they have to be found as, apart from hot springs, they are hidden below the surface. Indeed, some agencies refer to them as 'hidden' or 'undiscovered' sources (Fuergy 2019, USDOE n.d.). But how can these hidden sources be found? In the first instance, expert assessments using play fairway analysis (PFA) or similar, based on known geologic attributes of potential sites coupled with any available direct measurements of such factors as geothermal gradients, are used to identify sites worthy of further exploration (USDOE n.d., Vesselinov *et al* 2022) If the available measurements are not comprehensive then a small-bore hole can be drilled to determine where the resource is in terms of depth and temperature. It is very rare to encounter a 'dry hole', where no resource is detected, but, in some instances, the desired quality of the resource does not appear to meet a prescribed threshold (International Finance Corporation 2013). When the geological survey and test-drilling indicates a viable site, a production well is drilled.

According to the International Renewable Energy Agency (IRENA), about 78% of production wells prove to be 'successful' but this still indicates a measurable degree of uncertainty in the search for viable geothermal energy sources (IRENA and IGA 2023, IRENA 2023b). There may be certain intended applications for the identified geothermal source which are instrumental in making the decision as to whether or not the production well will be successful. However, as the geological characteristics will only be known after the drilling, it may be evident that opportunities for the utilization of the geothermal energy exist other than those originally planned. Indeed, there can be many potential applications of geothermal energy, both direct and indirect, depending upon the geothermal gradient, source temperature, heat flow rate and depth. The multiplicity of possible geothermal energy applications and their relationship to the subsurface temperature can be represented on a Lindal diagram, named after its creator, the Icelandic engineer Baldur Lindal (Energy Sector Management Assistance Program 2022b, NREL 2023a). There are many versions of the Lindal diagram, but all illustrate the range of applications associated with particular temperature bands.

The 'other' renewable energy sources appear to be peripheral in the present renewable energy scenarios, especially in terms of the maximum capacity to produce electricity and the actual electricity generated. Notwithstanding renewable hydropower, from the literature and the media, the impression can be gained that solar energy, for example, has long been one of the two dominant forms of renewable energy, the other being wind. However, this is not the case. Until 2007, geothermal energy capacity was greater than that of solar energy devices, especially solar photovoltaics, and it would be only in 2011 that global electricity generation from such solar devices surpassed that from geothermal generation (IRENA 2017). Moreover, the 'levelized cost of electricity' (LCOE), which can be defined as 'the average cost per unit of energy generated across the lifetime of a new power plant', for geothermal energy production compared with that from solar photovoltaics was significantly lower between 2010 and 2017, and only in 2021 did solar generation become cheaper (Ritchie *et al* 2023c).

Yet in 2021, a magnitude more, by a factor of almost 11 times, electricity was generated by solar photovoltaics than geothermal source generators (IRENA 2023a). However, a word of caution, the acronym LCOE is also used, unfortunately, to specify the 'level cost of energy' which takes into account a wider spectrum of economic factors, including subsidies, tax and usage incentives, and costs other than those associated with just the generating aspects, such as building, installation, maintenance, and other operating costs (Kabeyi and Olanrewaju 2023, USEIA 2022). The costing issue is just one of many factors involved in the choice of which renewable energy source to develop and utilize. The final decisions on choice or choices of renewable sources are normally based on an agreed multifaceted methodology developed at national or regional government levels (Shatnawi *et al* 2021). There is no universally accepted methodology so, arguably, the evolution of a globally optimum approach remains a work-in-progress, albeit that resolute governmental policies for favoured sources can prevail over other rationale.

A typical example of this latter situation is Germany's very strong solar culture, where, despite a very modest solar photovoltaic power generation, especially in the winter season, it leads the world in terms of installed solar capacity per capita (Energy Sector Management Assistance Program 2020, Limb 2024). The policy, no doubt, is one of Germany's responses to climate change concerns and the efforts to provide affordable electricity for all in harmony with Sustainable Development Goal 7 of Agenda 2030. Electricity prices for many years for households and small business have increased significantly and are the highest in Europe. Over the same period there has been a sizeable increase in the electricity generated from renewable sources, not just solar (Amelang 2023, Burger 2024, Statista 2024b). A correlation between increased renewable energy utilization and electricity prices to consumers is evident and the trend is hardly in harmony with the aspiration to produce affordable energy. Nevertheless, it may just be a case of 'short-term pain for long-term gain' since, although prices have increased, GHG emissions such as CO_2 per capita have been reduced by almost a third so far in this millennium, albeit the mean surface air temperature has increased by 1.62 °C (IEA 2022a, World Bank Group 2021).

As the momentum to replace traditional fossil-fuel energy sources has increased since the latter part of the twentieth century so have the efforts intensified to better exploit the potential use of known renewable sources. These endeavours have also given rise to research and development activities associated with newer or more novel forms of renewables, as outlined next.

1.2.2.3 Future renewable energy additions to the global mix

Although it can appear that renewable energy sources are a new phenomenon, most have a long history, as shown in table 1.4. In the nineteenth and just over halfway into the twentieth centuries the use of fossil fuels emerged and became ubiquitous in global energy mixes. The only new primary source during this period was nuclear energy. However, in the World Wars of the twentieth century countries without adequate access to fossil fuel oil made significant efforts to produce 'synthetic' oils, often from coal. In the second half of the twentieth century, and into the twenty-first, the concept of using plant matter, biomass, in the production of liquid biofuels such

Table 1.4. Past, present, and future developed energy sources.

Used prior to the eighteenth century	Added between the eighteenth and twentieth centuries	Recent developments this millennium
Muscle power[b]	Kerosene	Renewable hydrogen[a]
Solar[a]	Solid, liquid and gaseous fossil fuels	Nuclear fusion[b]
Traditional biomass[b]	Nuclear fission	Human generated[b]
Wind[a]	Non-traditional bioenergy[b]	Electrofuels[8] (e-fuels)[b]
Water[a]	Hydrogen[b]	Green ammonia[a]
Geothermal[a]	Ammonia	
	Ocean power[a]	

Key: [a]Renewable. [b]May be renewable.

as ethanol steadily increased. This outcome was facilitated in the United States by the 2007 Energy Independence and Security Act (EISA) which encouraged the development of renewable transportation fuels to reduce GHG emissions (USEPA 2024d). Under this Act, to qualify as a renewable fuel a biofuel has to have 'a life-cycle GHG profile at least 20 percent lower than the fossil fuel it replaces' (Rosenfeld *et al* 2018). While hardly qualifying as a future net-zero approach, the EISA was an early effort to reduce anthropogenic GHG emissions. Biofuels and bioenergy are modern manifestations of biomass energy and, like traditional biomass, their designation as carbon-neutral renewable energy sources is contentious (USEPA 2025).

The items listed in table 1.4 are energy sources that were harnessed through human ingenuity to generate power for countless practical applications. The underlying science on which many of the developed energy converters are based was generally discovered many years before such devices were shown to be technically and economically viable. For example, the scientific theory of nuclear fusion, i.e. the potential 'sun-in-a-box' method of producing energy was understood almost a century ago (Barbarino 2023). Globally, there are now over 100 experimental fusion devices, however, while some involved scientists, engineers and investors are confident that grid-electricity from nuclear fusion power plants will be available by 2030, such hopes are not generally shared (Schwägerl 2023, Sutter 2024, World Nuclear Association 2022). Nevertheless, there is optimism that towards the end of this century, or at least by the beginning of the next, nuclear fusion will become an established energy source (Sutter 2024). If nuclear fusion ever comes to fruition it would solve all the global energy and environmental issues as a renewable source, but not in time to meet the Paris Agreement targets or the net-zero scenario. However, while the energy density of nuclear fusion is at least 6×10^6 times that of natural gas, this astounding future potential of nuclear fusion, according to protagonists, cannot justify any lessening of the current focus on existing renewables (Schwägerl 2023, Stanford Energy 2023).

[8] Biofuels produced from clean renewable resources.

Until the unexpected discovery of naturally occurring 'white' hydrogen just over a decade ago, hydrogen had to be produced artificially from other sources such as natural gas or the electrolysis of water. Nevertheless, the use of hydrogen in power generation devices dates back to the early nineteenth century, but the methods used to produce the hydrogen emitted GHGs. It is only quite recently, as the cost of electrolysis was considerably reduced, that a renewable form of hydrogen, 'green' hydrogen, became available (Cordonnier and Saygin 2022, Norbeck *et al* 1996, Symons 2023). The colour coding of hydrogen to reflect the method of its production has become commonplace, but while the so-called 'rainbow' system is expanding, there is still no universally accepted code for matching the chosen colour with the specifics of the production technique, although there are some commonalities between the various codes. The colour system helps potential hydrogen users and producers to identify the types of hydrogen they may wish to pursue in their contributions to climate change mitigation. These provisions are the response to calls for an increased use of appropriate hydrogen products in transportation and heavier industry activities (IEA 2023b). Of the various published explanations of the hydrogen colour systems, the Mitsubishi colour system, in terms of completeness and clarity, in the author's opinion, has much to commend it. A version is illustrated in table 1.5 (H2 Bulletin 2024, National Grid 2023b, Willige 2022).

Ammonia, NH_3, is a naturally occurring chemical but can be produced, mainly by the German invented Harber–Bosch process, for widespread use in refrigeration and in the production of fertilizers, explosives, textiles, household cleaners, and pharmaceuticals (The Royal Society 2020). However, while fertilizers play a crucially important role in the global food supply, it is a toxic substance identified as an air pollutant in many countries such as the United States and the United Kingdom (APIS 2016, European Commission 2020, USDA Agricultural Air Quality Task Force 2014). Why then is it being considered as a renewable fuel source, especially for transportation? Ammonia can be burned or used as a zero-carbon fuel in internal combustion engines and fuel cells as well as a hydrogen carrier and a thermal storage medium (IRENA and AEA 2022). In these cases, the preferred type of ammonia would be 'green' ammonia, produced from green hydrogen, and would be both renewable and sustainable (The Royal Society 2020). The ammonia industry has largely adopted a very similar type of colour classification system as used for describing hydrogen production, as provided in table 1.5 (Tullo 2021). Grey or brown ammonia is the most common production method by which the steam reformation of methane or split natural gas yields ammonia and CO_2. If the generated CO_2 is captured, then the product is known as blue ammonia and, while not quite a zero-carbon source, it is considered to be low-carbon.

The production of conventional ammonia uses long-established processes. One of which, the conversion of atmospheric nitrogen to ammonia to a create a fertilizer, has played a dominant role in increasing agricultural yield, i.e. food, for almost a century (Penuelas *et al* 2023). A similar crucial role for ammonia in the production of zero- or low-carbon renewable fuels, especially for use in the transportation sector, has been identified in the scientific literature (Adeli *et al* 2023, David *et al* 2024). However,

Table 1.5. Hydrogen production method by colour.

Colour description	Feedstock	Production method—comments
Grey/grey	Natural gas	Catalytic chemical reaction between high temperature steam and methane from natural gas. CO_2 emissions generated.
Turquoise	Natural gas	Thermal splitting of natural gas methane into hydrogen and solid 'carbon black'. No CO_2 emissions.
Blue	Natural gas	Grey hydrogen plus CO_2 capture.
Green	Water	Electrolysis to split water into hydrogen and oxygen using renewable electricity.
Purple	Water	Thermochemical cycle using heat and chemical reactions coupled with electricity from nuclear energy.
Pink	Water	Electrolysis used to split water into hydrogen and oxygen using nuclear energy electricity.
Red	Water	Thermochemical cycle coupled with nuclear energy from a high temperature gas-cooled reactor.
Yellow	Water	Electrolytic method using electricity from solar power.
Gold	Hydrocarbons	Biological processes, i.e. fermenting microbes, to convert leftover hydrocarbons found in depleted oil wells to hydrogen plus CO_2 capture.
Brown	Coal	Gasification using brown coal. High CO_2 emissions.
Black	Coal	Gasification using black coal. High CO_2 emissions.
White	Natural hydrogen	Naturally occurring hydrogen found in underground deposits.

this eventuality would seem to be a futile endeavour as the internal combustion engines (ICEs) used extensively in all forms of transportation are to be banned in many countries over the next decade, at least in any new vehicles (Averna 2023). The implicit intentions of such legal sanctions are to encourage and enforce the manufacture of on-road electrical vehicles (EV) and by doing so to decarbonize the sector as part of the undertakings to combat climate change. In some jurisdictions almost unobtainable stringent emission regulations have been introduced to essentially ban small off-road vehicles, machinery, and equipment (Anandan 2023).

Indeed, governments are determining ways to replace or eliminate combustion engines in other situations such as 'any mobile machine, transportable equipment or vehicle not intended for the transport of goods or passengers on the road', referred to as non-road mobile machinery (NRMM) and so-called 'hard to electrify' situations, such as in the production of iron, steel, certain chemical and constructional materials (Hardisty *et al* 2023, Shaw *et al* 2024). It would then appear that most governments have decided that ICEs will no longer play a part in any form of energy conversion or generation by the 2040s. The close association of fossil fuels with both ICEs and external combustion engines by the general public is understandable but, arguably, more policymakers need to be aware that such engines are capable of operating with a variety of different fuels. Of the many examples, green ammonia and green hydrogen, as previously mentioned, are prime renewable fuel candidates along with blended e-fuels, such as ammonia–hydrogen. Industry, especially those involved in power generation for maritime transportation, such as the German company MAN and the Finnish corporation Wärtsilä, have developed ammonia fuelled engines and there are active research programs in many universities investigating the continued use of these fuels in combustion engines (Adeli *et al* 2023, David *et al* 2024, Lindstrand 2024, MAN Energy Solutions 2021).

Despite the growing interest in using green ammonia and green hydrogen as next generation combustion engine fuels, there are still many technical and economic challenges to overcome, as is normal with prototypes. But the technology is showing sufficient promise that its adoption, by mid-century, as a renewable energy source can likely be anticipated (Poore 2023, Tornatore *et al* 2022). However, will political environmental dogma ensure the demise of the internal combustion engine because of its current use of fossil fuels and, if so, could it prove to be a classic case of ‘throwing the baby out with the bathwater’ (Colin-Oesterlé 2023, Reitz *et al* 2022)? Perhaps with thoughtful reflection, particularly in the EU, the pathways to achieve ‘carbon neutrality’ in the energy and transportation sectors, will include ICEs (Berni *et al* 2024). This could lead to the idiom attributed to Mark Twain, ‘the report of my death was an exaggeration’, being more appropriate (Dictionary.com 2024).

1.2.3 Expanding menu of renewable sources

Whatever the rationales of the policy decisions to transition energy mixes to renewable sources and to focus on electrification, the momentum has increased to develop alternative sources and technologies in addition to those identified in the 1992 UNFCCC framework (United Nations 1992). Arguably, the unconstrained notion of a wholesale global transition to energy sources that are forever renewable and sustainable has become tempered to a more pragmatic vision enabling the possible inclusion of low carbon resources. The aim is still to achieve carbon neutrality, by CO_2 sequestration in some cases and in others by the source being inherently low carbon but not necessarily renewable, as defined by even the most doctrinaire environmental regulations. While this eventuality is providing an ever-growing number of choices for the replacement, or at least a radical reduction, in the use of fossil fuel resources, maybe this is detrimental to the desired timeline of the global energy transition? The core issue is the affordability of developing all these sources.

It has been estimated that over US$8 trillion will need to be invested in renewable energy deployment annually between 2023 and 2030 if global warming is to be kept below 2 °C by 2050, an increase from the annual US$6.9 trillion forecast in 2018 (Annex 2023, OECD 2018). However, there is a growing body of published work supporting the premise that not taking effective climate change mitigation measures to meet the Paris Agreement targets will further exacerbate global warming damage, causing greater socio-economic costs than have been suffered since the start of the millennium (Philip *et al* 2022, Waidelich *et al* 2024). For example, using a methodology known as 'extreme event attribution' (EEA) it has been possible to establish a causal link between anthropogenic GHG emissions and extreme weather events and quantify the relative contributions to such events from human activities and natural occurrences (Newman and Noy 2023). In addition to the frequency of extreme weather occurrences and their severity, other events such as wildfires, human and animal habitat loss, sea-level rises, the impact of changing precipitation patterns on agriculture and so on, all of which have been linked to global warming, could lead to the world-wide cost of damage to infrastructure, property, food production, and human health being between US$1.7 trillion to US$3.1 trillion annually by 2050 (Bennett 2023). In effect, the financial costs of not developing renewables could be comparable, or even greater, than those associated with renewable energy development.

However, forecasting the costs and benefits associated with climate change mitigation is a major challenge, relying heavily on the meticulousness of the selected economic model (Black 2022). Like all models the usual vagaries will be encountered, such as the acquisition of a sufficiency of verifiable data and the inclusion of all the pertinent factors and their contributions, and even then perhaps the best that can be expected is a perspective rather than a precise answer (Gillingham 2019, Gillingham and Stock 2018). Consequently, if exactness is desired then there are 'inherent uncertainties when predicting the likely impacts of climate change' and models should include some form of climate risk assessment and future economic growth impacts (Kotz *et al* 2024, Mullan and Ranger 2022). The exactness is necessary as signatories to the Paris Agreement have committed to align their finances with the Agreement's climate mitigation and resilience targets. Countries and states should then be able to identify where best to invest public and private funds in combating climate change, including the development of renewable energy sources. If such unambiguous models exist, they have yet to be reported in public domain publications.

What is clear is that, despite the trillions of dollars already invested in renewable energy, mainly solar and wind projects since 2013, these two sources only accounted for a 5.3% share of primary energy consumption in 2022 (Annex 2023, Ritchie and Rosado 2024b). This could appear to be a small energy return for a large financial investment but, if only electricity production is considered, the solar and wind global share rises to over 14% (Ritchie and Rosado 2024a). This difference highlights the need to be careful when quoting energy source statistics because the global energy mix is different from the global electricity mix. It can also indicate that particular sources of energy may be suited to specific applications. While global energy data provide an overall picture of which primary sources are used and what they are used for, at the individual national and local levels there are noticeable variances when compared with worldwide mixes. As the conversion to electricity of all primary energy sources is being increasingly nurtured, there is particular interest in how this can be or is being achieved globally, nationally, and locally as fossil-fuels are replaced. Current and likely future energy mixes are examined in section 1.3 together with the possible ramifications of forsaking traditional fossil-fuel sources in a timely manner.

1.3 Global energy mixes

With such a large selection of primary energy sources now available, albeit at different stages of technically maturity and commercial viability, perhaps it is not surprising that the type of energy consumed has changed in the first two decades or so of this century compared to the previous two decades. However, it is not just the types of energy use that have changed but also the global amounts of total energy consumption. Thus, although, as shown in figure 1.1, the world in general has increased the share of renewables in its primary energy consumption, some countries significantly more than others, the bulk of the additional energy demand between 1980 and 2022, some 74%, has been met by fossil fuel sources with just below 14% coming from renewables, not including hydropower energy (Ritchie and Rosado 2024b). The increase in electricity production, over a similar period, 1985–2023, has been even greater than the rise in overall energy demand but once again fossil fuel sources have accounted for almost 60% of the larger production levels, however, in this case, non-hydropower renewables have made more impact in the mix with an approximately 24% share (Ritchie and Rosado 2024a).

One of the key concerns about fossil fuels prior to the rising awareness of climate change issues was that the energy needs of the increasing global population could not be met by fossil fuel sources, see section 1.3.2.1. So, although renewables have contributed measurably to both the energy and electricity mixes' increased energy amounts, the largest contributor has been fossil fuels. However, some countries and regions consider hydropower as a renewable energy source, or at least a low-carbon source. If hydropower energy production is added to that of narrowly defined renewables, then their combined share of the additional energy demand rises from 14% to almost 21%, and if the other main low-carbon energy source, nuclear power, is also included then the share further increases to 26%. In 2023 the demand for electricity was over three times higher than in 1985, and 40% of the increased demand was met by a combination of renewables, hydropower, and nuclear power. Nevertheless, the amount of electricity produced from fossil-fuel sources increased by a factor of over 2.8 between 1985 and 2023. The increased use of fossil fuels may partly provide an explanation for why, despite the many UN environmental treaties (more than 20 since the 1992 Rio Convention) the recorded CO_2 levels have continued to rise, see figure 1.8 (Dunlea 2023).

Exactly how nations are going to replace their utilization of fossil fuels with alternatives, such as renewables (the most advocated by those seeking a complete solution to climate change mitigation), or a combination of renewable and low-carbon sources (perhaps the more pragmatic approach) is likely to depend on a country's natural resources, their provisions for importing energy, their ability to harvest global energy sources, such as solar and wind, and, especially, government policies. The latter have played increasingly significant roles since the advent of hydropower in the early twentieth century, the development of nuclear power in the late 1960s, and the rise of renewables towards the end of the first decade of the twenty-first century. One consequence of the expanding list of options of energy choices is that presently there is no global norm for how nations acquire and produce

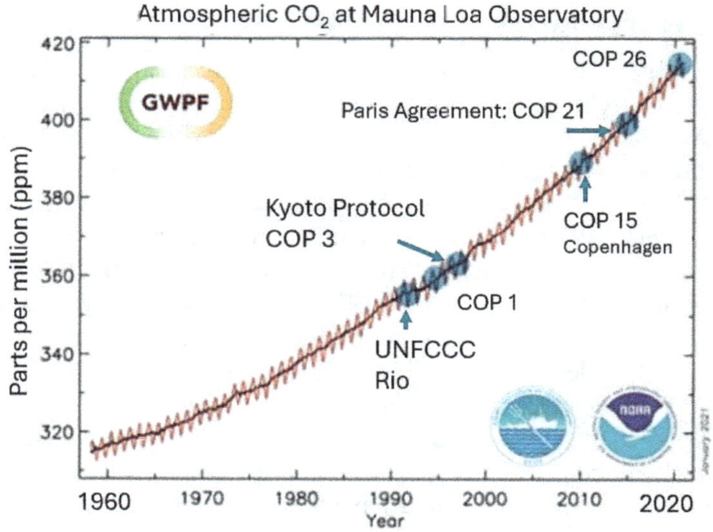

Figure 1.8. CO_2 trends and treaty impacts. (Modified from (Dunlea 2023). Image credit: The Global Warming Policy Foundation.)

their required primary energy and electricity. However, there are some commonalities in terms of energy mixes, particularly when electricity generation per capita is considered, as shown by the exemplars in the next section.

1.3.1 Present energy mixes

In discussing future energy mixes it would be useful to have knowledge of past and present mixes from respected and trustworthy sources. Fortunately, there is a wealth of such energy data as, for over 70 years, an annual statistical review of global energy was published, mainly by the British multinational company BP but, since 2023, this considerable undertaking was passed to the London, UK, based Energy Institute and its partners, with over 40 third-party contributors such as IRENA, United States Energy Information Administration (USEIA), the International Energy Agency (IEA), the Organization for Economic Co-operation and Development (OECD), BloombergNEF (BNEF), the UN, and Wood Mackenzie (BP 2022, Energy Institute 2023c, Rühl 2012). Since 2014, organizations such as the Oxford based Our World in Data have processed and analysed this energy data, presenting it on an open access website, and the energy think-tank Ember have now published five annual reports analysing global electricity generation (Roser 2019, Wiatros-Motyka *et al* 2024). Using data from these sources, especially those processed by Our World in Data, it can be readily appreciated that there are major differences between the amount of primary energy consumed and that of electricity generated, which are frequently ignored in media and in general climate rhetoric. This cannot be helpful in the efforts to convince the general public, especially the voters in democratic countries, of the needs to fully embrace the age of green electrification. Exemplars of this situation are given in table 1.6, highlighting the

Table 1.6. Present (2022) energy mixes—exemplars (Ritchie and Rosado 2024a, 2024b).

Country	Primary energy consumption kWh capita^{-1}	Renewable share of primary energy %	Electricity generation kWh capita^{-1}	Renewable share of electricity generation %	Electricity generation to primary energy ratio
Iceland	165 243	84	53 285	100	32.2
Norway	96 926	71.6	28 056	98.5	28.9
Brazil	17 300	48.7	3295	89	19.1
New Zealand	44 939	43.0	8516	87.3	19.0
Sweden	59 927	53.3	15 676	69	26.2
UK	30 098	19.3	4333	46	14.4
China	31 051	16.0	6635	31	21.4
USA	78 754	11.3	12 497	23	15.9
India	7143	10.4	1377	19.5	19.3
World	21 039	14.2	3664	30	17.4

difference in per capita energy consumption and electrical generation and the renewables' share of both. The chosen examples are countries with human populations, large and small, generally higher than average global per capita primary energy consumption rates, and many who generate electricity from renewables at a rate of twice the average world rate.

The heterogeneity of the national exemplars provides some insights into the challenges facing the adoption of renewable energy sources. Yet, overall, there are reasonable commonalities in the ratio of electricity generation compared to primary energy consumption, which is generally less than 1/5; Iceland, Norway, and Sweden are the exceptions but, even then, less than a third of the primary energy is used for electricity production. Arguably, the global share of primary energy consumption from renewable sources is discouragingly low. The situation improves for renewables in terms of electricity generation, but the high shares illustrated for Brazil and Sweden, in particular, and to a lesser extent the UK, are not proportionally reflected in primary energy consumption. This is not the case with Iceland and Norway whose primary energy consumption dominates electricity generation. However, as shown in figure 1.3, a United States example, about 65% of the primary energy supply used to generate electricity is 'lost' during the conversion processes. This underlines the issues of not only energy conversion efficiencies in general, but also which primary energy source can provide the highest in such efficiencies. These matters are discussed further in section 1.4.

As previously mentioned, fossil fuels have dominated energy mixes throughout the twentieth and, so far, the twenty-first century. Nevertheless, hydropower, as an energy source, has had a considerable impact on national mixes, particularly in terms of electricity generation during this period. Brazil and Sweden's present high shares of non-fossil fuel sources, especially in electricity generation, are the result of their policies to utilize hydropower. In the exemplars shown in table 1.7, the high

Table 1.7. Share of fossil, hydro, wind and solar energy sources; per capita exemplars.

	Fossil fuels %		Hydro %		Wind %		Solar %		Total population 2023
Country	PEC[a]	EEG[b]	PEC	EEG	PEC	EEG	PEC	EEG	Millions
Iceland	15.9	0.0	60	70.2	<0.1	<0.1	0.0	0.0	0.38
Norway	28.6	1.0	63.7	88.5	7.4	9.7	0.1	0.1	5.5
New Zealand	56.9	12.7	29.4	58.5	3.2	7.8	0.3	0.5	5.2
Sweden	26.3	1.8	29.7	39.7	13.9	20.9	1.1	1.5	10.6
Brazil	50.3	9.3	32.0	60.4	6.1	13.4	2.3	7.3	216
UK	74.7	40.0	0.7	1.8	10.4	28.1	1.8	4.5	68
USA	81.1	23.0	2.6	5.5	4.4	10.0	2.1	5.6	340
China	81.6	65.0	7.7	13.2	4.5	9.4	2.5	6.2	1426
India	88.5	78 0	4.5	7.6	1.8	4.1	2.5	5.7	1429
World	81.8	61 0	6.8	14.3	3.3	7.8	2.1	5.5	8045

Key: [a]Primary energy consumption; [b]electricity generation. Author analysis. Rounded up to 1 decimal place for energy sources. Data sources: (Ritchie and Rosado 2024a, 2024b, Worldometers 2023).

population countries of India and China, together with the world in general, use higher levels of hydropower for electricity generation relative to solar and wind, although the combined contributions of wind and solar are comparable and often greater. In the UK, fossil fuels still dominate, as in so many other countries, but wind now accounts for over 28% of electricity generation, once again driven by government policies in response to climate change and sustainability concerns (GOV.UK 2020).

Using the same databases, nuclear energy contributions globally are found to be just 4% of primary energy consumption, but over 9% of electricity generation. Nevertheless, in both cases, these are higher shares than wind or solar. As a baseload provider of electricity, and with no GHG emissions produced during the generation process, together with ultralow lifecycle emissions during powerplant construction, it should seem obvious that nuclear energy could replace fossil fuels in such applications (Addo *et al* 2023, Fisher and Liou 2021). However, nuclear energy was not identified as a renewable, sustainable, or environmentally sound energy source by the UNFCCC's Agenda 21 and as such was not seen as part of the solution to mitigate climate change. A status that has largely remained unchanged, somewhat inexplicably.

It will be noted from tables 1.6 and 1.7 that countries with high shares of renewables in their 'electricity mix' invariably achieve this share by the use of hydropower and, overall, hydropower is the favoured renewable for electricity generation globally, but whether all hydropower is a renewable source is contestable. In the UK hydropower plays a comparatively negligible role in primary energy consumption or renewable electricity generation with the wind power share overshadowing all other renewables, but not fossil fuel sources. Iceland is perhaps a unique case within the context of very high per capita rates or primary energy consumption and electricity generation. Here, the use of wind power is miniscule and solar power utilization is non-existent. However, in additional to high levels of hydropower and a modest contribution from fossil fuels for primary energy consumption, almost 30% of electricity generation and 25% of primary energy consumption comes from 'other renewables', mainly geothermal energy.

Another country that has a large proportion of geothermal energy (26.7%) and hydropower in its total energy supply is New Zealand which also uses almost the whole range of renewable sources in its energy mix, including bioenergy and waste-heat, as shown in figure 1.9, plus a very modest amount of solar (0.1%) (IEA 2023g). Nevertheless, fossil fuels still play a significant role in New Zealand's energy supply although, since 2023, all refined petroleum products are imported (Ministry of Business, Innovation and Employment, Hīkina Whakatutuki 2024).

As shown in table 1.7, there appears to be some correlation between a country's population size and its per capita use of fossil fuel energy, in that the larger the country the larger the fossil share. In contrast, the smaller population countries tend to use a bigger share of renewables in their electricity generation. An exception in the chosen examples being New Zealand, which uses twice as much fossil fuel energy per capita as the comparably sized country, Norway, in terms of population and population density. Nevertheless, New Zealand, whose land area is comparable to the UK's but whose population density is only 7% of the UK's, generates almost twice as much total electricity per capita and from almost double the renewable share, further highlighting the lack of similarity in countries' energy mixes. Nonetheless, it would seem that the popular renewable sources are wind, solar and hydropower, but in some cases such an observation could be considered specious as it depends how much bioenergy (traditional biomass and modern biofuels) is considered a renewable source of energy production. For example, in the EU, the regulations with some codicils post-2020, bioenergy and the use of 'woody' biomass are accepted as renewable, such that in 2021 it was claimed that 59% of renewable energy consumption was from bioenergy (Camia *et al* 2021, European Commission: Energy, Climate Change, Environment 2023).

Given the differing opinions and definitions about what constitutes 'carbon neutrality' and low-carbon sources, in essence how much primary energy

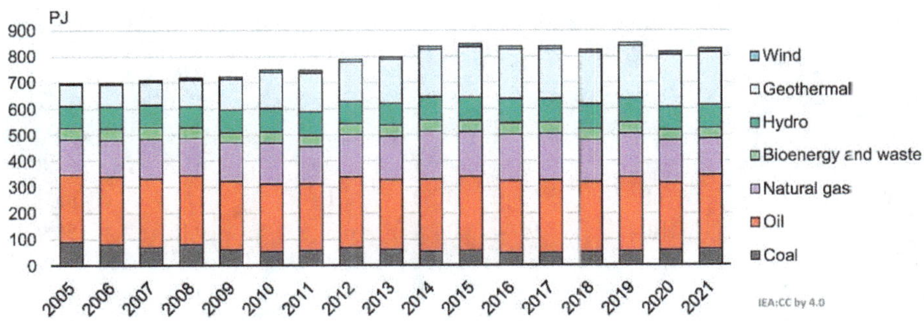

Figure 1.9. Total energy supply by source in New Zealand, 2005–21. (Reproduced from (IEA 2023g). CC BY 4.0.)

consumption from traditional biomass can be considered renewable, such data paradoxes are likely inevitable. In 2015, for instance, traditional biomass sources accounted for almost 7% of the global primary energy consumption, an order of magnitude more than from modern biofuels. Since that time no recent data regarding traditional biomass have been available (Ritchie and Rosado 2024b). Exactly which definitions to apply to the collected data can therefore be contentious. This is not entirely helpful in attempting to identify which energy sources could be part of future mixes and which should be. The latest (72nd) edition of the *Statistical Review of World Energy* provides very detailed information on the EU's energy mix, including modern biofuels, which is shown in table 1.8, as processed by Our World in Data, together with some European country examples (Energy Institute 2023a, Ritchie and Rosado 2024a).

As can be observed in table 1.8, the two main energy source providers for generating electricity are fossil fuels and nuclear power. Of the renewable sources, wind power is the largest contributor followed by hydropower. However, a smorgasbord of mixes is perceptible with the exemplars which represent over 70% of the total EU population, including five of the most populous countries. Electricity from fossil fuel dominates, especially in Poland and Italy, while nuclear power has the largest share in both France and Belgium. Wind power is the leading energy provider in Denmark, not just among renewables, while hydropower supplies almost 60% of Austria's electricity generation. While, as with the global energy scene, there appears to be no obvious common approach to how the energy transition away from fossil fuels can best be achieved, a close inspection indicates that many countries, in a financial position to do so, have developed their own natural resources. For example, a number of coastal countries who experience inherently high wind speeds and thus the potential for producing high wind power densities have developed such facilities, especially offshore.

Table 1.8. Energy sources in the EU for electricity generation—exemplars.

	% Source of electricity generation per capita						
Country	Fossil fuel	Hydro	Wind	Solar	Bioenergy	Nuclear	Population Millions
Austria	15.5	59.4	12.0	7.7	5.3	0.0	9.0
Belgium	26.4	0.5	18.5	9.3	4.8	40.6	11.7
Denmark	12.4	<0.05	57.7	9.2	20.6	0.0	5.9
France	8.4	10.3	9.5	4.5	1.9	65.3	64.8
Germany	26.8	3.8	27.2	12.2	9.1	1.7	85.3
Italy	56.3	14.4	8.9	11.8	6.3	0.0	58.9
Poland	72.9	1.4	13.7	7.2	4.8	0.0	36.7
Spain	28.9	7.4	23.8	16.7	2.2	21.1	47.5
EU	39.2	9.9	15.1	7.6	5.9	21.9	448

1.3.2 Further realities and knowledge resources to assist policymakers

But how do policymakers know about the quality and quantity of natural renewable resources? For wind, the answer is the increasingly sophisticated, and constantly updated, Global Wind Atlas, which was developed and launched in 2015 by the Technical University of Denmark (DUT) with support from IRENA and the Danish Government, and continues with funding support from organizations such as the World Bank Group (DUT Wind Energy 2024). A similar atlas and database are also available for solar power, the Global Solar Atlas, and IRENA has been developing its Global Atlas for Renewable Energy since 2012 (IRENA 2024, The World Bank Group 2024). How much attention the policymakers pay to these impressive knowledge sources in developing their policies remains to be seen. Most certainly the countries that have developed wind power appear to have installed their facilities in worthy locations but enthusiasm for solar power development in some countries can seem misplaced based on the quality of the radiance received and for how long. The locations for hydro powerplants can be artificially created, albeit using the natural features of the surrounding land, e.g. valleys, advantageous precipitation cycles, and associated water flow rates. The number of water reservoirs, natural and human-made, far outweigh those actually used for hydropower. In the United States, for example, only 2300 of over 90 000 water dams are involved in electricity generation or energy storage (US Department of Energy n.d.).

In 1992, 99.5% of primary energy consumption was sourced from fossil fuels, hydropower, nuclear energy, and traditional biomass. Three decades later this percentage has fallen to 92.6%, although the actual amount of energy consumption for these sources had risen by 55%. Including all sources, energy consumption has risen by 65% since 1992, so the greater part of the increase has not been met by renewables sources if hydropower and traditional biomass are ignored. However, if these two sources are included in the historical global mixes, then the presently somewhat unpopular sources, nuclear and fossil fuels, have only provided just over 37% of the increased per capita primary energy consumption since 1992. Nevertheless, even this more modest amount is hardly synonymous with the elimination of these 'out-of-favour' sources, particularly fossil fuels, by 2030 or even 2050. Indeed, a recent analysis of government policies[9], by the widely trusted Stockholm Environmental Institute (SEI), of the 19 major fossil fuel producers, who are also signatories to the Paris Agreement, found that these producers plan to double their production amounts by 2030 and that their share of global fossil fuel production, currently at 80%, will fall by a modest 5% by 2050 (SEI *et al* 2023).

It is worth remembering that such projections are extrapolations based on mathematical models. These can be useful in formulating future policies and implementation plans but they can also be used to draw attention to extant policies that, with hindsight, are deemed to be deficient or contradictory or both. Whatever the laudable intentions of the universal undertakings to simultaneously achieve sustainable global development, mitigation of global warming, and decarbonization

[9] Termed 'stated policies scenarios' or STEPS by organizations such as IEA.

of energy mixes largely by the elimination of fossil fuels and their replacement by renewable sources, the current policies are underperforming (World Economic Forum 2023). Arguably, if SEI's projections are accurate then those countries reaffirming their Paris Agreement commitments will need to change their energy transition implementation plans or acknowledge, for example, that there appears little likelihood that the Paris Agreement/IPCC global warming targets can be achieved in accordance with their ambitious global timetable.

However, such august organizations as the USEIA and the IEA appear more confident about future increases in the capacity of renewables, especially solar and wind power, in electricity generation (WEF 2023a). The USEIA are expecting that more than half of new electricity generating capacity in 2023 will be from solar sources. The IEA are forecasting that renewables plus contributions from nuclear will meet 90% of electricity demand from 2023–2025 and the share of renewables will increase to 35% by 2025 such that global rises in CO_2 emissions will not be significant and, indeed, will plateau by 2025 (IEA 2023c). The forecasts for the CO_2 maybe optimistic as the power sector emissions reached an all-time high in 2023, figure 1.10, according to a recent report (Wiatros-Motyka *et al* 2024). Nevertheless, the findings from this report are in harmony with the USEIA and IEA prognoses. They also stress that the growth in 'clean' electricity generation, particularly from solar and wind sources post-2013, has helped reduced the rate of growth of fossil fuel usage in the power generation sector by almost a third, thus avoiding many gigatons of CO_2 emissions.

The available data reveal that renewables have had a larger impact, i.e. percentage share, in the production of electricity than in overall primary energy consumption, accounting for 30% of the global electricity mix, and this is one of the reasons for the optimism surrounding the continued increase in their use. However, to put this in perspective, it needs to be emphasized that electricity accounts for only

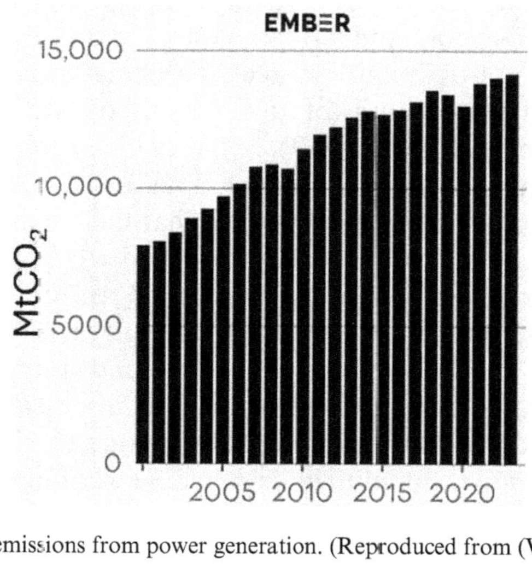

Figure 1.10. Historic CO_2 emissions from power generation. (Reproduced from (Wiatros-Motyka *et al* 2024). CC BY-SA 4.0.)

19%–20% of global primary energy consumption. There is a long way to go before renewables dominate the global energy mixes, but increasing the share of renewables in electricity production and consumption seems a sound strategy. Moreover, as previously discussed in section 1.1.1, the UNFCCC have set a target of 63% by 2030 for the renewable share of electricity generation. This is twice the share enjoyed today and to accomplish it represents an exceedingly difficult global task. It would no doubt be helpful to know the end-users of generated electricity as this could serve to identify applications on which to focus more attention. Governments and independent agencies categorize energy users into sectors such as residential, commercial, transportation, and industrial and may include electrical power (USEIA 2023b). The definitions for each category does vary depending on the reporting organization, although there are close similarities. Nevertheless, when interpreting sector data, it is necessary to be aware that care needs to be taken, since electrical power may be considered to be an energy-user sector, but not an end-user sector (USEIA 2024a). Fortunately, well-known energy reporting agencies usually provide detailed methodology guidelines explaining their definitions (Energy Institute 2023a). It is advisable to review these guidelines especially when comparing the same type of data, e.g. primary energy consumption, from the various sources.

As with all other energy data there can be significant differences in electricity consumption between countries, yet again demonstrating the global lack of homogeneity. Therefore, using world averages as benchmarks can be problematic. For example, in the United States in 2022 just 0.2% of the country's annual electricity consumption was associated with the transportation sector, whereas globally that share was 1.4%, down from 1.8% pre-COVID (IEA 2021, 2022b). Similarly, while the industrial sector dominated global electricity consumption with a 42% share, in the United States this sector share was 26%, residential usage at just over 38% being the largest consumption sector. Recent media headlines have trumpeted the fact that in the United States for the first two months of 2024 electricity consumption involving EVs has 'reached new highs' by consuming 50% more electricity than in a similar period in 2023 (Maguire 2024). However, it will only change the transportation sector annual proportion of electricity consumption by an infinitesimal amount. Moreover, the sector shares referred to above are for the end-use of electrical power and do not include the 'raw' primary energy consumption required to generate the electricity. Why is this important?

First of all, since energy-use definitions are not universal the inconsistencies may result in different data reporting. An accepted set of interconnected definitions of the form suggestion by Ritchie, shown as figure 1.11, could make data interpretation more straightforward. This approach could be used to identify opportunities for improving conversion efficiencies at various stages along the energy-chain pathways from raw resource to useful energy (Ritchie 2022b). Although not explicitly using the Ritchie technique, the US energy flow Sankey plot from the LLNL shown earlier in figure 1.3, is a good example of the value of energy pathway data. The largest user of 'raw' primary energy in the United States, for example, is electricity generation as shown on the LLNL figure and as reported by the USEIA (2024a). Once the primary energy source has been converted into transportable forms it will be used by

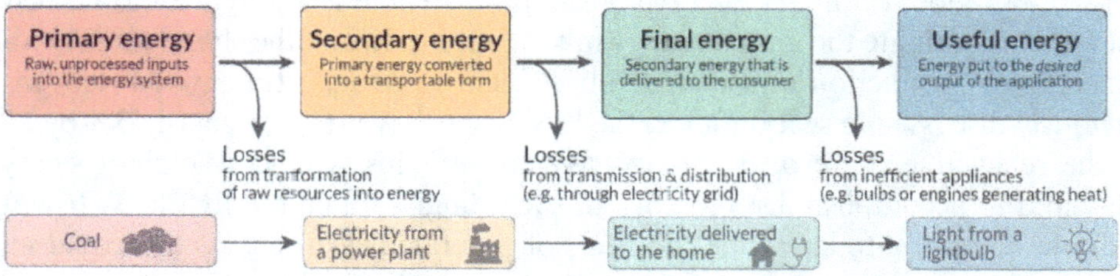

Figure 1.11. Ritchie's energy path definition. (Reproduced from (Ritchie 2022b). CC BY 4.0.)

consumers in the industrial, transportation, commercial and residential sectors, but the greater part of the processed primary energy, Ritchie's 'secondary energy', will be used for electricity production. Each sector, including that of power generation, will have its own energy mix and hence have different carbon footprints. Second, it is also important that the GHG emissions from the defined sectors are known to help policymakers determine where best to focus their mitigation strategies.

Determining global and national GHG emissions and their sources is a complex process and relies not only on high quality data gathering and analysis but also on the availability of high-resolution data. To date it is normally only from the world's major economies and high-income countries that completely *bona fide* emissions data are obtainable (Ritchie *et al* 2023a, USEPA 2024f). Nevertheless, the governance systems of all countries arguably share a common mantra of 'energy, environment and economy'. As a consequence, countries, who are able, collect GHG data on all their 'economic' sectors, not just the broadbrush categories of industry, transportation, commercial, and residential. A country's GHG inventory may also include the emissions generated in trade activities and, more problematically, emissions from marine and aviation transportation. The issue with the latter is to whom the emissions should be attributed—the country of origin, the destination country, or areas enroute? No international agreement exists on how such emissions should be allocated. In terms of policymaking, it appears that many countries do not have sufficiently detailed GHG emissions data to guide them into what economic sectors they should focus their attention on in their climate mitigation efforts. Invariably, the emissions data of the greatest interest is CO_2 as this colourless and odourless gas, at low concentrations, is considered to be the root cause of global warming and climate change, albeit, that it constitutes only about 0.04% of the Earth's atmosphere, although its share has been rising over the past two centuries.

However, there are countries where detailed GHG emissions data, both historic and current, is available, the most notable being the United States. This is not too surprising as the per capita consumption based CO_2 emissions from the United States are the second highest in the world, behind only Saudi Arabia, at 16.5 tonnes, which is 350% more than the global rate (Ritchie *et al* 2023a). In 2022, the shares of GHG emissions from the defined economic sectors in the United States were from three main sources, transportation (28%), electricity generation (25%), and industry (23%) (USEPA 2024c). If the GHGs of electricity generation are allocated to their end-use, the industrial share rises to 30%, slightly less than the combined commercial–residential sector, which increases from 13% to 31% because of the emissions from heating, cooling, ventilation, and other electrical appliances. Ironically, as the transportation sector uses so little of the United States' electricity generation, its shares of GHG emissions hardly changes, i.e. by just 1%, making it slightly above the global average for the sector. Does all this mean that the transportation sector is not the *bête noire* of GHG emissions? The answer is no, at least for the United States, which becomes clearer when it is considered that, overall, while 80% of United States' GHGs were in the form of CO_2, in the transportation sector this share rises to almost 98%. Hence, while for mitigation purposes all human generated GHG emissions need to be eliminated, those from the transportation sector are the most obvious target for urgent action, especially the emissions from on-road vehicles (European Environmental Agency 2023, USEPA 2024b).

Thus, although the present situation regarding energy mixes and the possible roles of renewable sources for individual nations can be bewildering, there is a growing sufficiency of data and knowledge tools available to policymakers to make sentient choices for the development of their future energy mixes to address climate change. The IEA, in their 2023 World Energy Outlook report, analysed the energy transition performance of selected individual countries and major geographical regions encompassing groups of countries and projected their energy mixes to 2050 using various pathways, including the already stated policies approach, STEPS[10], along with the pros and cons of those pathways (IEA 2023e). In all cases, the projections were for the global use of fossil fuel energy sources to peak in the mid-2020s, but oil and natural gas consumption would still be over 95% of the peak by 2050, while coal consumption would be about two-thirds of its peak rate. These projections can be expected to be unpalatable to those demanding or seeking fossil fuel utilization to cease without delay, and to the advocates for limiting global warming to 1.5 °C by 2030. Given the almost perennial concerns about the depletion of fossil fuel reserves, can the IEA's energy mix projections about their continued use be warranted?

1.3.2.1 Dwindling fossil fuels resources?

The overwhelming opinion of policymakers, the scientific community, and the public is that the elimination of the use of fossil fuels is viewed as the most crucial strategy for mitigating climate change. The scale of the measures that need to be taken are illustrated in table 1.7. Some countries who rely on sizeable shares of fossil fuels in their energy mixes, e.g. China, India, and Saudi Arabia, have announced targets for net-zero GHG emissions to be achieved between 2060 and 2070, others such as the United States and Australia have committed to 2050 (Climate Action Tracker 2023). Some countries have made no commitment, such as the Islamic Republic of Iran, while others, Norway being a prime example, have already met is own climate mitigation goals, albeit fossil fuels play a significant part in its economy

[10] See footnote 8.

given that large quantities of oil and gas are exported (International Trade Administration 2024).

By the end of 2023, 145 countries, generating 90% of the global GHG emissions, had announced substantive net-zero targets or were considering doing so (Climate Action Tracker 2023). However, while some countries had no formulated plans to realize their commitments, analyses of those that did found that many were not credible (Rogelj *et al* 2023). Indeed, if the yardstick of success is that the implementation of national plans is compatible with the 1.5 °C Paris Agreement/ IPCC goals then about 80% of the plans are considered to range from 'insufficient' to 'critically insufficient', including those of China, the United States, the Russian Federation, and India (Climate Action Tracker 2024). These four countries consume almost 54% of the global consumption of fossil fuels (Ritchie *et al* 2023b). The main point is that it seems unlikely that fossil fuels usage will cease in the next three or so decades regardless of the timely development of renewable sources, but will the present fossil fuel reserves be adequate?

Before the concerns of adverse climate change were raised about fossil fuel usage, the development of alternative energy sources was advocated because of the fears that the reserves of 'non-replenishable' fossil fuels, especially oil, would be insufficient to meet the increased energy demands of a growing population. Indeed, as oil became part of the global energy mix in the early twentieth century, especially in the United States, there were warnings that oil reserves would be exhausted by the mid-1920s, a timeline which has been periodically modified ever since, with the latest estimates being between 2070 for oil and gas, and 2130 for coal (Nordhauser 1973, Ritchie and Rosado 2024c). Perhaps the most famous of the end-of-oil predictions came from US President Jimmy Carter who, in 1977, stated that if the then current rate of oil consumption continued that all the *proven* reserves of oil in the entire world could be exhausted by the end of the 1980s (Carter 1977). But since the President's prediction was made, oil consumption has continued to rise and, in 2022, oil consumption was almost 52% higher than in 1977 and being used at a rate that could be globally maintained for the next 50 years. So, were all these predictions wrong?

The answer is both yes, and no, because they were based on so-called 'proven reserves'[11] which are *not* a measure of how much oil of all types[12] is present in the Earth's geological crust, but rather an economic assessment as described in a commonly used definition, i.e. '[r]eserves are those quantities of petroleum anticipated to be commercially recoverable by application of development projects to known accumulations from a given date forward under defined conditions' (Society of Petroleum Engineers 2022). Although not a universally generic definition, it could also be applied to other mineral resources, with lithium, cobalt, nickel, uranium, and so on replacing the term petroleum. The examples chosen are some of the critical minerals that will be essential if there is to be an energy transition to renewable

[11] Also known as P90 or proved reserves.

[12] Includes crude oil, shale oil, oil sands, condensates, and natural gas liquids, but excludes oil shales and kerogen extracted in solid form.

sources. As with fossil fuels, some countries have an abundance of these mineral resources, while others do not, so national import–export balances will become critical. For some countries whose natural mineral resources are considerable, e.g. Canada, the exploration, extraction, processing, manufacturing, and recycling of such minerals could be an economic growth game-changer, akin to the situation enjoyed by Norway with its oil and gas resources (Government of Canada 2023).

To operationalize the definition, the amount of a site's ratio of the proven reserves (*R*) divided by its production rate (*P*), i.e. the *R*/*P* ratio, provides an estimate of how many years the site will remain economically productive. At the time of President Carter's statement, the global *R*/*P* ratio for oil was 25–30 years so, while his 'end-of-oil' scenario timeline now seems premature, it was not unduly pessimistic at the time. However, since the Carter pronouncement, the value of the *R*/*P* ratio for oil has continually increased so that by 2022 it was twice its 1980 level. The *R*/*P* ratio can be determined for specific extraction sites, or for countries and regions, as well as on a global basis. For example, in 2022, the *R*/*P* ratio for oil for South and Central America was just over 150 years, whereas it was about 80 years for Middle East countries, and 30 years for North America, whilst the global average was just over 53 years (BP 2022, Energy Institute 2023a). The average *R*/*P* values for the other two fossil fuels were 139 years for coal, and 49 for gas. However, even among the top five countries in each of the three categories who have the largest global shares of proven reserves, there are significant disparities in the corresponding *R*/*P* measures, as shown in table 1.9 (Energy Institute 2023b). Moreover, these countries collectively account for the majority of reserves, 76% in the case of coal, 62% of oil, and 64% of gas.

The proven reserves of coal and oil have increased over the past five years, by 22% and 5.5%, respectively, while those for natural gas have diminished by 7%. However, there are known global resources of fossil fuels which could, in the future, be included in the inventory of proven reserves and 'undiscovered' resources, see

Table 1.9. Proven fossil reserves global shares and *R*/*P*—top five countries (Energy Institute 2023b).

Fossil fuels: coal, oil and gas sources—proven reserves								
Coal		Top five countries	Oil		Top five countries	Gas		Top five countries
%	*R*/*P*		%	*R*/*P*		%	*R*/*P*	
23.2	>500	USA	17.5	>500	Venezuela	19.9	58.6	Russian Federation
15.1	407	Russian Federation	17.2	73.6	Saudi Arabia	17.1	128	Iran
14	315	Australia	9.7	89.4	Canada	13.1	144	Qatar
13.3	37	China	9.1	139.8	Iran	7.2	230.7	Turkmenistan
10.3	147	India	8.4	96.3	Iraq	6.7	13.8	USA
100	139	World	100	53.5	World	100	48.8	World

section 1.2.2.2, whose amount is estimated from known geological surveys and other scientific data (Brownfield 2016, Schenk *et al* 2012). A few years ago, the ratios of resources to proven reserves were estimated to be from 2.8 to 58, depending upon the actual fossil fuel (Covert *et al* 2016, McGlade and Ekins 2015). Applied to present consumption rates oil could last for at least another 100 years and coal and gas for up to two millennium or more. The longevity of oil is probably an underestimate since new crude oil reservoirs are occasionally discovered, such as the recent find of a massive oil and gas deposit in British Antarctic territory, equivalent to 10 times the amount extracted from the North Sea oilfields over the past 50 years by Russia's largest geological exploration company Rosego (Offshore Staff 2024). Clearly, fossil fuels could play significant roles in future energy mixes, if needed. But will they?

1.3.3 Energy mixes to 2050

To attempt to predict how the energy transition will look in 2050 could be considered a futile exercise of the type attributed to the late, well-known management 'thinker' Peter F Druker, i.e.

> Trying to predict the future is like trying to drive down a country road at night with no lights while looking out the back window.

However, this assertion was made long before the advent of high-capacity supercomputers which allow increasingly sophisticated mathematical models to be explored. Nevertheless, there remain many uncertainties in the predictions, or 'projections', of climate models, which it is claimed can only be addressed by using a global network of the best supercomputers (Rees 2021, Tollefson 2023). This would enable even more advanced mathematical models to be rapidly explored using the techniques of the type being developed at institutions such as MIT (Zewe 2022). Whatever the quality of the answers obtained from such approaches they may not be to the liking of some policymakers, e.g. that the global warming temperature anomalies could be far higher than the 1.5 °C or 2 °C benchmarks of the Paris Agreement to which their country has committed. If this is the case policymakers, presumably, will need to know what policy changes are necessary and when to introduce and enforce them. But how?

To circumvent the implicit warning's in Drucker's observation, agencies such as USEIA, IEA, BP, IRENA, BNEF, Shell, ExxonMobil, Equinor[13], and Enerdata[14], who produce regular global energy mix 'outlooks' for projections over the next 20–30 years based on their collected data, use a defined set of possible scenarios or pathways, i.e. 'what if' notions, e.g. what if all energy was sourced from renewables? The IPCC currently use a similar approach to project how energy choice pathways could impact the scale of global warming anomalies by defining Shared Socioeconomic Pathways (SSPs), the successors of their Representative

[13] A Norwegian energy company.

[14] An independent research company based in France specializing in analyses of energy and climate data.

Concentration Pathways (RCPs) used in their assessment reports (ARs) prior to 2021 (Lee and Romero 2023). Unfortunately, as yet, because of some methodological differences and definition variations between the various organizations publishing 'energy outlooks', their projections are not easy to compare although efforts to harmonize them are being constantly developed and improved, especially by the Resources for the Future Research Institution (RFF) based in Washington, DC (Raimi and Newell 2023).

The agencies mentioned in the previous paragraphs invariably use two or three sets of different assumptions for their what-if scenarios which, according to RFF, can be categorized into three types: 'reference', 'evolving policies', and 'ambitious climate' (Raimi *et al* 2024). For example, in their 2023 global energy outlook report, BP, considered three scenarios that they refer to as 'new momentum', 'accelerated', and 'net-zero'. The assumptions in each are made to enable the global warming to be kept within the limits inherent in the IPCC's and Paris Agreement requirements of 2.0 °C and, preferably, 1.5 °C (BP 2023). Changes were also made to previous projections as a result of the impacts on energy supply chains caused by the Russia–Ukraine War coupled with the approval of the (US) IRA. The 2023 BP scenarios represent updated versions of previous analysis categories, such as 'new momentum' replaces the former 'business-as-usual' by taking account of evolving governmental policies, while the term 'accelerated' is a nuanced definition of the original 'rapid' scenario. All three address the issues of population and economic growth set against energy demand and the need to decarbonize. In the new momentum scenario, CO_2e[15] emissions are 30% below the 2019 levels, while they are 75% below in the accelerated version, and 95% in the net-zero approach (BP 2023). The decarbonization levels in the new momentum and accelerated scenarios for 2023 are more aggressive than in their former descriptions. For these and other ambitious climate scenarios since 2010 there has been no precedent for such rapid reductions in carbon emissions (Raimi *et al* 2024).

What would the energy mix look like in 2050, at least using the BP data and the 2023 outlook scenarios? As shown in figure 1.12, final energy consumption peaks in all BP scenarios before 2050 with 15%–30% overall reductions, compared to 2019, for the accelerated and net-zero scenarios after initial increases for the new momentum and accelerated contexts (BP 2023). Electricity final energy consumption increases by about 75% in all three scenarios while hydrogen is now a part of the final mix. The significant increase in electricity generation (figure 1.13) is dominated by the use of renewables and low-carbon sources, which would be considered as renewables by many jurisdictions, but natural gas is still in the mix as is coal (BP 2023).

The BP electrical generation projections are mirrored by the RFF's analyses of five other data agencies involving ten different scenarios with global electricity demand growing rapidly, especially in the ambitious climate scenarios, led by wind and solar sourced generation, but with natural gas, coal, and, to a far lesser extent, oil, still in the mixes (Raimi *et al* 2024). For global primary energy consumption, RFF had sufficient useable data to analyse 16 scenarios from nine data agencies.

[15] Carbon dioxide equivalent—a term for describing different greenhouse gases in a common unit.

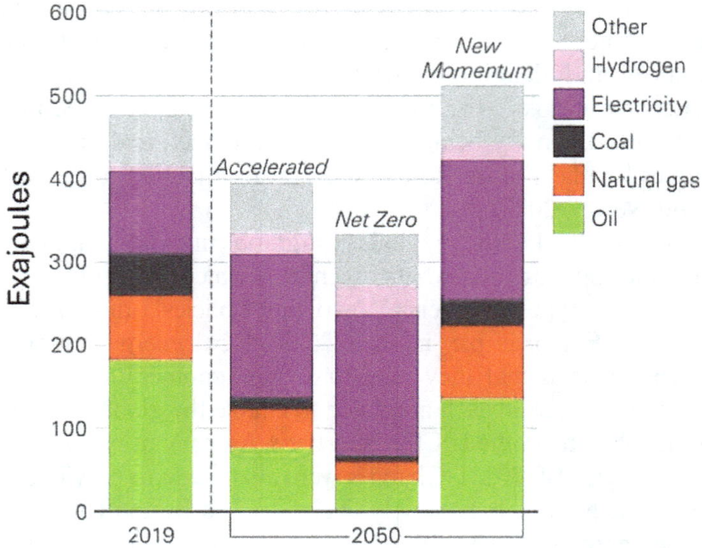

Figure 1.12. BP's energy mix projections—total final consumption. (Reproduced with permission from (BP 2023).)

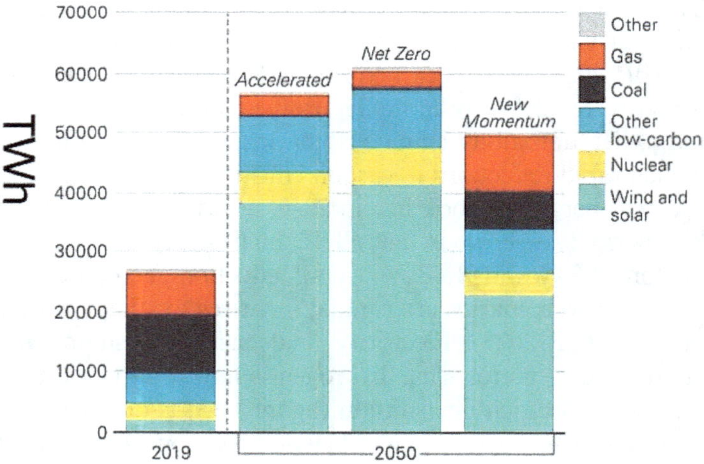

Figure 1.13. BP's energy mix projections—electricity generation. (Reproduced with permission from (BP 2023).)

In comparison with 2022, by 2050 50% of the scenarios were shown to lower consumption rates. However, in terms of scale, the differences between the scenarios were wide-ranging, with the highest global consumption being projected at 244 421 TWh, twice the lowest consumption.

Although the various outlooks vary in numerical details, the trends identified have many commonalities regarding specific issues surrounding energy transitions and compliance with the Paris Agreement targets. Given the physical size of the major outlook reports, such as those of BP and IEA, to discuss the congruences is outside the scope of this chapter. Thus, in table 1.10, a brief summary is provided of some of the key issues highlighted by three major reporting agencies involved in projecting future energy mixes. It should be noted that the summaries, while taken from the reports, are not direct quotes but are paraphrased by the chapter author (BP 2023, IEA 2023e, Raimi *et al* 2024).

Compared with the present energy mixes, those of the future (to 2050 and perhaps to the end of the century), it looks as if the analysts are in general agreement that, globally, there will be: (i) decreasing primary energy demand, (ii) significant increases in electricity generation, (iii) increased use of renewables, particularly wind and solar, (iv) fossil fuels will continue to be part of the energy role but in a diminishing role, (v) hydrogen production will increase especially for use in the transportation sector, and (vi) a momentous annual increase in financial investments in energy transition will be required up to 2050. However, there are still technical and economic impediments that need to be tackled if the desired pace of the transition to renewable dominated energy mixes is to be realized. These issues are discussed in the next section.

1.4 Enhancing the global share of renewable energy sources

The dominant forms of renewable energy are hydropower, wind and solar and, from the EU's point-of-view, biomass. The first three sources all have 'Achilles heels' due to weather influences. These issues are particularly acute with solar and wind sources because of the lack of energy storage capacity. Severe changes in precipitation patterns, both geographical and seasonal, can detrimentally impact hydropower facilities. Such pattern alterations are increasingly linked with climate change (USEPA 2023b). Another major challenge to the substitution of fossil fuels with renewables, and maybe low-carbon sources, is the seemingly insatiable demand for more useful energy by an increasing global population which needs to be quenched by improvements in the efficiency of primary energy and useful energy conversion, whatever the source or the converter. The conversion efficiency of dynamic heat converters using fossil fuels, nuclear, and geothermal sources are still higher than those of wind turbines and solar photovoltaic devices.

1.4.1 Improvements in energy efficiency

The efficiency of thermal conversion devices is governed by the universal laws of thermodynamics and physics. To increase a device's thermal efficiency requires that either a system's maximum temperature be increased, or its minimum temperature be decreased, and both are limited by the choice of device construction materials. The theoretical maximum conversion efficiency of wind turbines is limited by the Betz law[16], sometimes referred to as the Betz limit, a similar type of limit, the Shockley–Queisser limit applies to certain types of solar photovoltaic cells,

[16] Formulated by the British engineer Frederick W Lanchester five years before Albert Betz.

Table 1.10. Author's comparison of major issue projections taken from BP, IEA and RFF (BP 2023, IEA 2023e, Raimi *et al* 2024).

Issue	BP	IEA	RFF
Fossil fuels in energy mix	The importance of fossil fuel demand will decline but oil will continue to play a major role for the next 15–20 years. As natural base declines continuing investments will be required to meet future demands.	The end of the growth era for fossil fuels does not mean an end to fossil fuel investment, but it undercuts the rationale for any increase in spending. The global peaks in demand for each of the three fossil fuels mask important differences across economies at different stages of development.	Consumption remains high through 2050 in many scenarios but peaks before 2030, yet remaining substantial beyond the mid-century even under scenarios that limit warming to 1.5 °C.
Carbon dioxide removal	Carbon capture, use and storage will be needed to achieve deep and rapid global decarbonization.	Carbon capture, utilization and storage is making much needed progress. Over 400 million tonnes of CO_2 capture capacity are striving to be operational by 2030, but cost inflation could hamper progress.	CO_2 removal will be deployed rapidly and at a scale to meet global warming targets.
Critical metals and minerals	The growing requirement for a range of minerals associated with the energy transition has important implications for the sustainability of new and existing mining activities, especially lithium, with the rise in electrical vehicle use.	Copper, rare-earth elements, silicon, and various battery metals, notably lithium, are critical minerals for electrification. However, the mining and processing of certain critical minerals is heavily concentrated geographically, creating a security of supply risk.	Projected demand for energy-related metals and minerals grows rapidly, rising by orders of magnitude in some scenarios prompting various supply problems.
China impact	The use of coal is more persistent in 'new momentum', with a small increase in coal generation in China over the rest of this decade. But that rise is more than reversed by a sharp fall in Chinese coal generation in the final 20 years of the outlook. From the early 2030s onwards, natural gas demand declines but the growth in installed wind and solar capacity out to 2035 is dominated by China.	The total coal capacity will continue to increase to 2030, but output from coal-fired power plants in China peaks around 2025 and declines before 2030. Coal in China accounts for around one-quarter of total energy-related CO_2 emissions, cuts in its coal use have a major impact on the global outlook for emissions. CO_2 emissions from coal in China declines from 8.6 Gt in 2022 to 1.1 Gt in 2050.	A declining population and major economic headwinds will decrease energy demand contributing to lower coal use and CO_2 emissions.

especially those that are silicon based, but many researchers believe this limit can be exceeded with new materials and designs (Center for Sustainable Systems 2023, Office of Energy Efficiency and Renewable Energy 2019). None of the different types of energy conversion devices have achieved their maximum theoretical efficiencies, and probably never will, but individual device efficiency is only one element of the energy losses suffered in the conversion of an energy source to useful energy, as illustrated in figure 1.3 and figure 1.11 and discussed in the scientific literature (Cullen and Allwood 2010, Ritchie 2022b).

It has long been argued that improving energy efficiency can help reduce the levels of energy demand, reduce electricity prices, and be a principal component of climate change mitigation, but these claims have been contested (Allcott and Greenstone 2012, Gillingham *et al* 2009, Shove 2018). Among the concerns that have been expressed is that cheaper electricity could result in consumers using more, not less, electricity and therefore negate the declared advantages of efficiency improvements of agencies such as the IEA and the UN (IEA 2023d, 2023f, United Nations General Assembly 2015a). Notwithstanding, the concerns surrounding the drive to improve energy efficiency, the 193 signatories to the UN's Transforming Our World agenda adopted SDG target 7.3 which called for global progress on energy efficiency by 'doubling the rate of improvement in energy efficiency by 2030' as part of the goal to '[e]nsure access to affordable, reliable, sustainable and modern energy for all' (United Nations General Assembly 2015a).

However, to assess the progress towards target 7.3 it was agreed that 'energy intensity', would be the measure rather than energy efficiency, and this intensity would be determined by dividing the total primary energy supply by the gross domestic product (GDP), albeit it was acknowledged that it would be an 'imperfect proxy' (UN-UNStats 2024a, 2024b). Why then was 'energy intensity' chosen as an SDG indicator? The answer is that it could be determined on substantive energy data from the IEA and financial data from the World Bank, whereas to track changes in efficiency changes in everything from electrical appliances to all forms of transport, industrial machinery, and energy conversion equipment would be an overwhelming challenge. By definition the 'energy intensity' ratio is basically an indication of how much energy is used to produce a single unit of economic output. Thus, the lower the ratio the less energy is being used to produce one unit of financial output. Global and national results are published annually and usually expressed in MJ per US$, although other energy units, e.g. kWh $\$^{-1}$, are sometimes used. As shown in figure 1.14, energy intensity has been trending lower over the past two decades worldwide, but for some countries with highest ratios improvements have only been evident over the last ten years, while the performance of some has stalled, and others have mirrored the global benchmark (Our World in Data 2024).

Overall, it can appear that the trend in lowering energy intensity is encouraging, but the rate of lowering has not matched the UN's SDG 7.3 annual energy efficiency target, and the performance in recent years has deteriorated, such that only an extraordinary global response will correct the situation, but this is highly unlikely (Energy Sector Management Assistance Program 2022a). Does this mean that little of no effort is being made to improve individual source and conversion device

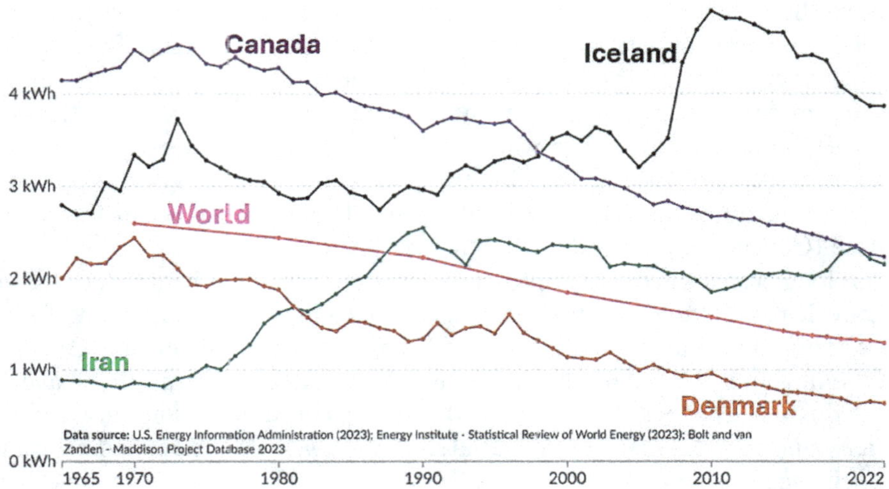

Figure 1.14. Global energy intensity trends since 2000 with exemplars. (Reproduced from (Our World in Data 2024). CC BY 4.0.)

efficiency? Not at all, but it has to be remembered that the devices used to convert fossil fuel sources into mechanical and electrical energy have been developed for well over a century whereas it is only relatively recently that the same development effort has gone into improving renewable sourced devices such as solar panels and wind turbines. It is only to be expected that such devices will need to overcome both anticipated and unanticipated obstacles. For example, in the case of solar photovoltaic cells, it is not always understood that, among many factors, as more radiance falls on the cell its temperature will rise causing its conversion efficiency to be reduced, so some form of cooling, water, air, or a combination, is required (Kurpaska *et al* 2018, US Department of Energy 2024).

The maximum efficiency of extracting kinetic energy from the wind to generate electricity is, as already cited, limited by the Betz law. However, in terms of performance, it is not just efficiency that is of interest, but also the amount of power that can be generated by a single wind turbine or a cluster of such turbines, i.e. a wind farm. Taller turbine towers at a particular location can take advantage of the higher winds found at increasing altitudes above the ground as natural and human-made obstacles which disrupt wind patterns can be avoided in addition to encountering lower aerodynamic drag forces. Consequently, locating the taller turbines offshore can be advantageous, albeit that for a variety of reasons, including legislative regulations, onshore wind turbine height is limited, although onshore installations have many practical advantages (National Grid 2022b). Nevertheless, building higher turbines with longer blades and using advanced control and sensor technologies so that the turbine can adjust to changes in wind-direction, i.e. wake-steering, do not necessarily improve extraction efficiency (Gupta 2015). On the other hand, the provision of the ability to optimize blade design aerodynamics for variable wind speeds can improve wind turbine efficiency (Fattal 2018). While, globally, wind power is projected to play an increasingly larger role in electricity generation, comparable to the share produced by solar sources, the energy transition and decarbonization policies of individual countries may influence how much wind power is to be used in their future energy mixes compared with modest increases in conversion efficiencies.

In the United States, where wind power enjoys a 10% share in electricity generation (table 1.7), it has been estimated that a five-fold increase in wind power installations will be needed to achieve the country's decarbonization goals (Eberle *et al* 2024). Obviously, improvements in efficiency would be an advantage in such endeavours, but there are a multitude of factors, from turbine location constraints, to mass production, to better control, that need to be addressed if the full potential of wind power is to be realized, especially in onshore applications (Roberts *et al* 2023). However, in terms of conversion efficiency from energy source to electricity, conventional hydropower is the clear front-runner. Hydropower is a technically mature technology whose development has a similar history to ICE and other dynamic heat converters. The individual devices in hydroelectric powertrains have efficiencies in the range of 80% and above giving overall conversion efficiencies in the same range. Traditional hydropower, until recently, has been associated with large scale utility size installations generating MW of power, but small, mini, and micro-size, e.g. 'vortex' water turbines, with outputs in the kW range are beginning to garner more attention (Maika *et al* 2023). These type of water turbines are not as efficient as the conventional hydroelectric versions, but nonetheless are still between 55% and 70% efficient (Quaranta *et al* 2020).

1.4.2 Energy storage

Energy storage is by no means a new concept, although its use with renewable sources is a more recent application stemming from the experience of using such sources, particularly solar energy, to produce electricity. The intermittent and variable nature of solar electricity generation means there are times when the supply of electricity from solar photovoltaics exceeds demand but, conversely, at other times that supply cannot meet demand. A phenomenon highlighted and illustrated in a 2008 NREL report has been since popularly labelled as the 'Duck Curve', or sometimes the 'Nessie Curve', although the suggestion of the calling it the 'Shark Curve' has not found favour as yet (Couture 2018, Denholm *et al* 2008). If the excess energy produced in periods of abundance could be saved to augment those times of insufficient production, then the habitual cyclic nature of electricity demand and supply could be advantageously addressed by helping to balance supply and demand at all times. To achieve this outcome a way to store the excess of the over generated electricity, and then release it at appropriate times, is needed. Fortunately, methods of energy storage have been developed over the past 12 millennia and particularly over the last two centuries and, even more recently, the rate of research and development of energy storage technologies has significantly increased.

Arguably, the first use of energy storage was prompted by the arrival of agriculture and more permanent human settlements, which resulted in stocks of wood being accrued. People living in colder winter climates would need sufficient stores of wood to survive until warmer weather arrived when foraging and tree-cutting would again be possible. Periodic community efforts to build up stocks under favourable conditions could also mean that the inhabitants would not need to forage daily for wood, or at least spend less time gathering wood. Storing energy in the form of wood continues today and the use of wood pellets, made from compressed wood fibre, has more than doubled since 2015 with almost 25 million tonnes being exported as replacements for coal in utility scale applications (Strauss 2024). These pellets are considered by many nations to be a renewable and sustainable biofuel and as such are an acceptable way of storing energy especially in Europe (Sikkema *et al* 2010). In the nineteenth century, energy stored in compressed gases, such as air, was developed to drive torpedoes, and mechanical energy stored in rotating masses, such as flywheels, were used for the same purpose (Burke 2017). These forms of energy storage are still in use for a variety of applications (Ibrahim *et al* 2007, Sterner and Stadler 2019).

It is easy to gain the impression that EVs are a very recent form of on-road vehicle, however, along with steam powered versions, they dominated the early automobile markets particularly in the United States. After the demise of steam driven automobile vehicles by the mid-1930s, battery electrical vehicles (BEVs) were only occasionally encountered, usually in small and slow vehicles, as the advent of Henry Ford's inexpensive gasoline powered cars soon became the vehicle type of choice (Melosi 2004, Wikipedia Contributors 2024a). Nevertheless, electrochemical secondary battery technology developments continued due to the advent of the electric starter, together with lighting and instrumentation services for ICE vehicles and the need for energy to power submarines when underwater. The battery of choice in almost all applications was the ubiquitous lead-acid type, although because of its low energy storage density alternatives were investigated, such as the nickel–cadmium and silver–zinc types, but the game-changer was the 1970s invention of the lithium-ion battery and its eventual commercial production by the Japanese company Sony (Blomgren 2017). Lithium-ion batteries now dominate the battery market from laptops to EV automobiles to utility-size electrical energy storage installations.

If they are to be used in energy transition applications, batteries must be rechargeable. Presently, there are 30+ main types of such batteries and an almost equal number of different versions. For example, there are at least 12 variants of the basic lithium-ion battery (Wikipedia Contributors 2024b). Why are so many forms being developed? Given that energy density and storage capacity are the two key criterion for battery choice, would it not be judicious to select the best performing battery versions and focus development on those? However, there are many other criteria which need to be accounted for in battery selection such as charging and discharging rates, cost, nominal voltage, and thermal runaway (Challa 2022). In particular, the latter needs to be avoided to ensure the safe operation of a battery system. All batteries, not just the lithium based ones, have to operate within a limited temperature range but if the top limit is exceeded then a phenomenon known as thermal runaway can occur causing a chemical chain reaction within the battery cell that can be very difficult to stop and, in extreme cases, can cause batteries to explode and start fires as battery temperatures can reach 400 °C (Dragonfly Energy 2022). Consequently, with so many factors involved, research into the so-called *super battery* is intensifying with the efforts focused on seeking safer, more powerful, longer-lasting, and less expensive batteries (Hall 2021).

Other forms of energy storage also began to be pursued in the second half of the twenty-first century, such as supercapacitors and thermal storage systems. Arguably, the latter is an old concept because ancient humans used heated rocks and stones to boil water for cooking. However, despite all the diverse efforts being made to provide energy storage systems suitable for use with intermittent and variable renewable sources, the dominant form is pumped-storage hydropower (PSH), developed in the United States and which, in 2022, accounted for 96% of utility-scale energy storage and almost the same percentage globally (International Hydropower Association 2022b, Uría-Martínez and Johnson 2023). PSH is based on a simple concept using two water reservoirs, one being higher than the other. Water is pumped to the upper reservoir during periods of low electricity demand and released to the lower reservoir to drive the water turbine systems to generate electricity when the demand is high. Low demand is usually associated with electricity prices being lowered, whereas at high demand they rise. The price differential offsets some of the costs of installation and operation of PSH systems. It may be possible to reduce the PSH costs and, simultaneously, enhance the use of solar and wind sources by hybrid combinations of the technologies. However, both nationally and globally, the use of PSH is dependent on geographical location, topographical features, and the availability of suitable electrical grids.

There are then a significant number of energy storage methods and devices available, as summarized in table 1.11 (Ibrahim *et al* 2007, Sterner and Stadler 2019). Perhaps there are too many choices to be fully developed in both a technical and commercial sense (Reader 2022)?

1.4.3 Other factors

Although efficiency and hence storage are universally agreed sustainable targets, there are many other factors involved in the pathways' wholesale adoption of renewable energy sources, such as the need to increase capacity factors, establish consumer affordable electricity, improve dispatchability, abate excessive land-use, reduce the use of unclean cooking fuels, and improve air quality. Each of these factors warrants an encyclopedic scale discussion but, in an abridged chapter such as this, only a brief synopses are provided.

1.4.3.1 Capacity factors

In much of the literature, especially when advocating for particular forms of energy conversion to generate electricity or other forms of useable energy such as wind and solar photovoltaics, emphasis is placed on the capacity of the installation. While the capacity of a generation system is a measure of the full-load sustained power output of the system or device, the use of the term can be somewhat cavalier. Moreover, the

Table 1.11. Forms of energy storage summarized from (Ibrahim *et al* 2007, Reader 2022, Sterner and Stadler 2019).

Energy storage classification	Basic forms
Mechanical	Pumped-storage hydro (PSH)
	Flywheel
	Compressed air (CAES)
Thermal (TES)	Sensible heat
	Latent heat
	Thermo-chemical reactions
Electrochemical	Primary batteries
	Secondary batteries
	Redox batteries
	Reversible fuel cells
	Bio-batteries
Electromagnetic	Supercapacitors
	Superconducting magnets (SMES)
Chemical	Ammonia
	Hydrogen
Various	Novel and innovative forms

term 'nameplate capacity' is frequently used as a synonym for installed capacity, however, for renewable sources such as wind and solar, it is the system's output under ideal conditions, such as maximum usable wind or solar radiance. 'Nameplate capacity' is also used to describe the legally registered theoretical output, or the maximum rated output under specific conditions, designated by the manufacturer. Other capacity descriptors are also frequently encountered, although all are attempts at specifying a power plant's generation capability. However, capacity specification is in units such as MW, and are a different measure from actual electricity generation, e.g. MWh. The amount of electricity will depend on how long the generator is operating and at what rate over a particular time period, usually in energy data statistics taken to be one year. No power plant operates 100% of the time, so power plants with identical capacities may generate noticeably different amounts of electricity annually, or over a specified time period. Consequently, to use the term capacity, in whatever form, to represent actual generation is Machiavellian. Thus, to highlight the crucial differences, a ratio known as the capacity factor (CF)[17] has been defined.

Although there are different variations of the phraseology of CF definitions, basically, it is the ratio, expressed as a percentage, of the actual energy produced by an energy generating system to the maximum amount of energy that could be generated at full rated power over a given time period without any interruption. CFs are published, like other energy statistics, by a number of different national and

[17] Sometimes referred to as the load factor (LF).

international agencies and the numerical details can be different depending upon the context. For example, IRENA provides the comparative CFs for renewables only while others include all types of energy conversion systems (IRENA 2023b, Statista 2024c, USEIA 2024b). While global average CFs provide a useful portrayal of the effectiveness of the various energy conversion systems within a particular category, e.g. solar, wind and coal, there can be marked performance differences, especially for electricity generation (Schernikau and Smith 2022). There are many reasons for these differences within each conversion category, such as the ages of particular systems and their technological stage of development (NREL 2023b).

A useful snapshot of the various CFs can be obtained from the latest extensive and reliable United States' data, as analysed by the independent multinational German company Statista, which specializes in collecting, consolidating, and analysing data on a wide-range of topics. According to their latest open-source report, renewables, such as solar and wind sources, have average CFs of about 23% and 34%, respectively, in the United States (Statista 2024c). Perhaps surprisingly, among acknowledged renewable sources, geothermal and biomass/wood sourced systems have the highest CFs at 70% and 52%–58%. Nevertheless, the source with the highest CF is nuclear power at 93%. The CFs for fossil-fuel electricity generating power plants vary between 40% and 60% but, globally, coal-fired CFs are closer to 70%. However, it has to be acknowledged that CFs are reliant not only on the fundamental technologies but also on government policies and the operational approaches of the end-use energy producers, private and public. Even so, the present much lower CF performances of the favoured renewables for electricity production, solar and wind, are a cause for concern in the efforts to address climate change mitigation.

1.4.3.2 Universal access to affordable electricity

One of the most persuasive political and scientific promises about the energy transition to renewables has been that energy, especially electricity, would be cheaper than the existing energy mixes. Indeed, there have been numerous announcements over the past few years of how solar fuelled systems now produce the cheapest electricity ever and the costs of wind producing electricity have been significantly reducing (Evans 2020). Subsequently, consumers could expect that electricity prices would also be reducing, but this has not been the case. Although the pricing of electricity is heavily contingent on government policies, which sometimes include additional taxation on energy use, the consumer cost rise has been dramatic. Globally, from over the past eight years, energy prices have increased by over 80% and, in some regions, e.g. India and Europe, electricity prices have risen between 80% and 100% (Federal Reserve Bank of St Louis 2024, IEA 2023c). However, of particular interest to the public are the costs of household electricity use and, whatever data sources are used, it is clear that there is a broad range of national charges.

In 2023, for example, an analysis of data from 29 countries, representing those mainly with the highest populations and largest economies, found that household prices varied from 0.002 to 0.53 US$ kWh^{-1} (Statista 2024a). Both the lowest (Iran) and highest (Ireland) rates were for countries whose electricity is generated from fossil-fuel sources. It appears that the situation in Ireland is an anomaly as the cost of electricity in countries relying on fossil-fuel is invariably much lower than for those aggressively pursuing the transition to renewables for electricity generation. Nevertheless, whatever the national price of electricity, perhaps a more relatable measure, to its population, is affordability. In some European countries, such as Poland, Hungary and Portugal, residents have to spend up to 8% of their income, whereas in Germany, where electricity prices are among the highest, residents spend only 5% of their annual income on such services because their salaries[18] are higher than the majority of other countries (ElectricRate 2024). Conversely, the residents of the United States have higher incomes than those of Germany and only spend 1.24% of their daily income on a day's supply of electricity.

As always there are exceptions to these trends such as Belgium, a country using significant amounts of nuclear energy, which has electricity prices higher than most of the other European countries that are using increasing amounts of renewable energy generation. Yet France, which has a similar mix as Belgium, charges households a third less for electricity. Norway also bucks the trends with its renewables strategy having low electricity charges of 0.11 US\$ kWh^{-1}, albeit almost twice as much as the majority of countries in the 29-group using fossil-fuels, but their residents only need less than 1% of their income to buy the electricity (ElectricRate 2024, Statista 2024a). So, the promise of cheaper electricity from renewable use has far from materialized and, in general, electricity has been increasingly expensive and unaffordable in sufficient quantities for many households. There seems to be a contradiction between the cheapness of generation and the costs of supply. Politicians have invariably espoused 'short-term pain for long-term gain' to acknowledging that electricity prices are driven by government policies, such as prices on pollution, i.e. carbon taxes, and strategies such as 'marginal cost pricing system', where the price is set by the most expensive method to meet demand (Stewart 2023).

In some electricity markets, known as 'capacity' markets, producers are paid for being available to meet peak electricity demand, not for the actual generation of electricity, the idea being to partially fund, 'the fixed costs of building and maintaining generating resources', with future requirements in mind (Goggin 2020). These costs are passed onto the consumer in a similar manner to the way incentives to install and operate household solar panels are underwritten by those consumers who do not, or cannot, take advantage of such inducements. The hoped for energy transition is then to be mostly funded by the consumers which governments likely believe to be a good investment and will ensure the eventual mitigation of climate change and cheap electricity for future generations, but the public may not be as convinced. However, it needs to be borne in mind that affordability is not a pressing issue for the 760 million people who do not have any access to electricity, or the almost 30% of the global population who cannot access clean cooking fuels

[18] Presumably based on GDP per capita.

(Ritchie *et al* 2024b). Many of these unfortunate populations live in lower income countries who do not have sufficient domestic financial resources to address these issues, and who will need access to international funding (Moses 2023). For many African countries, access to such funds is only available in their endeavours to address climate and transition issues if they utilize low carbon sources. Seemingly a 'carrot and stick' approach where the carrot is the promise of technical and financial aid, the stick being that poor countries must adopt the policies of the rich countries.

1.4.3.3 Dispatchability, land-use, unclean cooking fuels, and air quality

If the output of electrical generation devices within a grid system can be adjusted to meet demand fluctuations in real-time, maybe by turning some devices on or off, they are known as a 'dispatchable' power sources. The provision of such systems usually means that a reliable and uninterrupted supply of electricity can be assured, avoiding the likelihood of costly and damaging power outages for residential, commercial, and industrial consumers. However, bringing dispatchable systems online to scale production up or down is not as straightforward as may be thought. If a particular device, or set of devices, needs to become fully operational to allow the necessary adjustments, then this could take from a few minutes to several hours. It has been intimated that thermal power plants using fossil-fuels could take between 1 and 12 h to reach their operational states but, if using nuclear energy, the power plant could take more than 12 h, with the only systems capable of rapid responses of a few minutes being some hydroelectric power plants (Biddle 2024). Over the years, electricity providers and grid operators have accumulated sufficient temporal supply–demand data, for all weather conditions, to allow sophisticated software to be developed to enable them (i) to have dispatchable devices readily available and (ii) automatically monitor the real-time balancing of grids to lessen the impact of any unanticipated events. The latter may involve the import and export of electricity to and from other interconnected grids, of which, for instance, there are three major ones in the United States (USEPA 2024f).

The number of unanticipated events has increased in recent years, mainly because electricity from non-dispatchable renewable sources, such as of solar and wind, can also be fed into the grids. However, some renewable sources such as biomass, geothermal and PSH are dispatchable, as is single reservoir hydropower on a seasonal basis. The variability of the solar and wind generated electricity presents major challenges to grid management systems and many software providers are addressing these issues (Alotaibi *et al* 2020, Barth *et al* 2024). Clearly, these matters could be largely overcome if ample energy storage accompanied the non-dispatchable, making them, in essence, dispatchable. Other approaches could involve hybrids or combinations of renewable sourced electricity generating systems to generate dispatchable power outputs. Yet, it appears that such dispatchable renewable configurations may cost up to twice as much, if not more, than variable renewable systems (Lovegrove *et al* 2018). If solar and wind sources are to be the dominant providers of electricity in the future their dispatchability needs to be improved, considerably.

Not all the solar radiation from the Sun reaches the ground, but that which does is sufficient to meet in, one hour, the world's total annual power consumption, that is, if it could all be captured. Sunlight has many uses, such as vegetation growth including human driven agricultural activities and, together with the greenhouse-effect' helps maintain the Earth's temperature at habitable levels. Indeed, the use of solar energy to produce power, especially electricity, represents a relatively modest share of the total radiation received on Earth. Unmistakeably, the energy potential of solar radiation in power generation is gargantuan, but the problem is capturing the radiation in an efficient and economically viable manner. Two main methods of solar capture and its conversion to useable energy that have been developed are solar photovoltaics (PV) and concentrated solar power (CSP). In the former, the radiation is directly converted to electricity, whereas with CSP the radiation generates thermal energy which can then be used for several purposes including conversion to electricity. Both methods use conversion devices and equipment that require land space to be installed in the same way as all other power generation plants. An issue that has been raised over the past decade or more is exactly how much space would be needed to accommodate renewable energy devices such as solar (MacKay 2013, Ritchie 2022c).

One proposal, to completely cover the Sahara Desert with solar PV panels to meet global energy demand, estimated that this would represent only 1.5% of the Earth's land surface. A similar study suggested that, if all the UK's electrical energy were provided by solar, only 1% of their land would have to used, a seemingly modest amount, but it was pointed out that all the buildings in the UK only covered 1.2% of land space and for roads it was about 1.5% (MacKay 2013). It was further highlighted that if all the energy needs of the UK and countries in Europe, such as Belgium, Germany, and the Netherlands, were to be met by renewables then just about all the country's land area would be needed! In all probability it was these type of analyses that prompted the United Nations Economic Commission for Europe (UNECE) to undertake an in-depth investigation of the land use requirements for electricity generation, not only in Europe, but also in other world regions. The UNECE lifecycle assessments considered more than just the footprints of the various power plants and included the land used for the mining of materials for their construction, 'fuel inputs', eventually decommissioning and disposal of waste, expressed in m^2 per MWh (Gibon *et al* 2022). It was found that there are wide variations, not only between different types of power plants, but also within each category of power plant, figure 1.15 (Ritchie 2022c).

The determination of the land used by onshore wind power systems proved to be very challenging. Overall, it was found that nuclear energy power plants used the least amount of land but, based on medians, both on-ground CSP and solar PV systems used between 40 and 70 times more than nuclear, and small-to-medium hydroelectric plants of less than 360 MW required over 100 times that of nuclear. Fossil fuelled electricity generating plants using coal used more land than renewable systems in some cases but less in others, whereas gas powered plants required only 3–4 times more land than nuclear. Interestingly, it was determined that if solar systems were installed on roofs rather than on the ground, land use could be dramatically reduced. It is unlikely that the global electricity and energy demands will be met by roof-top installations but the concept of using them in localized

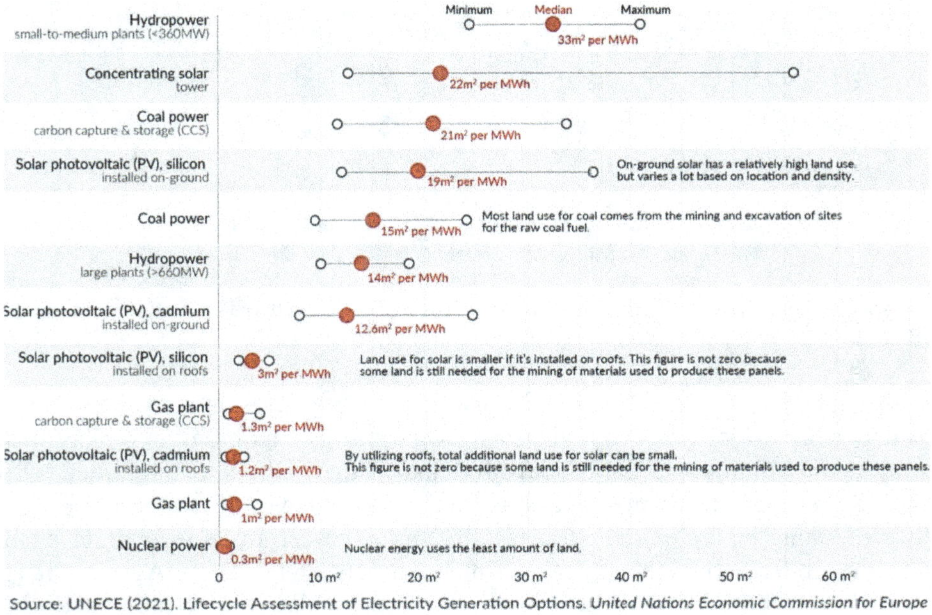

Figure 1.15. Land use for electricity generation systems. (Reproduced from (Ritchie 2022c). CC BY 4.0.)

micro-grids warrants further investigation. Nevertheless, the implications of having to use far more land when using renewables compared with nuclear and gaseous fossil fuelled electricity generation needs to be thoroughly documented.

To a large extent, the issues involved in no access to clean cooking fuels, i.e. indoor air pollution, and air quality are almost identical in many global regions. Of particular concern, is the millions of premature deaths attributed to household air pollution caused by the use of polluting fuels, especially in regions such as sub-Saharan Africa (SSA) (Roser 2021b). There are many socio-economic problems in this region, whose population has been increasing this millennium and will likely double to over 2 billion by the net-zero carbon deadline of 2050, by which time the largest global share of people using unclean cooking fuels, 44%, will live in SSA (figure 1.16, Oliver *et al* 2021). Moreover, as in some other regions, and especially in their rural communities, the cooking fuel of choice is wood or other forms of biomass. While biomass/wood is considered to be carbon-neutral and a renewable fuel it does emit particle matter (PM). The finer/smaller particles, called $PM_{2.5}$, can penetrate into human lungs and bloodstream contributing significantly to an assorted range of harmful impacts on human health and increased risk of premature death. Respiratory and cardiovascular diseases, lung cancer, and susceptibility to strokes, are among the cited detrimental impacts of biomass use. Although particle filters are readily available for such applications as diesel fuelled vehicles and generators, they are rarely, if ever, found in households that are burning biomass. In many regions, such as South-East Asia, the scale of the health problems linked to biomass use are now considered to constitute a crises (Amnuaylojaroen and Parasin 2023).

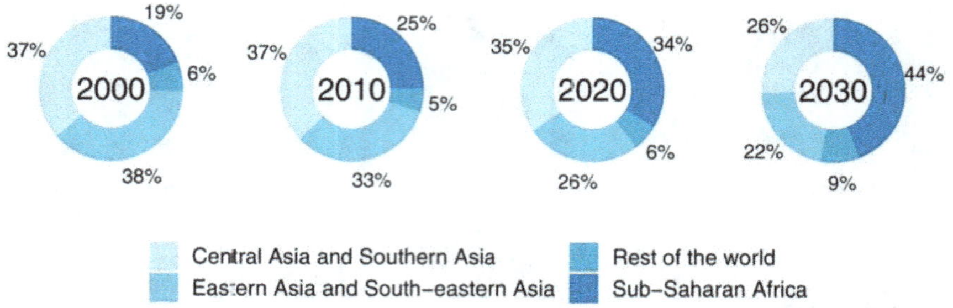

Figure 1.16. Regional shares of unclean fuel usage 2000–2050. (Reproduced from (Oliver *et al* 2021). CC BY 4.0.)

The countries where the use of unclean cooking fuels is prevalent initially considered the use of gaseous fuels, such as natural gas, as biomass replacements. However, the voluntary 2023 Global Methane Pledge (GMP), championed by the United States and the United Nations, and endorsed by 155 countries, seeks to reduce methane emissions by 30% by 2030 as a key strategy in climate change mitigation and meeting the global warming targets of the Paris Agreement (US Department of State 2023). So, any intentions to the replace a renewable fuel, whose use adversely effects human health and mortality, by a 'healthier' fossil source were, to all intents and purposes, abrogated by climate change concerns. The problem of unclean cooking fuel usage remains.

1.4.4 Future perspectives and directions: global versus national

There are a multitude of choices of energy sources available to political decision-makers worldwide. Moreover, the mitigation of climate change is considered to be a global issue which can only be addressed by a complete transition to renewable energy sources. To accomplish this transition will require global action, but there is no one-size-fits-all energy mix which can be implemented by every country. Consequently, nations, and regions within nations, will need to determine which mix of energy sources can best contribute to the collective global effort to avoid the feared consequences of rising global surface temperatures while simultaneously meeting the socioeconomic needs of their populations. Thus, while it can appear that decisions have already been made to focus on solar and wind power in all future global energy mixes, especially for electricity generation (figure 1.13), an individual country's renewable source utilization strategies need to be carefully and strategically formulated. For example, the choice to use solar energy is not as straightforward as may be imagined and many different criteria or factors will need to be investigated before the decision is made to transition to solar energy alone. Important considerations include the daily length of sunlight at a location, and this is very much dependent on its geographical latitude and season. Moreover, the intensity of the sunlight reaching the location will vary throughout the day and will depend on the prevailing weather conditions, especially with respect to the amount of cloud cover and the surrounding topography. However, solar energy is renewable and sustainable, and it is available everywhere, albeit not always on a wholly guaranteed basis either daily or throughout the day. Accordingly, as an example of the types of energy mix challenges facing the decision makers, some of the vagaries associated with use of solar energy are now considered.

At their higher latitudes, countries which are the closest to the Arctic and Antarctic poles experience shorter periods of daylight and sunlight during their autumn and winter months, which can also include days of complete darkness. Indeed, the Norwegian village of Rjukan did not enjoy sunlight for 6 months of the year until, in 2013, a giant sunlight tracking heliostat was installed with mirrors that directed sunlight onto the village from a nearby mountain top (Visit Norway 2024). Conversely, in the spring and summer months such countries contend with increasingly longer daily periods of daylight and sunlight and there are regular instances when the Sun never sets for a month or more. However, as always, when a location is affected by an unfortunate combination of natural weather and geographical phenomena, the results can be surprising. For instance, the Faroe Islands, in the North Atlantic between Scotland and Iceland some 400 km below the Arctic Circle, average less than 40 days and 1000 h of sunshine annually, far less than the world average of over 2300 h and less than 25% of that at the sunniest global location, i.e. Yuma, Arizona (The e face 2018). The Faroes are an anomaly but, in general, the inherent and problematic intermittency and variable intensity of solar energy can be greatly exaggerated as the latitude increases. Does this mean that at such locations solar would not be a source of choice?

A town like Utqiaġvik (formerly known as Barrow) in Alaska's North some 560 km above the Arctic Circle experiences an extended polar night of up to two months without any sunlight, usually beginning towards the end of November. Solar PVs would be of no use in these circumstances of darkness. But Utqiaġvik also experiences the midnight Sun phenomenon from about mid-May, when the Sun never goes below the horizon and which lasts longer than the polar night. In this regard Utqiaġvik is not unique as there are several other locations in the higher latitudes of the northern hemisphere, approximately 67° and above, where the polar-night effect occurs in the winter months and the midnight Sun happens in the summer months. Arguably, in such instances, it may be possible to use the 'excess-to-needs' electricity generation in the summer months to compensate for the 'no-generation' periods in the winter months, as the following example implies.

The city of Tromsø in northern Norway, some 300 km inside the Arctic Circle, enjoys more sunshine hours in the three months from May to July than either the French capital, Paris, or the Belgian capital Brussels (Wikipedia Contributors 2024c). However, from November to January the Norwegian city only receive a total of 9 h of sunshine, with none at all in December, which is two orders of magnitude less than that experienced by either Paris or Brussels. Nevertheless, on an annual basis Paris only receives about 30% more sunshine than Tromsø. The number and capacity of solar PV farms in Paris have been increasing in recent years, the latest one installed being a 36G Wh facility at Disneyland Paris which is sufficient to met the annual electricity consumption of a town of over 17 000 inhabitants (Ubrasolar 2024). Perhaps, a demonstration of the possibly of using solar PV technology in Tromsø, provided, of course, that suitable solar collector and energy storage capabilities were available. But how technically realistic would such a choice be for the largest city north of the Arctic Circle, i.e. Murmansk?

Based on the average national energy use per capita the annual primary energy needs of a highly populated Arctic City such as Murmansk, about 300 000 inhabitants, would require approximately 17 TWh. If this were to be provided by solar PV power alone then an installed capacity estimated at over 6.7 GW[19] would be needed, a third more than the Xinjiang solar farm in China, the world's largest as of June 2024 (Gill 2024). Even if only the electricity generation of such a city is considered then a solar-panel farm capacity of 770 MW would be required assuming a capacity factor of 30%. As a single facility this would still make it one of the world's twenty largest solar farms. However, single facilities of up to 440 MW are planned to be in commercial operation by 2025 in Russia including the Latgale Solar PV Project in the Magadan region about 200 km below the Arctic Circle (GlobalData 2024). These Norwegian and Russian examples demonstrate the use, or potential use, of solar PV energy at almost any geographical location, albeit the region between 25° and 35° latitudes from the equator in both hemispheres enjoys the most annual sunshine hours. Thus, it can be readily recognized that the global focus on developing solar power systems as a key source of renewable energy is unquestionably justified. However, not all countries and regions may decide to make solar energy their major renewable source as other forms may have better local characteristics. An example of such situations, and perhaps dilemmas, is the United States.

The United States is a country that is rich in a variety of renewable sources. In some locations within the conterminous 48 states and regions, both on land and in coastal areas, wind energy is more inherently available than solar and vice versa, while other regions may have significant geothermal resources, especially in the Western States, along with hydropower and biomass resources. Moreover, a combination of these sources, e.g. wind plus solar, solar plus hydropower, could also prove particularly attractive in other areas. This abundance of sources, and therefore choices of renewable energy mixes that are technically recoverable, is illustrated in figure 1.17 (Brooks 2022). It can be seen that solar, wind, and geothermal are the most abundant renewable energy resources nationwide, but hydropower also plays a prominent role in regions such as the Pacific northwest. It should be noted that not all renewable resources, such as biomass and ocean energy, are shown on this energy locations map but are included in the estimates of the technical potential of the country's renewable energy resources. In total these renewable resources have been determined to have the promise of being able to generate a hundred times more electricity annually than consumed in the United States in 2021 (Brooks 2022).

[19] Author's calculations.

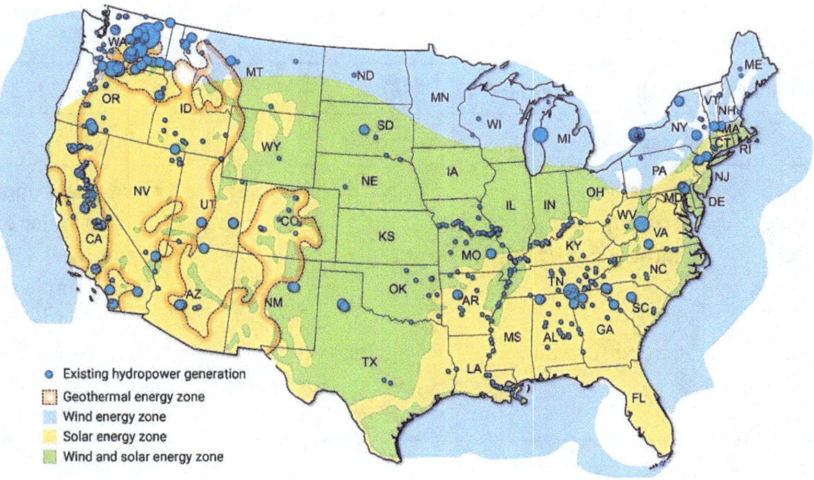

Figure 1.17. NREL renewable energy resource abundance. (Reproduced from (Brooks 2022). Image credit: DOE/NREL.)

1.4.4.1 Strategic decision making: SWOT

It is evident that globally, depending upon their geographical location, not all renewable resources can be developed to the same extent, so decision-makers will have to make choices about which form or forms to pursue. However, resource and technical potential are not the only criteria that need to be considered in framing strategic plans and policies for the increased future use of renewable sources. Among the other benchmarks, targets will be included not only for the overall contributions that renewables will eventually make to the energy mixes, but the shares of each of the available renewables. This is because there are several global and national regions, as illustrated in the figure 1.17 example, where several forms of renewable sources could be in competition for development and use. Account also needs to taken of the relative capital and operational costs of the competing sources as well as the prevailing political–legal realities which may favour particular sources. These factors can result in the type of selection framework, as summarized in table 1.12, which has been recommended for use in the United States by a series of NREL studies (Brown *et al* 2016, Cox *et al* 2018).

Quantifying the generic criteria provided in table 1.12 can provide the basis for a strengths, weaknesses, opportunities and threats (SWOT) analysis, made popular in the business community for assessing strategic planning and forecasting market penetration (Namugenyi *et al* 2019). This technique can be applied to the comparative assessment of renewables in general, complete energy mixes, i.e. renewable versus non-renewable, and to explicit aspects of the sector such as electricity generation and transportation. In all cases the approach will be used to compare options and choices. Thus, the system is also useful to compare specific renewable and low-carbon energy sources pertinent to particular individual regions and countries (Bayraktar *et al* 2023, Shaikh *et al* 2022, SpendEdge 2024). More recently an additional form of energy source selection analysis has been developed

Table 1.12. Recommended criterion framework for assessing energy generation from renewable sources (Brown *et al* 2016).

Types of renewable generation potential			
Resources	Technical	Economic	Market
Energy content of resource	System and topographical and land-use constrains	Projected technology costs	Regional competition with other energy sources Policy implementation and impacts
Theoretical physical potential	System performance	Available versus required revenue for energy project	Regulatory limits Investor response

termed PESTEL[20], an acronym for an approach embracing 'political, economic, social, technological, environmental and legal' aspects of renewable energy development (Kansongue *et al* 2023). All three frameworks, the NREL framework, SWOT and PESTEL, have the same objective, which is to identify the main factors that need to be considered in assessing how the various types of renewable energy sources could meet future energy production and consumption requirements while achieving the goals and targets of the Paris Agreement and the Sustainable Development Agenda.

In these type of frameworks, it is usual to attribute the 'strengths and weaknesses' to internal factors and the 'opportunities and threats' to external factors. So, in table 1.12, what NREL refer to as 'market' correlates approximately with threats, and 'economic' with opportunities, although there will be some overlap, e.g. the investor response factor. In all the frameworks mentioned the factors are usually formulated by sizeable groups of experts whose collective work then passes through various levels of peer review and scrutiny (Benzaghta *et al* 2021, Haan 2024). To help the expert groups decide the elements of the particular SWOT category both internal and external data are required, much of which will be obtained through published sources, commission reports and individual expert experiences. However, whatever the level of expertise and state of knowledge of these groups the outcomes will still be somewhat subjective and therefore liable to what the business community refers to as 'cognitive bias', i.e. differing individual interpretation of the same facts. Published SWOT analyses on individual energy sources display differences depending upon the region that is being considered although they also contain similarities. For example, one SWOT analysis of solar energy on a global scale considers the scale of the amount of land needed for solar PV as a weakness whereas a similar analysis, but only for a region on the African continent, identifies the abundance of

[20] Also referred to as PESTLE in some literature.

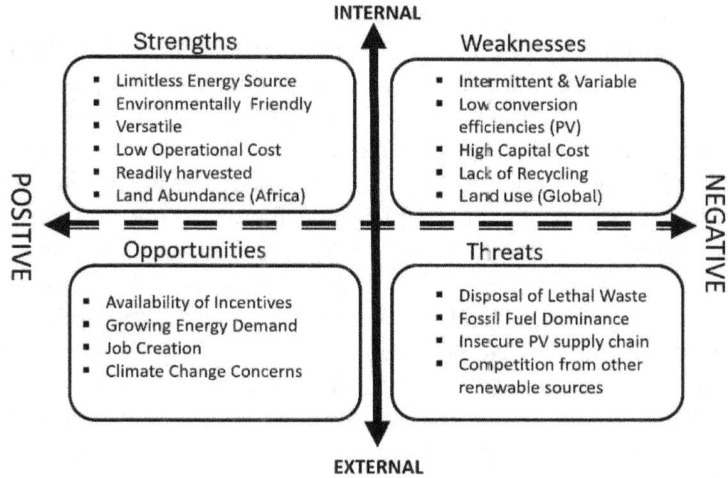

Figure 1.18. SWOT analysis of solar power. (Modified from (Guangul and Chala 2019, Lei *et al* 2019).)

land on the continent as a strength, albeit there are common factors evident (Guangul and Chala 2019, Lei *et al* 2019). This type of situation can be illustrated on a conventional SWOT analysis quadrant map as shown in figure 1.18, showing that land-use can be considered both as a weakness and strength depending upon the context. The main point is that a SWOT analysis needs to be formulated with appropriate regard to the prevailing circumstances.

It should also be noted that SWOT analyses need to be updated regularly since the rates of technical and economic development of all energy sources, especially renewable forms, have continually gained pace. One of the results of such progress is that electricity from solar sources, once considered to be amongst the most expensive methods, is now the cheapest in many regions (Okoye *et al* 2017, Ritchie 2024b). Moreover, account needs to be taken of the changing categorization of factors, e.g. from threats to opportunities and vice versa, particularly those stemming from changes associated with significant adjustments in governmental energy and environmental policies. For example, when governments first introduced 'incentives', perhaps a less than subtle euphemism for subsidies, to encourage on-road vehicular users to purchase EVs, they were an obvious opportunity for the automobile and secondary battery industries, but subsequent removals of such incentives by some countries, whether total or partial, such as in the UK, led to opportunities becoming threats to the financial viability of industrial strategies resulting in such headlines as 'UK BEV sales plummeted from 186% in 2020 to just 15% in 2023 due to incentive removals' (Destribats 2023). Furthermore, assumptions used in the original expert interpretations of data related to some SWOT factors may, over time, be found to be unsatisfactory as more reliable data and contemporary observations become available (Haan 2024). Nevertheless, despite the limitations in determining strategic directions, SWOT analysis has more advantages than shortcomings.

1.4.4.2 Strategic decision making: hierarchy processes

Identifying the various SWOT categories involves factors such as those given in table 1.12 and figure 1.18, but does not provide comparative weighting and priority of the factors or the categories. These issues become increasingly important if, regionally, there are several candidate renewable energy sources, as illustrated in figure 1.17. For instance, should more weight be placed on strengths rather threats, and, within a category, which of the identified factors is the most important? To answer such questions, methods have been developed which enable category and factor priorities to be determined to elaborate on the initial SWOT analyses, such as the analytic hierarchy process (AHP), whereby a numerical scale of 1–9, the 'Saaty scale', is used to indicate the relative importance of the various criteria (Saaty 1987, 2008, Saaty and Vargas 2012). Although the assignments of the scores are again collected from expert groups, their responses are subjected to a decision-making statistical evaluation using an approach referred to by mathematicians and management specialists as 'pairwise comparisons' or 'pairwise ranking' (Kyne 2024).

With the AHP system the so-called importance intensities are determined on a scale of 0 to 1 for the SWOT groups which, collectively, will add-up to 1, e.g. strengths = 0.498, weakness = 0.096, opportunities = 0.301, and threats = 0.015. Obviously in this example case, strengths are the group priority to address. However, within each group there will be factors, and maybe subfactors, that also need to be ranked in terms of priorities using the same approach. Once ranked, an overall priority ranking is established by multiplying the group priority by the factor/subfactor priority, i.e. in the preceding example it may be that keeping the operational cost low is the highest priority, equal to say 0.25, then the overall ranking would be $0.498 \times 0.25 = 0.125$. This calculation is carried out for each group and set of factors in the group. The results are then compared, and it could be that highest strategic priorities do not necessarily come from the highest priority group. An AHP–SWOT study of a manufacturing company found that although the strengths group achieved the highest group priority, it was one of the factors in the opportunities group had the highest overall priority, providing a key focus for the company's strategic planning process of future directions (Görener *et al* 2012).

Combining a ranking system with a SWOT analysis has become more prevalent in the scholarly literature in recent years, including studies using different 'multi-criteria decision-making' techniques other than AHP such as VIKOR[21] (Kumar and Samuel 2017). In all cases, the identification of the ranking criteria becomes crucial. So, while illustrations such as figure 1.18 and table 1.12 highlight the need to identify criteria within a SWOT group, they are not necessarily comprehensive. For example, in a recent in-depth study the obstacles, i.e. threats, to increasing the rate of renewable source adoption, especially with regard to meeting the climate targets, identified nine key threats associated with five key sources, solar, wind, bioenergy, geothermal and offshore wind (De *et al* 2023). The nature of the barriers, e.g. energy storage, critical minerals, supply chain issues, etc, were not of equal intensity for the

[21] An acronym developed from a Serbian description of the method.

assessed renewable sources. For offshore wind it was found that access to capital investment was an 'acute' challenge, whereas, among the sources considered, only geothermal had a lower intensity of supply chain issues. The results from this type of study could be used in a hybrid ranking-SWOT analysis to identify the critical factors that need to be addressed. The combination of the qualitative ranking and quantitative SWOT systems enable more erudite analyses to be undertaken with regard to existing energy mixes and the possible ways in how the pace of renewable sources can be positively impacted.

Nevertheless, if the elected decision makers are to obtain a clearer picture of what the constituent parts of future energy mixes could be, then important societal and environmental factors need to included in their policy assessments. These additional factors should take account of population growth, increasing energy consumption, economic growth rates, the impact of current and stated policies regarding climate change mitigation, poverty elimination, shelter and health care affordability, sustainable development progress plus a myriad of other issues of public concern. However, such an approach would require the acquisition of an overwhelming amount of reliable data and analysis. With regard to future energy mixes, international, national, governmental and non-governmental agencies have extended the framework systems discussed to include mathematical models of future population trends along with similar approaches to climate change assessments and final energy consumptions together with other inclusions of statistical projections not widely publicized. Normally, such institutions, e.g. the UN, IEA, USDOE, BP, IRENA, etc, provide the results of their analyses including explanations of their approaches in the open literature but not the precise details of the individual models used or how they are connected to produce their projections (Brinkman *et al* 2021).

Just how factors are included in energy mix transition projections by the various agencies named in the last paragraph is not known, but their projections for future electricity generation all identify solar and wind as the major candidates among renewable sources and with the greatest combined shares of all primary energy demand sources (BP 2023, Raimi *et al* 2024). However, a review of the bulk of agency projections and their associated scenarios found that, by 2050, all the current energy sources in use today will still be in the global mixes including fossil fuels (IEA 2023e, Raimi *et al* 2024, Raimi and Newell 2023). While SWOT and hybrid analyses provide, at the very least, talking points among scholarly researchers, entrepreneurs, the media and industrial lobbyists, the political decision makers have to factor in a much broader set of criteria in planning the way ahead for a future renewables-alone energy mix. Notwithstanding this schism in action advocacy and action pragmatism, SWOT and similar analyses should provide insights into how best to tackle energy and environmental issues globally, nationally and locally in the pursuit of a sustainable and renewable future. Although it has always to be borne in mind that, to paraphrase a quote attributed to the French author-philosopher Voltaire[22], 'the best can be the enemy of the good'.

[22] The *nom de plume* of François-Marie Arouet, the eighteenth century author and philosopher.

In section 1.5, some concluding remarks are provided with regard to the development of renewable energy sources since their identification at the 1992 UN Rio Conference as the cornerstone of a future energy transition, away from fossil-fuels. After three decades the concept has made much progress largely because of government investments driven by global fears of climate change. The overall aim is to decarbonize the Earth's energy systems but the choices of how to achieve this state of affairs are extremely challenging for most nations.

1.5 Concluding remarks

The scale of the transition to renewables in such a short time has been remarkable, albeit achieved less by persuasion and choice and more by massive financial investments and incentives, governmental policies, and enforceable regulations, but most with honest intentions to mitigate climate change. No particularly new sources of renewable energy, with the exception of white hydrogen, have been discovered, but the technical development of the associated energy converters has been rapid as has the declining cost of the many devices. Solar panels and wind turbines are now familiar sights, whether on a roof or clustered on farmland. The profound fears of the potential catastrophic effects of climate change have led to accelerated efforts to have a complete energy transition to renewable sources as soon as possible. Although there are differing opinions about what constitutes a renewable source, solar and wind power are the favoured choice for deployment and development, probably because the source is considered free, requiring no costly penetrations of the planet's crust or the removal of its natural resources. Unfortunately, this a somewhat whimsical scenario as, even for solar and wind, devices have to be manufactured to capture the energy source, convert it to usable forms, and supply it in a timely fashion to consumers through distribution facilities. The capture, conversion, and distribution processes are not free, and all require the use of natural resources largely obtained from mining activities.

Despite perennial warnings about the finiteness and growing scarcity of the energy sources that have dominated the global energy mix for almost three centuries, i.e. fossil fuels, there are sufficient amounts to maintain that position for at least another century and likely much longer. Their long history of use has been accompanied by an equal period of technical development and continual improvement together with the establishment of immense infrastructure capacities. They are reliable, with higher energy densities than any practical alternative sources, apart from nuclear energy, have continually provided competitively priced electricity and, for transportation applications, are the cheapest, if taxes are excluded. Why then are alternatives being sought? The answer is their carbon content, and resulting emissions, when combusted. These are considered to be harming both the health of the planet and the health of humans. Moreover, whatever the scale of fossil fuel reserves and resources, at sometime in the future new sources will be required. Why now? The global effects of fossil fuel usage are considered to have accelerated since the late nineteenth century causing atmospheric and ocean temperatures to increase, glaciers and ice-caps to melt, and sea-levels to rise. To limit, or even eliminate, the adverse effects and maybe even reverse them, the decision has been made to eliminate fossil fuels from global energy mixes without delay.

To correct all that is wrong with fossil fuels requires using energy sources that can never be exhausted or produce carbon emissions, especially CO_2 and CH_4. These requirements provide the recipes for renewable energy. But in the endeavours to identify and develop renewable sources there are some contradictions, a prime example being biomass. Biomass is only renewable, if human activities ensure it is renewable, and its use generates not only GHGs but also particulate matter. The arguments in favour of its use suggest that any emissions will eventually be sequestrated and that, in any case, the emissions are lower than those from fossil fuels, a somewhat contestable claim. However, if lower carbon emissions are acceptable, why is nuclear energy frowned upon, and the use of large-scale hydropower neglected? Regardless of these ambiguities, the development and adoption of renewable energy systems is projected to increase throughout the remainder of this century and play a major role in establishing electricity as the energy form of choice. Yet, as the internationally acknowledged energy expert Vaclav Smil has succinctly intimated in his assessment of the fossil fuel transition, 'The quest for abandoning these abundant, affordable and reliable sources of energy [fossil fuels] is driven overwhelmingly by a single objective, the eventual elimination of fossil carbon from the global energy supply' (Smil 2020).

There is now an expanding menu of technically viable renewable energy systems each having distinct national and global advantages and disadvantages, not only when comparing particular performance criteria, but also inherent resources. Nevertheless, while the overwhelming attention paid to solar and wind systems may be too blinkered, attempts to deliver too many different systems may be counterproductive. Moreover, the development pathways of renewable systems are in their infancy and, as discussed in the previous section, there are still many obstacles to be overcome and performances to be improved, i.e. conversion efficiencies, energy storage methods and capabilities, capacity factors, affordability, access, dispatchability, land-use, provision of clean cooking facilities, and improved air quality for all. The development of other energy sources has encountered similar bumps on their path to full technical and economic maturity as well societal acceptance, so solutions to the problems accompanying the increased use of renewables are likely to emerge. But it is to be expected that some of the solutions, if not all, will necessitate trade-offs, but should conform to the words attributed to the legendary and influential economist J M Keynes, that '[i]t is better to be roughly right than precisely wrong'.

The projections of the energy mix by 2030 and 2050 suggest that it will still involve fossil fuels and other non-renewables along with increasing shares of renewable sources especially, but not exclusively, solar and wind. It could be argued that the continued use of fossil fuels will enable sufficient time for the challenges associated with the energy transition to renewables to be better addressed. It could also be the case that the forecasters are 'hedging their bets' as, underpinning the transition, is the assertion that the elimination of fossil fuels and their GHG emissions are the only way to solve future climate change issues, real and perceived, and to placate all the political and scientific climate concerns. Notionally a simple solution to the complex problem of climate change, yet:

> For every complex problem there is an answer that is clear, simple, and wrong.
>
> H L Mencken

References

Acosta L 2015 Legal effect of United Nations resolutions under international and domestic law *Report* LL File No. 2015-012099 Global Legal Research Directorate, The Law Library of Congress, Washington, DC https://tile.loc.gov/storage-services/service/ll/llglrd/2019669646/2019669646.pdf

Adams H 2021 Biomass energy can power the planet from animal droppings *Sustainability Magazine* https://sustainabilitymag.com/renewable-energy/biomass-energy-can-power-planet-animal-droppings

Addo E K, Kabo-bah A T, Diawuo F A and Debrah S K 2023 The role of nuclear energy in reducing greenhouse gas (GHG) emissions and energy security: a systematic review *Int. J. Energy Res.* **2023** 31

Adeli K, Nachtane M, Faik A, Saifaoui D and Boulzehar A 2023 How green hydrogen and ammonia are revolutionizing the future of energy production: a comprehensive review of the latest developments and future prospects *Appl. Sci.* **13** 43

Al-Habaibeh A 2022 Want to burn less fossil fuel? Use body heat to warm buildings instead *FastCompany* https://fastcompany.com/90746028/want-to-burn-less-fossil-fuel-use-body-heat-to-warm-buildings-instead

Allcott H and Greenstone M 2012 Is there an energy efficiency gap? *J. Econ. Perspect.* **26** 3–28

Alotaibi I, Abido M A, Khalid M and Savkin A V 2020 A comprehensive review of recent advances in smart grids: a sustainable future with renewable energy resources *Energies* **13** 41

Amelang S 2023 Renewables cover more than half of Germany's electricity demand for first time this year *Clean Energy Wire* https://cleanenergywire.org/news/renewables-cover-more-half-germanys-electricity-demand-first-time-year

Amnuaylojaroen T and Parasin N 2023 Perspective on particulate matter: from biomass burning to the health crisis in mainland Southeast Asia *Toxics* **11** 14

Anandan V 2023 Internal combustion engines: emissions regulations for small off-road engines *delta-q Technologies* https://delta-q.com/industry-news/internal-combustion-engines-emissions-regulations-for-small-off-road-engines/

Annex M 2023 Renewable energy investment hits record-breaking $358 billion in 1H 2023 *BloombergNEF* https://about.bnef.com/blog/renewable-energy-investment-hits-record-breaking-358-billion-in-1h-2023/

APIS 2016 Ammonia *Air Pollution Information System* https://www.apis.ac.uk/overview/pollutants/overview_nh3.htm

Averna 2023 Is this the end of the internal combustion engine? *The Averna Blog e-learning series* https://insight.averna.com/en/resources/blog/the-end-of-the-internal-combustion-engine

Barbarino M 2023 What is nuclear fusion? *IAEA* https://iaea.org/newscenter/news/what-is-nuclear-fusion

Barth A *et al* 2024 How grid operators can integrate the coming wave of renewable energy *McKinsey* https://mckinsey.com/industries/electric-power-and-natural-gas/our-insights/how-grid-operators-can-integrate-the-coming-wave-of-renewable-energy#/

Bayraktar M P, Sokukcu M and Yuksel O 2023 A SWOT-AHP analysis on biodiesel as an alternative future marine fuel *Clean Techn. Environ. Policy* **25** 2233–48

Bennett P 2023 Climate change is costing the world $16 million per hour: study *World Economic Forum: Climate Action* https://weforum.org/agenda/2023/10/climate-loss-and-damage-cost-16-million-per-hour/

Benzaghta M A, Elwalda A, Mousa M M, Erkan I and Rahman M 2021 SWOT analysis applications: an integrative literature review *J. Global Business Insights* **6** 55–73

Berni F, Pessina V, Teodosio L, d'Adamo A, Borghi M and Fontanesi S 2024 An integrated 0D/1D/3D numerical framework to predict performance, emissions, knock and heat transfer in ICEs fueled with NH_3–H_2 mixtures: the conversion of a marine diesel engine as case study *Int. J. Hydrogen Energy* **50** 908–38

Biddle C J 2024 Understanding the differences between non-dispatchable and dispatchable generation *PCI Energy Solutions* https://pcienergysolutions.com/2024/05/01/understanding-the-differences-between-non-dispatchable-and-dispatchable-generation/

Black R 2022 Climate economics—costs and benefits *Energy and Climate Intelligence Unit* https://eciu.net/analysis/briefings/climate-impacts/climate-economics-costs-and-benefits

Blomgren G E 2017 The development and future of lithium ion batteries *J. Electrochem. Soc.* **164** A5019–25

BP 2022 *BP Statistical Review 2022* 71st edn (London: BP) https://bp.com/content/dam/bp/business-sites/en/global/corporate/pdfs/energy-economics/statistical-review/bp-stats-review-2022-full-report.pdf

BP 2023 *BP Energy Outlook: 2023* (London: BP) https://bp.com/content/dam/bp/business-sites/en/global/corporate/pdfs/energy-economics/energy-outlook/bp-energy-outlook-2023.pdf

Brake W H 1975 Air pollution and fuel crises in preindustrial London, 1250–1650 *Technol. Cult.* **16** 337–59

Brinkman G, Bain D, Buster G, Draxl C, Das P, Ho J and Avi P 2021 *The North American Renewable Integration Study: A US Perspective* (Golden, CO: NREL) https://nrel.gov/docs/fy21osti/79224.pdf

Brooks A 2022 *Renewable Energy Resource Assessment Information for the United States* (Washington, DC: Office of Energy Efficiency and Renewable Energy, USDOE) https://energy.gov/sites/default/files/2022-03/Renewable%20Energy%20Resource%20Assessment%20Information%20for%20the%20United%20States.pdf

Brown A, Beiter P, Heimiller D, Davidson C, Denholm P, Melius J and Porro G 2016 Estimating renewable energy economic potential in the United States: methodology and initial results *Technical Report* NREL/TP-6A20-64503 National Renewable Energy Laboratory, Golden, CO https://nrel.gov/docs/fy15osti/64503.pdf

Brownfield M E 2016 Assessment of undiscovered oil and gas resources of the Orange River Coastal Province, Southwest Africa *Geologic Assessment of Undiscovered Hydrocarbon Resources of Sub-Saharan Africa: US Geological Survey Digital Data Series 69–GG* (Reston, VA: USGS) ch 18 https://pubs.usgs.gov/dds/dds-069/dds-069-gg/REPORTS/69_GG_CH_8.pdf

Brundtland G H and Mansour K 1987 *Our Common Future: Report of the World Commission on Environment and Development* (Oxford: Oxford University Press) https://sustainabledevelopment.un.org/content/documents/5987our-common-future.pdf

Bryson B 2003 Getting the lead out *A Short History of Nearly Everything* (New York: Broadway—Crown) ch 10

Burger B 2024 Electricity generation in Germany 2023 *Report* Fraunhofer ISE, Freiburg https://www.ise.fraunhofer.de/content/dam/ise/en/documents/downloads/electricity_generation_germany_2023.pdf

Burke A E 2017 *Torpedoes and Their Impact on Naval Warfare* 1st edn (Newport: Defense Technical Information Center) https://apps.dtic.mil/sti/tr/pdf/AD1033484.pdf

California Energy Commission 2024 Hydroelectric power *California Energy Commission* https://www.energy.ca.gov/data-reports/california-power-generation-and-power-sources/hydroelectric-power

Camia A *et al* 2021 *The Use of Woody Biomass for Energy Production in the EU* (Luxembourg: Joint Research Centre (European Commission))

Carrington D 2021 How much of the world's oil needs to stay in the ground? *The Guardian* www.theguardian.com/environmenthttps://theguardian.com/environment/2021/sep/08/climate-crisis-fossil-fuels-ground

Carter J E 1977 Address to the Nation on Energy *Miller Center, University of Virginia* https://millercenter.org/the-presidency/presidential-speeches/april-18-1977-address-nation-energy

Center for Sustainable Systems 2023 Wind energy factsheet *Center for Sustainable Systems, University of Michigan* https://css.umich.edu/publications/factsheets/energy/wind-energy-factsheet

Challa V 2022 Selecting the right battery for your application, part 1: important battery metrics *ANSYS Blog* https://ansys.com/blog/important-battery-metrics

Chandrasekaran S, Khan F and Abbassi R 2022 *Wave Energy Devices: Design, Development, and Experimental Studies* 1st edn (Boca Raton, FL: CRC Press—Routledge) https://www.routledge.com/Wave-Energy-Devices-Design-Development-and-Experimental-Studies/Chandrasekaran-Khan-Abbassi/p/book/9781032250779

Climate Action Tracker 2023 CAT net zero target evaluations *Climate Action Tracker* https://climateactiontracker.org/global/cat-net-zero-target-evaluations/

Climate Action Tracker 2024 Countries *Climate Action Tracker* https://climateactiontracker.org/countries/

Climate Champions 2023 2030 Climate Solutions Implementation Roadmap *Report* UNFCCC, Bonn https://www.climatechampions.net/media/vknpzpxo/2030-climate-solutions-publication-implementation-roadmap.pdf

Colin-Oesterlé N 2023 The end of combustion engines: the cart before the horse! *Policy paper* No. 670 Fondation Robert Schuman: European Issues p 3 https://server.www.robert-schuman.eu/storage/en/doc/questions-d-europe/qe-670-en.pdf

Cordonnier J and Saygin D 2022 *Green Hydrogen Opportunities for Emerging and Developing Economies: Identifying Success Factors for Market Development and Building Enabling Conditions* (Paris: OECD)

Couture T D 2018 Enter the Shark Curve *Analytical Brief* E3 Anlaytics, Berlin https://e3analytics.eu/wp-content/uploads/2019/11/Analytical-Brief_May-2018_FINAL.pdf

Covert T, Greenstone M and Knittel C R 2016 Will we ever stop using fossil fuels? *J. Econ. Perspect.* **30** 117–38

Cox S, Lopez A, Watson A, Grue N and Leisch J E 2018 Renewable energy data, analysis, and decisions: a guide for practitioners *Technical Report* NREL/TP-6A20-68913 National Renewable Energy Laboratory, Golden, CO https://nrel.gov/docs/fy18osti/68913.pdf

Crowley T J 2000 Causes of climate change over the past 1000 years *Science* **289** 270–7

Cullen J M and Allwood J M 2010 Theoretical efficiency limits for energy conversion devices *Energy* **35** 2059–69

David W I *et al* 2024 2023 roadmap on ammonia as a carbon-free fuel *J. Phys.: Energy* **6** 57

De A, Hayes M, Jodas F and Virley S 2023 Turning the tide in scaling renewables *Report* KPMG, London https://assets.kpmg.com/content/dam/kpmg/gr/pdf/2024/01/gr-turning-the-tide-in-scaling-renewables.pdf

Dell M, Jones B F and Olken B A 2008 Climate change and economic growth: evidence from the last half-century *Working Paper* 14132 National Bureau of Economic Research, Cambridge, MA https://nber.org/system/files/working_papers/w14132/w14132.pdf

Denholm P, Margolis R and Milford J 2008 Production cost modeling for high levels of photovoltaics penetration *Technical Report* NREL/TP-581-42305 National Renewable Energy Laboratory, Golden, CO https://nrel.gov/docs/fy08osti/42305.pdf

Destribats J 2023 The European countries doing the most for the EV transition *Motor Finance Online News* https://motorfinanceonline.com/news/incentives/? cf-view

Dictionary.com 2024 The reports of my death are greatly exaggerated *Dictionary.com* www.dictionary.com/browsehttps://dictionary.com/browse/the-reports-of-my-death-are-greatly-exaggerated

Divadkar Y, Gutta S and Vyas A 2021 Mission 2070: A Green New Deal for a net zero India *White Paper* World Economic Forum, Geneva https://www3.weforum.org/docs/WEF_Mission_2070_A_Green_New_Deal_for_a_Net_Zero_India_2021.pdf

Dodds J 2021 The psychology of climate anxiety *BJPsych Bull.* **45** 222–6

Dragonfly Energy 2022 What is thermal runaway in batteries? *Dragonfly Energy* https://dragonflyenergy.com/thermal-runaway/

Dunlea M A 2023 *Putting Out The Planetary Fire* (Poestenkill, NY: Big Toad) https://gelfny.org/wp-content/uploads/2023/02/Putting-Out-The-Planetary-Fire-book-pdf.pdf

DUT Wind Energy 2024 Introduction *The Global Wind Atlas* https://globalwindatlas.info/en/about/introduction

Dyke J, Watson R and Knorr W 2021 Climate scientists: concept of net zero is a dangerous trap *The Conversation* https://theconversation.com/climate-scientists-concept-of-net-zero-is-a-dangerous-trap-157368

Earle S and Panchuk K 2019 *Physical Geology* 2nd edn (Victoria, BC: BC Open Collection) https://opentextbc.ca/physicalgeology2ed/

Eberle A, Mai T, Roberts O, Williams T, Pinchuk P, Lopez A and Lantz E 2024 Incorporating wind turbine choice in high-resolution geospatial supply curve and capacity expansion models *Technical Report* NREL/TP-6A20-87161 National Renewable Energy Laboratory, Golden, CO https://nrel.gov/docs/fy24osti/87161.pdf

Ediger V and Akar S 2023 Historical pattern analysis of global geothermal power capacity development *Conference Paper* NREL/CP-7A40-86996 National Renewable Energy Laboratory, Golden, CO https://nrel.gov/docs/fy24osti/86996.pdf

ElectricRate 2024 Pricing of electricity by country *ElectricRate* https://electricrate.com/data-center/electricity-prices-by-country/

Energy Education 2024 Geothermal gradient *Energy Education* https://energyeducation.ca/encyclopedia/Geothermal_gradient

Energy Institute 2023a *Statistical Review of World Energy 2023* (London: Energy Institute) https://energyinst.org/statistical-review

Energy Institute 2023b Statistical Review of World Energy *Data* Energy Institute, London https://www.energyinst.org/__data/assets/excel_doc/0020/1540550/EI-Stats-Review-All-Data.xlsx

Energy Institute 2023c About the statistical review *Energy Institute* https://energyinst.org/statistical-review/about

Energy Sector Management Assistance Program 2020 *Global Photovoltaic Power Potential by Country* (Washington, DC: World Bank) https://documents1.worldbank.org/curated/en/466331592817725242/pdf/Global-Photovoltaic-Power-Potential-by-Country.pdf

Energy Sector Management Assistance Program 2022a *Regulatory Indicators for Sustainable Energy (RISE)* (Washington, DC: The World Bank) https://rise.esmap.org/data/files/reports/2022/RISE%202022%20Report%20Building%20Resilience.pdf

Energy Sector Management Assistance Program 2022b Direct utilization of geothermal resources *Technical Report* 21/22 ESMAP and World Bank, Washington, DC https://esmap.org/sites/default/files/esmap-files/16103-WB_ESMAP%20Direct%20Use-WEB.pdf

European Commission 2020 On the progress made on the implementation of directive (EU) 2016/2284 on the reduction of national emissions of certain atmospheric pollutants *Report* European Commission, Brussels https://eur-lex.europa.eu/resource.html? uri = cellar: 7199e9c2-b7bf-11ea-811c-01aa75ed71a1.0007.02/DOC_1&format = PDF

European Commission, Directorate-General for Energy 2024 Union Bioenergy Sustainability Report—Study to support reporting under Article 35 of Regulation (EU) 2018/1999 *Final Report* Publications Office of the European Union, Brussels

European Commission: Energy, Climate Change, Environment 2023 Biomass *European Commission* https://energy.ec.europa.eu/topics/renewable-energy/bioenergy/biomass_en

European Environmental Agency 2023 Digitalisation in the mobility system : challenges and opportunities *Transport and Environment Report 2022* EEA, Copenhagen

European Environmental Agency 2024 *European Climate Risk Assessment* (Luxembourg: Publications Office of the European Union)

European Union 2023 *Directive (EU) 2023/2413 of the European Parliament and of the Council* Document 32023L2413 Official Journal of the European Union, EUR-Lex, Brussels http://data.europa.eu/eli/dir/2023/2413/oj

Evans S 2020 Solar is now 'cheapest electricity in history', confirms IEA *CarbonBrief* https://carbonbrief.org/solar-is-now-cheapest-electricity-in-history-confirms-iea/

Fattal T N 2018 Increasing wind turbine efficiency *Int. J. Technol. Eng. Stud.* **4** 120–31

Federal Reserve Bank of St Louis 2024 Global price of Energy index *FRED* https://fred.stlouisfed.org/series/PNRGINDEXM

Feigin S V *et al* 2023 Proposed solutions to anthropogenic climate change: a systematic literature review and a new way forward *Heliyon* **9** 26

Fisher M and Liou J 2021 How can nuclear replace coal as part of the clean energy transition? *IAEA Newscenter* https://iaea.org/newscenter/news/how-can-nuclear-replace-coal-as-part-of-the-clean-energy-transition

Fuergy 2019 Geothermal energy: the clean renewable energy hidden inside our planet *Fuergy* https://fuergy.com/blog/geothermal-energy-the-clean-renewable-energy-hidden-inside-our-planet

Gibon T, Menacho Á H and Guirton M G 2022 *Carbon Neutrality in the UNECE Region: Integrated Life-cycle Assessment of Electricity Sources* (Geneva: United Nations Econoomic Commisson for Europe) https://unece.org/sites/default/files/2022-04/LCA_3_FINAL%20March%202022.pdf

Gill T 2024 The world's biggest solar farms *the eco experts* https://theecoexperts.co.uk/solar-panels/biggest-solar-farms

Gillingham K 2019 Carbon calculus *International Monetary Fund: Finance and Development* https://imf.org/en/Publications/fandd/issues/2019/12/the-true-cost-of-reducing-greenhouse-gas-emissions-gillingham

Gillingham K and Stock J H 2018 The cost of reducing greenhouse gas emissions *J. Econ. Perspect.* **32** 53–72

Gillingham K, Newell R G and Palmer K 2009 Energy efficiency economics and policy *Annu. Rev. Resour. Econ.* **1** 597–620

GlobalData 2024 Top five solar PV plants in development in Russia *Power Technology* https://power-technology.com/data-insights/top-5-solar-pv-plants-in-development-in-russia/

Goggin M 2020 Capacity markets: the way of the future or the way of the past? *ESIG Energy Systems Integration Group* https://esig.energy/capacity-markets-the-way-of-the-future-or-the-way-of-the-past/

Görener A, Toker K and Uluçay K 2012 Application of combined SWOT and AHP: a case study for a manufacturing firm *Procedia—Soc. Behav. Sci.* **58** 1525–34

GOV.UK 2020 Strategic environmental assessment and sustainability appraisal *Guidance* Department for Levelling Up, Housing and Communities; Ministry of Housing, Communities and Local Government https://gov.uk/guidance/strategic-environmental-assessment-and-sustainability-appraisal

Government of Canada 2023 *The Canadian Critical Minerals Strategy* (Ottawa: Natural Resources Canada) https://canada.ca/content/dam/nrcan-rncan/site/critical-minerals/Critical-minerals-strategyDec09.pdf

Government of Malta/UN 1988 Conservation of climate as part of the common heritage of mankind *Draft Resolution* United Nations, New York http://digitallibrary.un.org/record/49293

Guangul F M and Chala G T 2019 Solar energy as renewable energy source: SWOT analysis *4th MEC Int. Conf. on Big Data and Smart City (ICBDSC) (Muscat, Oman)* (Piscataway, NJ: IEEE) p 1–5

Gupta A K 2015 Efficient wind energy conversion: evolution to modern design *J. Energy Resour. Technol.* **135** 10

H2 Bulletin 2024 Hydrogen colours codes *H2 Bulletin* https://h2bulletin.com/knowledge/hydrogen-colours-codes/

Haan S 2024 SWOT analysis; learn about the SWOT framework, the process of a SWOT analysis, and its advantages and disadvantages *The Chartered Institute of Personnel and Development* https://cipd.org/en/knowledge/factsheets/swot-analysis-factsheet/#the-swot-process

Haeberlin A, Rosch Y, Tholl M V, Gugler Y, Okle J, Heinisch P P and Zurbuchen A 2020 Intracardiac turbines suitable for catheter-based implantation-an approach to power battery and leadless cardiac pacemakers? *IEEE Trans. Biomed. Eng.* **67** 1159–66

Halawa E 2024 Sustainable energy: concept and definition in the context of the energy transition—a critical review *Sustainability* **16** 14

Hall C 2021 Future batteries, coming soon: charge in seconds, last months and power over the air *Pocket-Lint* https://pocket-lint.com/gadgets/news/130380-future-batteries-coming-soon-charge-in-seconds-last-months-and-power-over-the-air/

Hardisty A, Pearce-Higgins R and Hadaya N 2023 Industrial non-road mobile machinery decarbonisation options: techno-economic *Final Report* Project No.: 0671307 The Department for Energy Security and Net Zero, Energy Resources Management, London https://assets.publishing.service.gov.uk/media/658443f3ed3c3400133bfd4d/nrmm-decarbonisation-options-feasibility-report.pdf

Harvey C and Heikkinen N 2018 Congress says biomass is carbon-neutral, but scientists disagree *Scientific American* https://scientificamerican.com/article/congress-says-biomass-is-carbon-neutral-but-scientists-disagree/

Hawken P 2018 *Drawdown: The Most Comprehensive Plan Ever Proposed to Reverse Global Warming* (New York: Penguin Random House)

Hegeral G C, Brönnimann S, Cowan T, Friedman A R, Hawkins E, Iles C and Undorf S 2019 Causes of climate change over the historical record *Environ. Res. Lett.* **14** 26

Hong J, Chae C, Kim H, Kwon H, Kim J and Kim I 2023 Investigation to enhance solid fuel quality in torrefaction of cow manure *Energies* **16** 4505

Ibrahim H, Ilinca A and Perron J 2007 Energy storage systems—characteristics and comparisons *Renew. Sustain. Energy Rev.* **12** 1221–50

IEA 2021 Electricity information: overview *IEA* https://iea.org/reports/electricity-information-overview

IEA 2022a How much CO_2 does Germany emit? *IEA* https://iea.org/countries/germany/emissions

IEA 2022b Energy consumption in transport by fuel in the Net Zero Scenario, 2000–2030 *IEA* https://iea.org/data-and-statistics/charts/energy-consumption-in-transport-by-fuel-in-the-net-zero-scenario-2000-2030

IEA 2023a A vision for clean cooking access for all *World Energy Outlook Special Report* IEA, Paris https://iea.org/reports/a-vision-for-clean-cooking-access-for-all

IEA 2023b *Global Hydrogen Review 2023* (Paris: IEA) https://iea.blob.core.windows.net/assets/ecdfc3bb-d212-4a4c-9ff7-6ce5b1e19cef/GlobalHydrogenReview2023.pdf

IEA 2023c *Electricity Market Report 2023* IEA, Paris https://iea.blob.core.windows.net/assets/255e9cba-da84-4681-8c1f-458ca1a3d9ca/ElectricityMarketReport2023.pdf

IEA 2023d Energy efficiency 2023 *Report* IEA, Paris https://iea.org/reports/energy-efficiency-2023

IEA 2023e *World Energy Outlook 2023* (Paris: IEA) https://iea.blob.core.windows.net/assets/86ede39e-4436-42d7-ba2a-edf61467e070/WorldEnergyOutlook2023.pdf

IEA 2023f Energy end-uses and efficiency indicators, highlights *IEA* https://iea.org/data-and-statistics/data-product/energy-efficiency-indicators-highlights

IEA 2023g *New Zealand 2023 Energy Policy Review* (Paris: IEA) https://iea.org/reports/new-zealand-2023

International Finance Corporation 2013 *Success of Geothermal Wells: A Global Study* (Washington, DC: World Bank) https://documents1.worldbank.org/curated/en/305681468168834775/pdf/782300WP0Succe00Box0377330B0PUBLIC0.pdf

International Hydropower Association 2022a Hydropower's carbon footprint: factsheet *IHA* https://hydropower.org/factsheets/greenhouse-gas-emissions

International Hydropower Association 2022b Pumped storage hydropower; the world's oldest battery *Report* International Hydropower Association—ESMAP, London https://esmap.org/sites/default/files/ESP/WB_PSH_16Jun22.pdf

International Renewable Energy Agency 2024 Ocean energy *IRENA* https://www.irena.org/Energy-Transition/Technology/Ocean-energy

International Trade Administration 2024 Norway: offshore energy—oil, gas and renewables *International Trade Administration* https://trade.gov/country-commercial-guides/norway-off-shore-energy-oil-gas-and-renewables

IPCC 2018 *Global Warming of 1.5°C* (Cambridge: Cambridge University Press)

Ireland R 2022 The rise of utility wood pellet energy in the era of climate change *Working Paper* ID-088 US International Trade Commission, Washington, DC https://usitc.gov/publications/332/working_papers/wood_pellets_final_060622.pdf

IRENA 2017 Renewable energy statistics 2017 *Report* International Renewable Energy Agency, Abu Dhabi https://www.irena.org/Publications/2017/Jul/Renewable-Energy-Statistics-2017

IRENA 2023a Renewable energy statistics 2023 *Report* International Renewable Energy Agency, Abu Dhabi https://www.irena.org/Publications/2023/Jul/Renewable-energy-statistics-2023

IRENA 2023b Renewable capacity statistics 2023 *Report* International Renewable Energy Agency, Abu Dhabi https://www.irena.org/Publications/2023/Mar/Renewable-capacity-statistics-2023

IRENA 2023c Renewable power generation costs in 2022 *Report* International Renewable Energy Agency, Abu Dhabi https://www.irena.org/Publications/2023/Aug/Renewable-Power-Generation-Costs-in-2022

IRENA 2024 *The Global Atlas for Renewable Energy: A Decade in the Making* (Abu Dhabi: International Renewable Energy Agency) https://www.irena.org/Publications/2024/Apr/The-Global-Atlas-for-Renewable-Energy-A-decade-in-the-making

IRENA and AEA 2022 Innovation outlook: renewable ammonia *Report* International Renewable Energy Agency and Ammonia Energy Association, Abu Dhabi and Brooklyn https://www.irena.org/publications/2022/May/Innovation-Outlook-Renewable-Ammonia

IRENA and CPI 2023 *Global Landscape of Renewable Energy Finance* (Abu Dhabi: International Renewable Energy Agency)

IRENA and IGA 2023 *Global Geothermal Market and Technology Assessment* (Abu Dhabi and The Hague: International Renewable Energy Agency and International Geothermal Association) https://www.irena.org/Publications/2023/Feb/Global-geothermal-market-and-technology-assessment

Jaremko D 2023 A matter of fact: the IEA's upated net zero scenario is still unrealistic *Canadian Energy Centre* https://www.canadianenergycentre.ca/a-matter-of-fact-the-ieas-updated-net-zero-scenario-is-still-unrealistic/

Johns Hopkins School of Advanced International Studies 2022 The inflation reduction act and renewable energy policy *Johns Hopkins School of Advanced International Studies* https://energy.sais.jhu.edu/articles/renewable-energy-policy/

Johnson E 2009 Goodbye to carbon neutral: getting biomass footprints right *Environ. Impact. Asses. Rev.* **29** 165–8

Kabeyi M J and Olanrewaju O A 2023 The levelized cost of energy and modifications for use in electricity generation planning *Energy Rep.* **9** 495–534

KamLAND Collaboration 2011 Partial radiogenic heat model for Earth revealed by geoneutrino measurements *Nat. Geosci.* **4** 647–51

Kansongue N, Njuguna J and Vertigans S 2023 A PESTEL and SWOT impact analysis on renewable energy development in Togo *Front. Sustain.* **3** 19

Kilajian A and Mercier-Blais S 2019 Carbon emissions from hydropower reservoirs: facts and myths *IHA* https://hydropower.org/blog/carbon-emissions-from-hydropower-reservoirs-facts-and-myths

Kilcher L, Fogarty M and Lawson M 2021 Marine energy in the United States: an overview of opportunities *Technical Report* NREL/TP-5700-78773 National Renewable Energy Laboratory, Golden, CO https://nrel.gov/docs/fy21osti/78773.pdf

Kotz M, Levermann A and Wenz L 2024 The economic commitment of climate change *Nature* **628** 551–7

Kovarik W 2005 Ethyl-leaded gasoline: how a classic occupational disease became an international public health disaster *Int. J. Occup. Environ. Health* **11** 384–97 https://cdn.toxicdocs.org/gD/gDY3Qqag1rRVG877ydd6e5r0V/gDY3Qqag1rRVG877ydd6e5r0V.pdf

Kumar A, Schei T, Ahenkorah A, Caceres Rodriguez R, Devernay J-M, Freitas M and Killingtveit Å 2012 Hydropower *IPCC Special Report on Renewable Energy Sources and Climate Change* (Cambridge and New York: Cambridge University Press and IPCC) https://ipcc.ch/site/assets/uploads/2018/03/Chapter-5-Hydropower-1.pdf

Kumar M and Samuel C 2017 Selection of best renewable energy source by using VIKOR method *Technol. Econ. Smart Grids Sustain. Energy* **2** 10

Kurpaska S, Knaga J, Latała H, Sikora J and Tomczyk W 2018 Efficiency of solar radiation conversion in photovoltaic panels *BIO Web Conf.: Contemporary Research Trends in Agricultural Engineering* **10** 02014

Kurth C and Pihkala P 2023 Eco-anxiety: what it is and why it matters *Front. Psychol.* **13** 981814

Kyne D 2024 Pairwise comparison (definition, methods, examples, tools) *OpinionX* https://opinionx.co/blog/pairwise-comparison

L J 2023 Decoding the climate giants: top carbon emitters race to net zero and renewable future *Carbon Credits* https://carboncredits.com/carbon-emission-country-decoding-the-climate-giants-china-u-s-and-indias-race-to-net-zero-and-renewable-future/

Lai C 2022 Tidal energy: advantages, disadvantages, and future trends *Earth.Org* https://earth.org/what-is-tidal-energy/

Lanphear B P, Rauch S, Auinger P, Allen R W and Hornung R W 2018 Low-level lead exposure and mortality in US adults: a population-based cohort study *Lancet: Public Health* **3** e177–84

Lattanzio R K 2022 Clean Air Act: a summary of the Act and its major requirements *Report* Congressional Research Service, Washington, DC https://crsreports.congress.gov/product/pdf/RL/RL30853

Lawrence Livermore National Laboratory 2023 US energy consumption in 2022 *Energy Flow Charts* https://flowcharts.llnl.gov/commodities/energy

Lee C 2020 The buildings heated by human warmth *BBC Future* https://bbc.com/future/article/20200908-the-buildings-warmed-by-the-human-body

Lee H and Romero J 2023 Climate Change 2023: Contribution of Working Groups I, II and III to the Sixth Assessment Report *AR6 Synthesis Report* IPCC, Geneva

Lei Y, Lu X, Shi M, Wang L, Lv H, Chen S and da Silveira S D 2019 SWOT analysis for the development of photovoltaic solar power in Africa in comparison with China *Environ. Impact Assess. Rev.* **77** 122–7

Levin K, Fransen T, Schumer C, Davis C and Boehm S 2023 What does 'net-zero emissions' mean? 8 common questions, answered *World Resources Institute* https://wri.org/insights/net-zero-ghg-emissions-questions-answered

Li K, He Q, Wang J, Zhou Z and Li X 2018 Wearable energy harvesters generating electricity from low-frequency human limb movement *Microsyst. Nanoeng.* **4** 13

Limb L 2024 Solar balconies are booming in Germany. Here's what you need to know about the popular home tech *euro news* https://www.euronews.com/green/2024/07/23/solar-balconies-are-booming-in-germany-heres-what-you-need-to-know-about-the-popular-home-

Lindstrand N 2024 Unlocking ammonia's potential for shipping *MAN Energy Solutions* https://man-es.com/discover/two-stroke-ammonia-engine

Lindwall C 2022 What are the effects of climate change? *Natural Resources Defense Council* https://nrdc.org/stories/what-are-effects-climate-change#weather

Lo J 2023 UN tells governments to 'fast forward' net zero targets *Climate Home News* https://climatechangenews.com/2023/03/20/un-tells-governments-to-fast-forward-net-zero-targets/

Lovegrove K, James G, Leitch D, Milczarek A, Ngo A, Rutovitz J and Wyder J 2018 Comparison of dispatchable renewable electricity options *Report* Australian Renewable Energy Agency, Canberra https://arena.gov.au/knowledge-bank/comparison-of-dispatchable-renewable-electricity-options/

Lund J W 2024 Geothermal energy *Encyclopedia Britannica* https://britannica.com/science/geothermal-energy (Accessed: 22 April 2024)

Lund J W and Toth A N 2020 Direct utilization of geothermal energy worldwide review *Geothermics* **90** 40

MacKay D J 2013 Solar energy in the context of energy use, energy transportation and energy storage *Phil. Trans. R. Soc.* A*371 1–24*

Magalhães P 2020 Climate as a concern or a heritage? Addressing the legal structural roots of climate emergency *Rev. Electron. Dir.* **21** 36

Maguire G 2024 US electricity demand from EVs jumps to new highs in early 2024 *Reuters* https://reuters.com/business/energy/us-electricity-demand-evs-jumps-new-highs-early-2024-maguire-2024-05-22/

Maika N, Lin W and Khatamifar M 2023 A review of gravitational water vortex hydro turbine systems for hydropower generation *Energies* **16** 39

MAN Energy Solutions 2021 *MAN B&W Two-Stroke Engine Operating on Ammonia* (Munich: MAN-es) https://man-es.com/docs/default-source/marine/tools/man-b-w-two-stroke-engine-operating-on-ammonia.pdf

Mardani A, Jusoh A, Zavadskas E K, Cavallaro F and Khalifah Z 2015 Sustainable and renewable energy: an overview of the application of multiple criteria decision making techniques and approaches *Sustainability* **7** 13947–84

Martinez J 2024 Great Smog of London *Encyclopedia Britannica* https://britannica.com/event/Great-Smog-of-London (Accessed: 3 March 2024)

McGlade C and Ekins P 2015 The geographical distribution of fossil fuels unused when limiting global warming to 2°C *Nature* **517** 187–90

Mehetre S A, Panwar N L, Sharma D and Kumar H 2017 Improved biomass cookstoves for sustainable development: a review *Renew. Sustain. Energy Rev.* **73** 672–87

Melosi M V 2004 The automobile and the environment in American history *Automobile in American Life and Society* http://autolife.umd.umich.edu/Environment/E_Overview/E_Overview.htm

Ministry of Business, Innovation and Employment, Hīkina Whakatutuki 2024 Energy in New Zealand 2024 *Report* Government of New Zealand, Wellington https://mbie.govt.nz/assets/energy-in-nz-2024.pdf

Moses O 2023 Who finances energy projects in Africa? *Working Paper* Carnegie Endowment for International Peace, Washington, DC https://carnegie-production-assets.s3.amazonaws.com/static/files/Moses_Energy_Finance_1.pdf

Mullan M and Ranger N 2022 Climate-resilient finance and investment: framing paper *Environment Working Paper* #136 OECD, Environment Directorate, Paris https://one.oecd.org/document/env/wkp(2022)8/en/pdf

Namugenyi C, Nimmagadda S L and Reiners T 2019 Design of a SWOT analysis model and its evaluation in diverse digital business ecosystem contexts *Procedia Comp. Sci.* **159** 1145–54

National Academy of Sciences 2020 *Climate Change: Evidence and Causes: Update 2020* (Washington, DC: The National Academies Press)

National Geographic Society 2024a Renewable energy *National Geographic Education* https://education.nationalgeographic.org/resource/renewable-energy/

National Geographic Society 2024b Core: Earth's core is the very hot, very dense center of our planet *National Geographic Education* https://education.nationalgeographic.org/resource/core/

National Grid 2022a What are the different types of renewable energy? *National Grid Energy Explained* https://nationalgrid.com/stories/energy-explained/what-are-different-types-renewable-energy

National Grid 2022b Onshore vs offshore wind energy: what's the difference? *National Grid Energy Explained* https://nationalgrid.com/stories/energy-explained/onshore-vs-offshore-wind-energy

National Grid 2023a What is green energy? *National Grid Energy Explained* https://nationalgrid.com/stories/energy-explained/what-is-green-energy

National Grid 2023b The hydrogen colour spectrum *National Grid* https://nationalgrid.com/stories/energy-explained/hydrogen-colour-spectrum

Needleman H L 2000 The removal of lead from gasoline: historical and personal reflections *Environ. Res.* **84** 20–35

Newman R and Noy I 2023 The global costs of extreme weather that are attributable to climate change *Nat. Commun.* **14** 13

Norbeck J M, Heffel J W, Durbin T D, Tabbara B, Bowden J M and Montano M C 1996 *Hydrogen Fuel for Surface Transportation* (Warrendale, PA: Society of Automotive Engineers)

Nordhauser N 1973 Origins of federal oil regulation in the 1920s *Bus. Hist. Rev.* **47** 53–71

NREL 2023a Geothermal *Annual Technology Baseline* https://atb.nrel.gov/electricity/2023/geothermal

NREL 2023b Utility-scale PV *Annual Technology Baseline* https://atb.nrel.gov/electricity/2023/utility-scale_pv

NREL 2024 Geothermal electricity production basics *NREL* https://nrel.gov/research/re-geo-elec-production.html (Accessed: 22 April 2024)

Nunes L J, Matias J C and Catalao J P 2017 *Torrefaction of Biomass for Energy Applications: From Fundamentals to Industrial Scale* (Amsterdam: Academic Press—Elsevier)

OECD 2018 *Financing Climate Futures: Rethinking Infrastructure, Policy Highlights* (Paris: OECD/The World Bank/UN Environment) https://www.oecd.org/en/publications/financing-climate-futures_9789264308114-en.html

Office of Energy Efficiency and Renewable Energy 2019 PV Cells 101, part 2: solar photovoltaic cell research directions *Solar Energy Technologies Office* https://energy.gov/eere/solar/articles/pv-cells-101-part-2-solar-photovoltaic-cell-research-directions

Office of Energy Efficiency and Renewable Energy 2024 Enhanced geothermal systems *US Department of Energy* https://energy.gov/eere/geothermal/enhanced-geothermal-systems

Offshore staff 2024 Russia reportedly finds vast oil and gas reserves in British Antarctic territory *Offshore* https://offshore-mag.com/geosciences/article/55039736/russia-reportedly-finds-vast-oil-and-gas-reserves-in-british-antarctic-territory

Okoye P U, Ezeokonkwo J U and Nworji G C 2017 Sustainability of renewable sources of energy for rural communities in Anambra State *Adv. Energy Power* **5** 37–47

Oliver S, Jessica L, Martínez I L, Sophie G, Economou T and Adair-Rohani H 2021 Household cooking fuel estimates at global and country level for 1990 to 2030 *Nat. Commun.* **12** 8

Our World in Data 2024 Data page: energy intensity *Our World in Data* https://ourworldindata.org/grapher/energy-intensity-of-economies (Accessed: 2 June 2024)

Penuelas J, Coello F and Sardans J 2023 A better use of fertilizers is needed for global food security and environmental sustainability *Agric. Food Secur.* **12** 5 9

Peterson T C, Connolley W M and Fleck J 2008 The myth of the 1970s global cooling scientific consensus *Bull. Am. Meteorol. Soc.* **89** 1325–38

Philip P, Ibrahim C and Hodges C 2022 *The Turning Point: A Global Summary* (London: Deloitte Economics Institute) https://deloitte.com/an/en/issues/climate/global-turning-point.html

Planete Energies 2021 Very low-temperature geothermal energy *Planete Energies* https://planete-energies.com/en/media/article/very-low-temperature-geothermal-energy

Poore C 2023 Ammonia fuel offers great benefits but demands careful action *Princeton Engineering* https://engineering.princeton.edu/news/2023/11/07/ammonia-fuel-offers-great-benefits-demands-careful-action

Quaranta E *et al* 2020 Hydropower case study collection: innovative low head and ecologically improved turbines, hydropower in existing infrastructures, hydropeaking reduction, digitalization and governing systems *Sustainability* **12** 78

Quaschning V V 2019 *Renewable Energy and Climate Change* 2nd edn (Hoboken, NJ: Wiley)

Raimi D and Newell R G 2023 Global energy outlook comparison methods: 2023 update *Report* 23-02 Resources for the Future, Washington, DC https://media.rff.org/documents/Methodology_for_Report_23-02.pdf

Raimi D, Zhu Y, Newell R G and Prest B C 2024 Global energy outlook 2024: peaks or plateaus? *Report* Resources for the Future, Washington, DC https://rff.org/publications/reports/global-energy-outlook-2024/

Reader G T 2022 Renewable energy storage: too many options, not enough time? *Sustainable Energy Storage for Furthering Renewable Energy* ed D S-K Ting and J A Stagner (Danbury, CT: Begell House) p 41

Rees A 2021 History of predicting the future *Wired* https://wired.com/story/history-predicting-future/

Reitz R D *et al* 2022 IJER editorial: The future of the internal combustion engine *Int. J. Engine Res.* **21** 3–10

Ritchie H 2022a How the world eliminated lead from gasoline *Our World in Data* https://ourworldindata.org/leaded-gasoline-phase-out (Accessed: 9 March 2024)

Ritchie H 2022b Primary, secondary, final, and useful energy: why are there different ways of measuring energy? *Our World in Data* https://ourworldindata.org/energy-definitions

Ritchie H 2022c How does the land use of different electricity sources compare? *Our World in Data* https://ourworldindata.org/land-use-per-energy-source

Ritchie H 2024a More people care about climate change than you think *Our World in Data* https://ourworldindata.org/climate-change-support

Ritchie H 2024b Solar panel prices have fallen by around 20% every time global capacity doubled *Our World in Data* https://ourworldindata.org/data-insights/solar-panel-prices-have-fallen-by-around-20-every-time-global-capacity-doubled

Ritchie H and Rosado P 2024a Electricity mix *Our World in Data* https://ourworldindata.org/electricity-mix

Ritchie H and Rosado P 2024b Energy mix *Our World in Data* https://ourworldindata.org/energy-mix

Ritchie H and Rosado P 2024c Fossil fuels *Our World in Data* https://ourworldindata.org/fossil-fuels

Ritchie H, Rosado P and Roser M 2023a CO_2 and greenhouse gas emissions *Our World in Data* https://ourworldindata.org/co2-and-greenhouse-gas-emissions

Ritchie H, Rosado P and Roser M 2023b Data page: primary energy consumption from fossil fuels *Our World in Data* https://ourworldindata.org/grapher/fossil-fuel-primary-energy

Ritchie H, Rosado P and Roser M 2023c Levelized cost of energy by technology *Our World in Data* https://ourworldindata.org/grapher/levelized-cost-of-energy

Ritchie H, Rosado P and Roser M 2024a Energy production and consumption *Our World in Data* https://ourworldindata.org/energy-production-consumption (Accessed: 2 March 2024)

Ritchie H, Rosado P and Roser M 2024b Energy access *Our World in Data* https://ourworldindata.org/energy-access

Ritchie H, Roser M and Rosado P 2024c Renewable energy *Our World in Data* https://ourworldindata.org/renewable-energy

Roberts O, Williams T, Lopez A, Maclaurin G and Eberle A 2023 Exploring the impact of near-term innovations on the technical potential of land-based wind energy *Technical Report* NREL/TP-5000-81664 National Renewable Energy Laboratory, Golden, CO https://nrel.gov/docs/fy23osti/81664.pdf

Rogelj J, Fransen T, Elzen M G, Lamboli R D, Schumer C, Kuramochi T and Portugal-Pereira J 2023 Credibility gap in net-zero climate targets leaves world at high risk: supplementary materials *Science* **380** 26

Rogers P G 2016 EPA History: Clean Air Act of 1970/1977 *EPA* https://epa.gov/archive/epa/aboutepa/epa-history-clean-air-act-1970.html

Rosenfeld J, Lewandrowski L, Hendrickson T, Jaglo K, Moffroid K and Pape D 2018 A life-cycle analysis of the greenhouse gas emissions from corn-based ethanol *Report* ICF, Washington, DC https://usda.gov/sites/default/files/documents/LCA_of_Corn_Ethanol_2018_Report.pdf

Roser M 2019 History of Our World in Data *Our World in Data* https://ourworldindata.org/history-of-our-world-in-data

Roser M 2021a Data review: how many people die from air pollution? *Our World in Data* https://ourworldindata.org/data-review-air-pollution-deaths

Roser M 2021b Energy poverty and indoor air pollution: a problem as old as humanity that we can end within our lifetime *Our World in Data* https://ourworldindata.org/energy-poverty-air-pollution

Rosner D and Markowitz G 1985 A 'Gift of God'?: The public health controversy over leaded gasoline during the 1920s *Am. J. Public Health* **75** 344–52

Rühl C 2012 60 years: BP statistical review of world energy 1951–2011 *Report* BP, London https://bp.com/content/dam/bp/business-sites/en/global/corporate/pdfs/energy-economics/statistical-review/bp-statistical-review-of-world-energy-60-anniversary.pdf

Saaty R W 1987 The analytic hierarchy process—what it is and how it is used *Math. Model.* **9** 161–76

Saaty T L 2008 Decision making with the analytic hierarchy process *Int. J. Serv. Sci.* **1** 83–98

Saaty T L and Vargas L G 2012 *Models, Methods, Concepts and Applications of the Analytic Hierarchy Process* 2nd edn (New York: Springer)

Schenk C J, Brownfield M E, Charpentier R R, Cook T A, Gautier D L, Higley D K and Whidden K J 2012 An estimate of undiscovered conventional oil and gas resources of the world *Fact Sheet* 2012–3042 USGS, Denver, CO https://pubs.usgs.gov/fs/2012/3042/fs2012-3042.pdf

Schernikau L and Smith W H 2022 *The Unpopular Truth about Electricity and the Future of Energy* (Singapore: Energeia) https://unpopular-truth.com/

Schwägerl C 2023 Will tech breakthroughs bring fusion energy closer to reality? *Yale Environment 360* https://e360.yale.edu/features/nuclear-fusion-research-startups

SEIClimate AnalyticsE3GIISDUNEP 2023 Phasing down or phasing up? Top fossil fuel producers plan even more extraction despite climate promises *Online Appendix to the 2023 Production Gap Report* Stockholm Environment Institute, Climate Analytics, E3G, Stockholm https://productiongap.org/wp-content/uploads/2023/11/PGR2023_Appendix.pdf

Shaikh A, Shaikh P H, Kumar L, Mirjat N H, Memon Z A, Assad M E and Eskandarpoor B 2022 A SWOT analysis for a roadmap towards sustainable electric power generation *Int. Trans. Electr. Energy Syst.* **2022** 15

Shatnawi N, Abu-Qdais H and Abu Qdais F 2021 Selecting renewable energy options: an application of multi-criteria decision making for Jordan *Sustain.: Sci., Pract. Policy* **17** 209–19

Shaw W *et al* 2024 A US perspective on closing the carbon cycle to defossilize difficult-to-electrify segments of our economy *Nat. Rev. Chem.* **8** 376–400

Shove E 2018 What is wrong with energy efficiency? *Build. Res. Inform.* **46** 779–89

Shu F H 2019 Stopping and reversing climate change *Resonance* **24** 181–200

Sikkema R, Junginger M, Pichler W, Hayes S and Faaij A P 2010 The international logistics of wood pellets for heating and power production in Europe: costs, energy-input and greenhouse gas balances of pellet consumption in Italy, Sweden and the Netherlands *Biofuels, Bioprod. Biorefin.* **4** 132–53

Smil V 2010 *Energy Transitions: History, Requirements, Prospects* (Oxford: Praeger)

Smil V 2017 *Energy and Civilization: A History* (Cambridge, MA: MIT Press)

Smil V 2020 Energy transitions: fundamentals in six points *Papeles de Energia—Energy Papers* **8** 20 https://www.funcas.es/wp-content/uploads/Migracion/Articulos/FUNCAS_PE/009artC3.pdf

Society of Petroleum Engineers 2001 *Guidelines for the Evaluation of Petroleum Reserves and Resources* (Richardson, TX: Society of Petroleum Engineers) https://spe.org/industry/docs/Guidelines-Evaluation-Reserves-Resources-2001.pdf

Society of Petroleum Engineers 2022 *Petroleum Resources Management System* (Richardson, TX: Society of Petroleum Engineers) https://spe.org/media/filer_public/0c/83/0c835db9-501f-4ce7-97f1-a1d6bb4e3331/prmgmtsystem_v103.pdf

SpendEdge 2024 A comprehensive SWOT analysis of the power sector *SpendEdge Smarter Procurement* https://spendedge.com/blogs/swot-analysis-power-sector/

Stanford Energy 2023 Nuclear fusion *The Understand Energy Learning Hub* https://understand-energy.stanford.edu/energy-resources/nuclear-energy/nuclear-fusion

Statista 2024a Household electricity prices worldwide in September 2023, by select country *Statista* https://statista.com/statistics/263492/electricity-prices-in-selected-countries/

Statista 2024b Average electricity bill for a 3-person household in Germany from 1998 to 2024 *Statista* https://statista.com/statistics/1346248/electricity-bill-average-household-germany/

Statista 2024c Capacity factors for selected energy sources in the United States in 2023 *Statista* https://statista.com/statistics/183680/us-average-capacity-factors-by-selected-energy-source-since-1998/

Sterner M and Stadler I 2019 *Handbook of Energy Storage* 2nd edn (Heidelberg: Springer Nature)

Stewart I 2023 Why is cheap renewable electricity so expensive on the wholesale market? *House of Commons Library* https://commonslibrary.parliament.uk/why-is-cheap-renewable-electricity-so-expensive/

Stori V 2020 The role of hydropower in state clean energy policy *Report* Clean Energy States Alliance, Montpelier, VT https://cesa.org/wp-content/uploads/Role-of-Hydropower.pdf

Strauss W 2024 Global wood pellet markets: 2023 in review and why industrial wood pellets are key for the future *Canadian Biomass Magazine* https://canadianbiomassmagazine.ca/global-wood-pellet-markets-2023-in-review-and-why-industrial-wood-pellets-are-key-for-the-future/

Subsidiary Body for Scientific and Technological Advice 2023 Technical dialogue of the first global stocktake *Synthesis Report* FCCC/SB/2023/9 UN and UNFCCC, Washington, DC https://unfccc.int/sites/default/files/resource/sb2023_09E.pdf

Sutter P 2024 We've been 'close' to achieving fusion power for 50 years. When will it actually happen? *Space.com* https://space.com/when-will-we-achieve-fusion-power

Symons A 2023 What is 'white hydrogen'? The pros and cons of Europe's latest clean energy source *euro news* https://euronews.com/green/2023/11/05/what-is-white-hydrogen-the-pros-and-cons-of-europes-latest-clean-energy-source

Thant U 1968 Activities of United Nations organizations and programmes relevant to the human environment *Report of the Secretary-General* E/4553 United Nations, New York https://digitallibrary.un.org/record/729430? ln = en&v = pdf#files

The e face 2018 The cities of the world with the most and fewest daylight hours *endesa* https://endesa.com/en/the-e-face/energy-efficiency/cities-hours-sunlight

The European Commission 2022 *Commission Implementing Decision* 2022/1657 Official Journal of the European Union, Brussels https://eur-lex.europa.eu/legal-content/EN/TXT/PDF/?uri = CELEX:32022D1657

The Royal Society 2020 Ammonia: zero-carbon fertiliser, fuel and energy store *Policy Briefing* The Royal Society, London https://royalsociety.org/-/media/policy/projects/green-ammonia/green-ammonia-policy-briefing.pdf

The White House 2023 *Building a Clean Energy Economy: A Guidebook to the Inflation Reduction Act's Investments in Clean Energy and Climate Action* (Washington, DC: CleanEnergy.gov) https://case.house.gov/uploadedfiles/inflation-reduction-act-guidebook.pdf

The World Bank Group 2024 Welcome to the Global Solar Atlas *Global Solar Atlas* https://globalsolaratlas.info/map

Tollefson J 2023 Climate scientists push for access to world's biggest supercomputers to build better Earth models *Nature News* **6** 11

Tornatore C, Marchitto L, Sabia P and De Joannon M 2022 Ammonia as green fuel in internal combustion engines: state-of-the-art and future perspectives *Front. Mech. Eng.* **8** 16

Tso K 2021 Has there been climate change before? *Climate Portal* https://climate.mit.edu/ask-mit/has-there-been-climate-change

Tullo A H 2021 Is ammonia the fuel of the future? *Chemical and Engineering News* https://cen.acs.org/business/petrochemicals/ammonia-fuel-future/99/i8

Ubrasolar 2024 Disneyland Paris and Urbasolar commission Europe's largest photovoltaic parking canopies power plant *Ubrasolar* https://urbasolar.com/disneyland-paris-and-urbasolar-commission-europes-largest-photovoltaic-shadow-power-plant/

UNEP 2024 UNEP: 50 years of environmental milestones *UNEP* https://unep.org/environmental-moments-unep50-timeline

United Nations 1945 Charter of the United Nations *United Nations* https://refworld.org/legal/constinstr/un/1945/en/27654 (Accessed: 17 March 2024)

United Nations 1968 Question of convening an international conference on the human environment *Resolution* E/RES/1346(XLV) United Nations, New York and Geneva https://digitallibrary.un.org/record/214491?ln=en&v=pdf

United Nations 1973 Report of the United Nations Conference on the Human Environment *Report* A/CONF.48/14/Rev.1 United Nations, New York https://docs.un.org/en/A/CONF.48/14/Rev.1

United Nations 1987 International co-operation in the field of the environment *Resolution* A/RES/42/184 United Nations General Assembly, New York https://documents.un.org/doc/resolution/gen/nr0/514/16/img/nr051416.pdf? token = QvKRwTJKlrmaQYyuDF&fe = true

United Nations 1988 Protection of Global Climate for Present and Future Generations of Mankind *Resolution adopted by the General Assembly* United Nations, New York https://digitallibrary.un.org/record/54234? ln = en&v = pdf

United Nations 1992 *United Nations Framework Convention on Climate Change* (New York: United Nations) https://unfccc.int/resource/docs/convkp/conveng.pdf

United Nations 1993 Report of the United Nations Conference on Environment and Development *Report* A/CONF.151/26/Rev.1(Vol.II) United Nations, New York https://digitallibrary.un.org/record/168679?ln=en&v=pdf

United Nations 2015a *Paris Agreement* (New York: United Nations) https://treaties.un.org/doc/Treaties/2016/02/20160215%2006-03%20PM/Ch_XXVII-7-d.pdf

United Nations 2015b United Nations Charter (full text) https://un.org/en/about-us/un-charter/full-text (Accessed: 17 March 2024)

United Nations 2023 Renewable energy—powering a safer future *United Nations Climate Action* https://un.org/en/climatechange/raising-ambition/renewable-energy (Accessed: 12 March 2024)

United Nations 2024 Climate action fast facts *United Nations Climate Action* https://www.un.org/en/climatechange/science/key-findings

United Nations Climate Action 2023a For a livable climate: net-zero commitments must be backed by credible action *United Nations Climate Action* https://un.org/en/climatechange/net-zero-coalition

United Nations Climate Action 2023b COP28 signals beginning of the end of the fossil fuel era *United Nations Climate Action* https://un.org/en/climatechange/cop28

United Nations Climate Action 2024 What is renewable energy? *United Nations Climate Action* https://un.org/en/climatechange/what-is-renewable-energy (Accessed: 6 April 2024)

United Nations Economic and Social Council 2016 *Specifications for the Application of the United Nations Framework Classification for Fossil Energy and Mineral Reserves and Resources 2009 to Renewable Energy Resources* (Geneva and New York: United Nations) https://unece.org/DAM/energy/se/pdfs/comm25/ECE_ENERGY_2016_4.pdf

United Nations General Assembly 2015a *Transforming Our World: The 2030 Agenda for Sustainable Development* (New York: United Nations) https://sdgs.un.org/2030agenda

United Nations: Sustainable Development 1992 Agenda 21 *United Nations Conference on Environment and Development (Rio de Janerio, Brazil,, 3–14 June 1992)* (New York: United Nations) https //sustainabledevelopment.un.org/content/documents/Agenda21.pdf

United Nations–UNDRR 2015 *Sendai Framework for Disaster Risk Reduction 2015–2030* (Geneva: United Nations) https://preventionweb.net/files/43291_sendaiframeworkfordrren.pdf

UN-UNStats 2024a *Global Indicator Framework for the Sustainable Development Goals and Targets of the 2030 Agenda for Sustainable Development* United Nations, New York https://unstats.un.org/sdgs/indicators/Global-Indicator-Framework-after-2024-refinement-English.pdf

UN-UNStats 2024b *SDG Indicator Metadata* 07-03-01 United Nations, New York https://unstats.un.org/sdgs/metadata/files/Metadata-07-03-01.pdf

Uppal A 2018 Tidal energy: an overview of Indian scenario *Int. J. Adv. Res. Innov.* **6** 76–80

Uría-Martínez R and Johnson M M 2023 *US Hydropower Market Report* (Oak Ridge, TN: US Department of Energy) https://www.energy.gov/sites/default/files/2023-09/U.S.%20Hydropower%20Market%20Report%202023%20Edition.pdf

USDA Agricultural Air Quality Task Force 2014 Ammonia emissions: what to know before you regulate *White Paper* USDA, Washington, DC https://nrcs.usda.gov/sites/default/files/2022-10/AAQTF-Accomplishments-Ammonia-White-Paper.pdf

US Department of Energy n.d. Types of hydropower plants *USDOE Water Power Technologies Office* https://energy.gov/eere/water/types-hydropower-plants

US Department of Energy 2022 *The Inflation Reduction Act Drives Significant Emissions Reductions and Positions America to Reach Our Climate Goals* (Washinton, DC: US Department of Energy Office of Policy) https://www.ourenergypolicy.org/resources/the-inflation-reduction-act-drives-significant-emissions-reductions-and-positions-america-to-reach-our-climate-goals/

US Department of Energy 2024 Solar performance and efficiency *Solar Energy Technologies Office* https://energy.gov/eere/solar/solar-performance-and-efficiency

US Department of State 2023 Highlights from 2023 global methane pledge ministerial *Fact Sheet* US Department of State https://2021-2025.state.gov/highlights-from-2023-global-methane-pledge-ministerial/

USDOE n.d. Hidden systems *Office of Energy Efficiency and Renewable Energy—Geothermal Technologies Office* https://energy.gov/eere/geothermal/hidden-systems

USDOE 2023 Geothermal Energy *Fact Sheet* US Department of Energy, Office of Energy Efficiency and Renewable Energy, Washington, DC https://energy.gov/sites/default/files/2023-02/DOE-EE-2680-GTO-Geothermal-Overview-8-5x11.pdf

USDOI 2024 Old Faithful *National Park Service* https://nps.gov/yell/planyourvisit/exploreold-faithful.htm

USEIA 2022 Levelized costs of new generation resources in the Annual Energy Outlook 2022 *Report* USEIA, Washington, DC https://eia.gov/outlooks/aeo/pdf/electricity_generation.pdf

USEIA 2023a What is renewable energy? Renewable energy explained *USEIA* https://eia.gov/energyexplained/renewable-sources/

USEIA 2023b The United States uses a mix of energy sources. US energy facts explained *USEIA* https://eia.gov/energyexplained/us-energy-facts/

USEIA 2024a *Monthly Energy Review April 2024* (Washington, DC: USEIA Office of Energy Statistics) https://eia.gov/totalenergy/data/monthly/pdf/mer.pdf

USEIA 2024b Electric power monthly *USEIA* https://www.eia.gov/electricity/monthly/

USEPA 2021 Seasonality and climate change: a review of observed evidence in the United States *Report* EPA 430-R-21-002. USEPA, Washington, DC https://epa.gov/system/files/documents/2021-12/30339_epa_report_climate_change_and_seasonality_v12_release_508.pdf

USEPA 2023a Climate-indicators/view-indicators *USEPA* https://epa.gov/climate-indicators/view-indicators

USEPA 2023b Climate change indicators: US and global precipitation *USEPA* https://epa.gov/climate-indicators/climate-change-indicators-us-and-global-precipitation

USEPA 2024a What is green power? *EPA Green Power Markets* https://epa.gov/green-power-markets/what-green-power

USEPA 2024b Fast Facts: US transportation sector greenhouse gas emissions 1990–2022 *Fact Sheet* United States Environmental Protection Agency Office of Transportation and Air Quality, Washington, DC https://climateprogramportal.org/wp-content/uploads/2025/02/Fast-Facts-US-Transportaton-Sector-GHG-Emissions-1990-2022.pdf

USEPA 2024c Green power equivalency calculator—calculations and references *EPA Green Power Markets* https://epa.gov/green-power-markets/green-power-equivalency-calculator-calculations-and-references

USEPA 2024d Summary of the Energy Independence and Security Act: Public Law 110-140 (2007) *USEPA* https://epa.gov/laws-regulations/summary-energy-independence-and-security-act

USEPA 2024e Inventory of US greenhouse gas emissions and sinks: 1990–2022 *Report* EPA 430-R-24-004 US Environmental Protection Agency, Washington, DC https://epa.gov/system/files/documents/2024-04/us-ghg-inventory-2024-main-text_04-18-2024.pdf

USEPA 2024f US electricity grid and markets *EPA Green Power Markets* https://epa.gov/green-power-markets/us-electricity-grid-markets

USEPA 2025 Biofuels and the Environment: Third Triennial Report to Congress (Final Report) United States Environmental Protection Agency, Washington, DC https://assessments.epa.gov/biofuels/documents/&deid=363940

Vesselinov V V, Ahmmed B, Mudunuru M K, Pepin J D, Burns E R, Siler D L and Middleton R S 2022 Discovering hidden geothermal signatures using non-negative matrix factorization with customized *k*-means clustering *Geothermics* **106** 15

Visit Norway 2024 The giant Sun mirrors in Rjukan *Visit Norway* https://visitnorway.com/listings/the-giant-sun-mirrors-in-rjukan/3632/

Waidelich P, Batibeniz F, Rising J, Kikstra J S and Seneviratne S I 2024 Climate damage projections beyond annual temperature *Nat. Clim. Chang.* **14** 592–9

WEF 2023a IEA: more than a third of the world's electricity will come from renewables in 2025 *WEF Energy Transition* https://weforum.org/agenda/2023/03/electricity-generation-renewables-power-iea/

Welsby D, Price J, Pye S and Ekins P 2021 Unextractable fossil fuels in a 1.5°C world *Nature* **597** 230–4

Wiatros-Motyka M, Fulghum N and Jones D 2024 Global Electricity Review 2024 *Report* Ember, London https://ember-climate.org/app/uploads/2024/05/Report-Global-Electricity-Review-2024.pdf

Wikipedia Contributors 2024a History of the automobile *Wikipedia, The Free Encyclopedia* https://en.wikipedia.org/w/index.php?title = History_of_the_automobile&oldid = 1226405487

Wikipedia Contributors 2024b List of battery types *Wikipedia, The Free Encyclopedia* https://en.wikipedia.org/w/index.php? title = List_of_battery_types&oldid = 1216632171

Wikipedia Contributors 2024c List of cities by sunshine duration *Wikipedia, The Free Encyclopedia* https://en.wikipedia.org/w/index.php? title=List_of_cities_by_sunshine_duration&oldid=1230742878

Wikipedia Contributors 2024d Tetraethyllead *Wikipedia, The Free Encyclopedia* https://en.wikipedia.org/w/index.php? title=Tetraethyllead&oldid=1211606822

Willige A 2022 The colors of hydrogen: expanding ways of decarbonization *Spectra* https://spectra.mhi.com/the-colors-of-hydrogen-expanding-ways-of-decarbonization

World Bank Group 2021 Germany—climatology *Climate Change Knowledge Portal* https://climateknowledgeportal.worldbank.org/country/germany/climate-data-historical

World Economic Forum 2023 Fostering effective energy transition *Insight Report* World Economic Forum, Geneva https://www3.weforum.org/docs/WEF_Fostering_Effective_Energy_Transition_2023.pdf

World Nuclear Association 2022 Nuclear fusion power *World Nuclear Association* https://world-nuclear.org/information-library/current-and-future-generation/nuclear-fusion-power

Worldometers 2023 Countries in the world by population (2024) *Worldometers* https://worldometers.info/world-population/population-by-country/

Worldometers 2024 World population by year *Worldometers* https://worldometers.info/world-population/world-population-by-year/

Zewe A 2022 A tool for predicting the future *MIT News Office* https://news.mit.edu/2022/tensor-predicting-future-0328

IOP Publishing

Renewable Energy Systems
The way forward
David S-K Ting and Jacqueline A Stagner

Chapter 2

A methodological framework for modeling passive solar greenhouses

Gurpreet Khanuja, Rajeev Ruparathna and David S-K Ting

The high energy footprint of traditional greenhouses calls for a transition towards passive solar greenhouses, designed to suit the locational constraints, and this requirement is amplified, especially in cold regions. Greenhouse energy modeling predicts the energy requirements and evaluates the economic viability and profitability of these greenhouses in advance. There is a lack of comprehensive energy modeling guidelines and literature to model these greenhouses accurately. This research aims to conceptualize a general framework for greenhouse energy modeling to simulate actual operational and environmental conditions. The framework of the proposed methodology was demonstrated using a Chinese solar greenhouse in Elie, Manitoba. A significant challenge while modeling this greenhouse was the unavailability of precise information related to greenhouse ventilation, weather inputs, and evapotranspiration, which led to making some assumptions for greenhouse modeling. Despite these data gaps, the validation results reveal that the greenhouse energy model could accurately predict the hourly internal air temperature and heating load, with a root mean square error of 3.1 °C and 1.4 kW, respectively. Additionally, the investigation estimated the ground temperature and the north wall temperature, with root mean square errors of 6.4 °C and 4.3 °C, respectively. The findings suggest that this study can be used as a basis to develop the modeling approach for further greenhouse simulations to replicate dynamic and environmental conditions.

Abbreviations

ACH	air changes per hour
ASHRAE	American Society of Heating, Refrigerating and Air-Conditioning Engineers
BPS	building performance simulations
CA	canopy area
CFD	computational fluid dynamics

doi:10.1088/978-0-7503-6179-8ch2

CGSS	corrugated galvanized steel sheet
COMIS	conjunction of multi-zone infiltration specialists
CSG	Chinese solar greenhouse/Chinese mono-slope solar greenhouse
CWEC	Canadian weather for energy calculations
EPW	energy plus format
EVT	evapotranspiration
GLSR	ground-level solar radiation
IWEC	international weather for energy calculations
LAI	leaf area index
LSSVM	least-squares support vector machine
MAE	mean absolute error
PSG	passive solar greenhouse
RH	relative humidity
RMSE	root mean square error
rRMSE	relative root mean square error
TMY	typical meteorological year
TMY2	typical meteorological year version 2
TMY3	typical meteorological year version 3
TRNSYS	Transient System Simulation Program
VB	Visual Basic
VPD	vapor pressure deficit

2.1 Introduction

Conventional greenhouses are known for their extensive energy consumption, primarily relying on electricity or fossil fuels, contributing significantly to greenhouse gas emissions (Gorjian *et al* 2021). The energy from electricity and natural gas accounts for about 28% of global greenhouse gas emissions (Greer *et al* 2024). This massive energy demand not only poses environmental threats but also leads to higher operating costs for growers, thereby reducing the cost-effectiveness of conventional greenhouses. Energy consumption costs represent approximately 50% of greenhouse production expenses, making them the second largest operating cost (Acosta-Silva *et al* 2019, Golzar *et al* 2018). In northern latitudes, the heating expense can account for 70%–85% of total greenhouse operating costs (Rorabaugh *et al* 2002). This underscores the necessity for enhancing the energy efficiency of greenhouses to substantially decrease energy demands. Researchers have explored energy-efficient approaches, such as using renewable energy, to enhance greenhouse sustainability. Solar energy offers an economical solution for greenhouse heating (Beshada *et al* 2006), leading to a transition from traditional greenhouses to modern solar greenhouses (Wang *et al* 2017). Modern solar greenhouses, classified into passive and active, have the potential to reduce energy demand and greenhouse gas emissions. Active solar greenhouses are integrated with solar energy technologies such as photovoltaic and thermal solar collectors, while passive solar greenhouses (PSGs) are designed to increase solar energy capture (Gorjian *et al* 2021).

PSGs are specifically designed to enhance solar energy capture and minimize energy losses. Studies have been conducted to enhance the solar capturing efficiency of the greenhouse by modifying the greenhouse design and orientation (Mobtaker *et al* 2016, Sethi 2009). As a result, PSGs are oriented to ensure that the maximum possible surface area is exposed to the sun. The south side of a PSG typically consists of a thin and transparent surface that transmits solar radiation to enter during the daytime. A thermal curtain is employed on the south side at night to minimize heat loss. Despite this, a significant amount of heat, approximately 54%–68% during winter, still escapes to the external environment through the south roof (Xu *et al* 2017, Zhao *et al* 2019). The north wall and north roof provide heat storage, thermal insulation, and structural stability. The north wall absorbs solar radiation for passive heat storage during the day and then releases the stored heat through conduction and convection to warm the greenhouse air (Liu *et al* 2022). Hence, solar energy interactions were modeled within greenhouses to estimate energy consumption and analyse the reduction in energy costs.

The high energy demand of greenhouses particularly in cold regions, is a prime concern for determining their feasibility, enabling growers to make informed decisions about their establishment. Building performance simulations (BPSs) are essential because they generate valuable data to enhance the efficiency of the greenhouse without the significant time and expense associated with conducting real-life experiments (Beaulac *et al* 2024). Previous research by Vadiee and Martin (2013), Dong *et al* (2021), Ahamed *et al* (2018), and Choab *et al* (2021) performed BPSs to estimate the heating and cooling needs of greenhouses under different climatic conditions. Many studies eliminated the effect of plants inside the greenhouse or assumed constant evapotranspiration rates (Vadiee and Martin 2013, Semple *et al* 2017), as well as ventilation, supplemental heating, or cooling (Guo *et al* 1994, Yu *et al* 2016, Imafidon *et al* 2023) which significantly alters the greenhouse's energy requirements. Moreover, a major obstacle in performing greenhouse energy modeling was the absence of specific guidelines or standards. Existing standards, such as the International Building Code, National Building Code, or the National Energy Code, are used for designing and constructing different buildings but do not encompass greenhouse facilities. These facilities do not meet the criteria outlined in these codes and standards. Thus, this study aims to provide a methodological framework to perform energy modeling in greenhouses. While the greenhouse energy modeling process may vary slightly based on the simulation tool used, the fundamental framework to model the dynamic interactions within the greenhouse remains consistent. This framework was applied to simulate the actual operational and environmental conditions of an experimental mono-slope solar greenhouse in Elie, Manitoba. The simulation results were then compared with measured data to identify the accuracy of the greenhouse energy model. This approach offers a comprehensive understanding of the energy modeling process to accurately replicate real experimental conditions in a greenhouse.

2.2 Literature review

The absence of standardized guidelines for greenhouse modeling has resulted in a gap where no study certainly outlines critical considerations for modeling a greenhouse. Consequently, many studies simulated the thermal performance of greenhouses and predicted their heating and cooling requirements using MATLAB, CFD, FORTRAN, TRNSYS, and EnergyPlus. Table 2.1 summarizes the literature reviewed for this study, illustrating different modeling approaches used over the years. Some studies focused on analysing the indoor air temperature including the temperatures of the north wall, back roof, and soil. It includes different greenhouse models such as a mathematical model TEMP developed by Guo *et al* (1994), a computational fluid dynamics (CFD) study by Tong *et al* (2007), and the thermal environment simulation model using MATLAB and Visual Basic (VB) by Meng *et al* (2009). A greenhouse simulation model using a finite difference numerical approach was studied by Ma *et al* (2010) to predict and evaluate the thermal environment of solar greenhouses. This model has limitations for use outside China. Vadiee and Martin (2013) utilized TRNSYS software to study closed greenhouses

Table 2.1. Summary of the published literature on simulating greenhouses with different approaches.

Author	Modeling strategies	Studied parameters
Guo *et al* (1994)	Mathematical model—TEMP using FORTRAN programming	Indoor air temperature Surface temperature
Tong *et al* (2007)	Computational fluid dynamics (CFD)	Indoor air temperature
Meng *et al* (2009)	MATLAB and Visual Basic (VB)	Indoor air temperature Surface temperature Soil temperature Back roof temperature
Ma *et al* (2010)	Simulation model using finite difference numerical method	Comparison of the thermal performance of different solar greenhouses.
Vadiee and Martin (2013)	TRNSYS software	Heating and cooling load
Yu *et al* (2016)	Prediction model using a least-squares support vector machine (LSSVM)	Temperature variation in the CSG
Dong (2018)	Simulation model—SOGREEN using the finite difference numerical method	Energy consumption
Ahamed *et al* (2018)	CSGHEAT model using MATLAB	Ground temperature North wall temperature Hourly heating requirements
Ahamed *et al* (2019)	TRNSYS software	Heating requirements
Choab *et al* (2021)	TRNSYS software	Heating and cooling energy needs
Imafidon *et al* (2023)	TRNSYS software	Effects of ground parameters Solar-to-air fraction

with thermal seasonal storage to calculate their heating and cooling load. Yu *et al* (2016) developed a predictive model based on a least square support vector machine (LSSVM) to predict the occurrence of temperatures several hours early to reduce financial losses. Dong (2018) modified the original greenhouse model, developed by Chengwei Ma in China (Ma 2015), to create the SOGREEN model suitable for the cold climate in Saskatchewan. The model was validated using field data from a solar greenhouse in Elie, Manitoba. An average error of 1.9 °C for indoor air temperature and 7% for relative humidity was observed in the model. Another simulation model, CSGHEAT, developed by Ahamed *et al* (2018), estimated hourly heating requirements for a Chinese solar greenhouse (CSG) with a relative root mean square error (rRMSE) of 11.5%. Ahamed *et al* (2019) used TRNSYS software to calculate the heating requirements and compared these results with the CSGHEAT model's results. The findings indicate that the monthly average difference in the heating simulation load between the two models was about 5% when excluding thermal blankets and plants from the simulation. Choab *et al* (2021) used TRNSYS software to investigate the key design parameters affecting a greenhouse's thermal behavior along with heating and cooling energy needs. Evapotranspiration affected the greenhouse's thermal behavior, yielding a relative error of 1.66% for the annual heating demand. Imafidon *et al* (2023) utilized TRNSYS with a detailed radiation model to simulate a net-zero passive solar greenhouse in Alberta, Canada. The study investigated the effects of ground parameters and the solar-to-air fraction on the simulation results. The drawbacks of the above studies involve the exclusion of parameters such as evapotranspiration, ventilation, supplemental heating, and cooling (Guo *et al* 1994, Meng *et al* 2009, Yu *et al* 2016, Ahamed *et al* 2018, Imafidon *et al* 2023) leading to inaccuracies in estimating the energy demands whereas other involved complex estimation because of the simulation software (Tong *et al* 2007, Yu *et al* 2016, Ahamed *et al* 2019).

Most research focused on the methods to model greenhouses with different simulation tools to estimate the heating and cooling requirements or thermal performance of a greenhouse. The modeling strategies provided in the above studies remained specific to the software used, however, they lack a detailed, unified framework for the modeling process.

2.3 Guidelines for simulating a passive solar greenhouse

The challenges encountered during the greenhouse energy modeling include complex greenhouse interactions because of the greenhouse locations, shapes, orientation, construction materials, crops, and weather conditions (Chen *et al* 2016, Sethi *et al* 2013). The energy modeling employed to simulate greenhouse heating and cooling requirements was developed using a heat balance approach for greenhouse air. All heat gains to the greenhouse air were considered positive, while all heat losses were considered negative. Thus, the greenhouse heating and cooling demand can be expressed as the difference between all heat gains and heat losses. Figure 2.1 depicts the greenhouse volume indicating the heat interactions between the greenhouse air and its surroundings. The greenhouse air temperature is primarily influenced by the

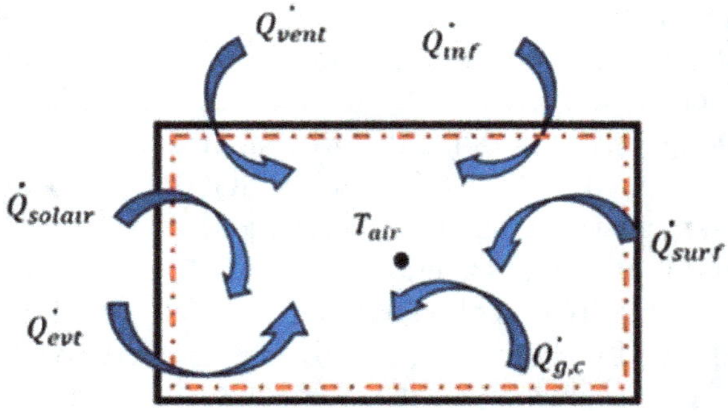

Figure 2.1. Heat interactions between greenhouse air and its surroundings. (Note. This figure is re-created from the TRNSYS-18 (2021).)

heat gain from solar radiation entering the greenhouse through glazing ($\dot{Q}_{\text{solair}}$). A portion of this heat is added to the greenhouse air, while another fraction is absorbed by the greenhouse surfaces. Furthermore, heat is gained and lost via conduction, convection, and radiation between the internal and external air and the different greenhouse surfaces ($\dot{Q}_{\text{surf}}$). Heat loss from the greenhouse air also occurs due to infiltration ($\dot{Q}_{\text{inf}}$) and ventilation ($\dot{Q}_{\text{vent}}$). Environmental control systems such as motors and lights may be installed in the greenhouse and contribute additional heat to the air ($\dot{Q}_{\text{g,c}}$). Additionally, evapotranspiration heat flux ($\dot{Q}_{\text{evt}}$) creates a cooling effect, thereby reducing the greenhouse temperature.

Sensible heat flux to the volume of the greenhouse air is presented by $\dot{Q}_{\text{i}}$ in kJ h^{-1} and is given by the following equation (TRNSYS-18 2021):

$$\dot{Q}_i = \dot{Q}_{\text{surf}} + \dot{\text{Q}}_{\text{inf}} + \dot{Q}_{\text{vent}} + \dot{Q}_{\text{g, c}} - \dot{Q}_{\text{solair}} + \dot{Q}_{\text{evt}}, \tag{2.1}$$

where

$\dot{Q}_{\text{surf}}$ is the convective heat gain from surfaces in kJ h^{-1},

$\dot{Q}_{\text{inf}}$ is the infiltration gain in kJ h^{-1},

$\dot{Q}_{\text{vent}}$ is the ventilation gain in kJ h^{-1},

$\dot{Q}_{\text{g, c}}$ is the internal convective gain (by people, equipment, and illumination) in kJ h^{-1},

$\dot{Q}_{\text{solair}}$ is the fraction of solar radiation entering the greenhouse through external windows in kJ h^{-1}, and

$\dot{Q}_{\text{evt}}$ is the evapotranspiration heat loss due to the plants in kJ h^{-1}.

A negative Q_i indicates that greenhouse heat losses outweigh heat gains, which means heating is required in the greenhouse, whereas positive Q_i signifies more heat gains in comparison to heat loss, requiring cooling inside the greenhouse.

Based on the heat interactions, the greenhouse energy modeling process was proposed to include the following important considerations, which are covered in the respective sections: greenhouse model (section 2.3.1); weather and initial data

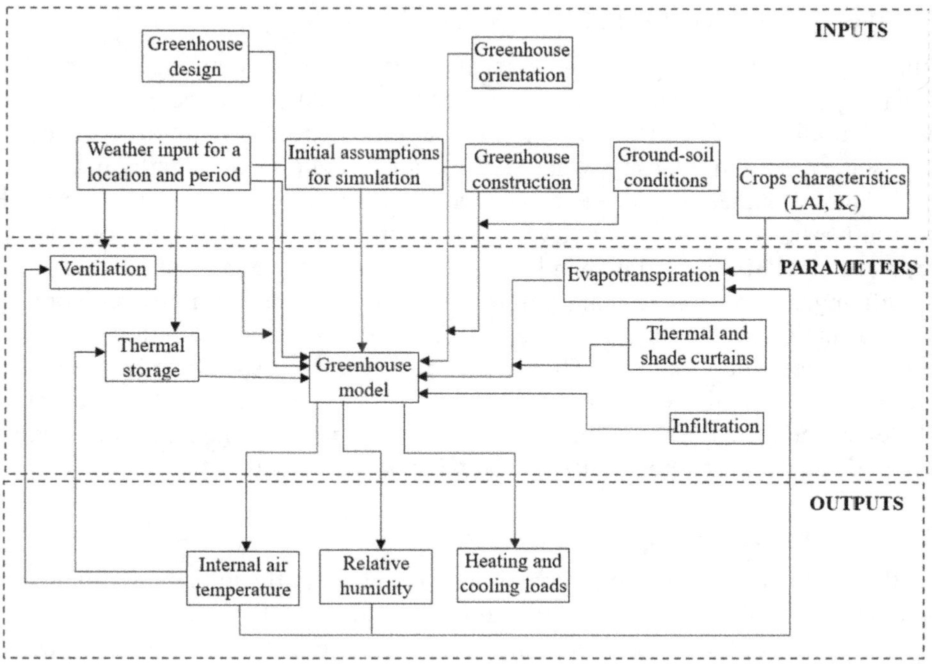

Figure 2.2. Flow chart explaining the main processes involved in greenhouse energy modeling.

for simulation (section 2.3.2); greenhouse construction material (section 2.3.3); solar radiation distribution and interactions within the greenhouse (section 2.3.4); evapotranspiration (section 2.3.5); infiltration (section 2.3.6); and energy-efficiency features such as thermal and shade curtains, supplemental heating and cooling, thermal storage, and ventilation (section 2.3.7).

A flow chart explaining the general framework for greenhouse energy modeling is shown in figure 2.2. The specific steps and parameters may vary depending on the unique characteristics and requirements of each greenhouse.

2.3.1 Greenhouse model

The first step in the energy modeling process is to develop the greenhouse model, which provides the physical presence of the greenhouse. The greenhouse model can be created either by providing the geometric details of the greenhouse manually or using design software, depending upon the simulation software being used. The greenhouse model should define comprehensive information about the greenhouse such as its orientation, design, dimensions, volume, inclination, and orientation for each surface. The shape and inclination of the surfaces are the significant cross-sectional parameters that affect the reflection and transmission of solar radiation inside the greenhouse (Tong *et al* 2013). Hence, greenhouse design and orientation are key factors that influence the thermal behavior and energy demand of a greenhouse.

2.3.2 Weather and initial data for simulation

Hourly weather inputs specific to a location and time of year are important for simulating the thermal environment, as greenhouses interact directly with external weather conditions. The greenhouse air temperature depends on ambient air temperature, solar radiation intensity, wind velocity, and wind direction (Sethi and Sharma 2007). These weather parameters are connected as inputs to the greenhouse model. The weather file must contain hourly weather outputs such as solar radiation, dry bulb temperature, relative humidity, wind velocity, wind direction, solar zenith angle, solar azimuth angle, and other relevant parameters. Energy modeling may also require an initial estimation of parameters such as temperature and relative humidity to calculate the hourly greenhouse outputs. These estimations can be assumed to be the average measured internal air temperature and relative humidity obtained from sensors installed in the greenhouse. If measured parameters are unavailable, average temperature and relative humidity values can be taken for the weather file.

2.3.3 Greenhouse construction material

The thermal performance of a greenhouse is significantly influenced by its construction materials as it varies the absorbed, transmitted, and reflected solar radiation entering the greenhouse. The modeling of a greenhouse includes thermal properties of materials such as density, heat capacity, conductivity, absorptance, and emissivity. In addition, the g value, U value, and transmittance value are also necessary for glazing materials as they significantly impact a greenhouse's heating and cooling requirements. The thermal properties of the materials can be found in the *ASHRAE Handbook* under the non-residential heating and cooling chapter (ASHRAE 2013). The details of the construction material can be added to the greenhouse model while developing the model or to the simulation interface depending on the simulation software used.

2.3.4 Solar radiation distribution and interactions within the greenhouse

The solar radiation that falls on the surfaces comprises beam solar radiation and diffuse solar radiation. The amount of solar radiation available inside the greenhouse depends upon the transmissivity and orientation of glazing surfaces (Ahamed *et al* 2017). Solar radiation heat gain $\dot{Q}_{\text{solair}}$ entering the greenhouse from the glazing is given by equation (2.2) (Ahamed *et al* 2018) and equation (2.3) (Liu and Jordan 1963):

$$\dot{Q}_{\text{solair}} = \sum_{i=1}^{i=n} \tau_{g,i} \, A_{g,i} \, I_{g,i} \tag{2.2}$$

$$I_{g,i} = I_b \frac{\cos\theta_{g,i}}{\cos\theta_z} + \left[I_d\left(\frac{1+\cos\beta}{2}\right) + (I_b + I_d)\rho_r\left(\frac{1-\cos\beta}{2}\right)\right], \tag{2.3}$$

where

$\tau_{g,i}$ is the solar transmissivity of glazing i,
$A_{g,i}$ is the glazing area i in m^2,

$I_{g,i}$ is the total solar radiation through inclined glazing i in kJ h^{-1} m^2,
I_b, I_d is the beam and diffuse radiation on a horizontal surface in kJ h^{-1} m^2,
$\Theta_{g,i}$ is the angle of incidence for glazing i in degrees,
Θ_z is the zenith angle of the sun in degrees,
β is the angle of the inclined surface with horizontal in degrees, and
ρ_r is the reflectivity of outdoor ground.

The glazing area and its inclination can be estimated using the greenhouse model. Beam and diffuse solar radiation on a horizontal surface, zenith angle, angle of incidence, and reflectivity of outdoor ground can be collected from the weather data. The transmittance is influenced by the angle of incidence and decreases as the angle of incidence increases. The transmittance of a material varies from 0 to 1 and can be identified for different materials from the *ASHRAE Handbook* (ASHRAE 2013). The transmitted solar radiation is then absorbed by the internal air, inside surfaces, and plants. This results in various heat interactions within the greenhouse including the net radiative heat transfer with all surfaces inside and outside the greenhouse, the convection heat flux from the inside surfaces to the greenhouse air, the convection heat flux from the outside surface to the ambient environment, the conduction heat flux from the walls at the inside surfaces and the conduction heat flux into the wall at the outside surfaces (TRNSYS-18 2021). The convective heat gain $\dot{Q}_{surf}$ within the greenhouse can be expressed by the following equations (Ahamed *et al* 2018):

$$\dot{Q}_{surf} = Q_t + Q_r \tag{2.4}$$

$$Q_t = (A_g U_g + \sum A_w U_w).\ (T_{air} - T_{outside}) \tag{2.5}$$

$$Q_r = \sigma\varepsilon_g A_g F_g\left(T_{air}^4 - T_{glazing}^4\right) + \sigma\varepsilon_i\tau_l A_f F_{sk}\left(T_{air}^4 - T_{sk}^4\right), \tag{2.6}$$

where
Q_t is the conduction and convection heat transfer in kJ h^{-1},
Q_r is the radiative heat transfer in kJ h^{-1},
A_g is the glazing area in m^2,
U_g is the combined conduction and convection heat transfer coefficient for the glazing in kJ h^{-1} m^2 °C,
A_w is the area of each wall or roof in m^2,
U_w is the combined conduction and convection heat transfer coefficient for each wall and roof in kJ h^{-1} m^2 °C,
T_{air} is the internal air temperature in °C,
$T_{outside}$ is the ambient air temperature in °C,
$T_{glazing}$ is the glazing temperature in °C,
σ is the Stefan–Boltzmann constant in kJ h^{-1} m^2 °C ^{4},
ε_g, ε_i is the emissivity of the glazing and indoor components,
F_g, F_{sk} is the glazing view factor and sky view factor,
τ_l is the transmissivity of glazing to longwave radiation,
A_f is the floor area in m^2, and
T_{sk} is the sky temperature in °C.

The values of U_g and U_w can be calculated using the relation given by Tiwari (2003). The emissivity and transmissivity to longwave radiation of a surface can be selected for different materials from the *ASHRAE Handbook* (ASHRAE 2013). T_{sk} and $T_{outside}$ can be estimated from the weather data, whereas T_{air} and $T_{glazing}$ will be received from the greenhouse model simulation. The view factor between the greenhouse and the sky can be considered as 1 as the greenhouse is completely enclosed by the sky (Vadiee 2011) whereas the view factor for the glazing can be calculated using the relation given by Liu and Jordan (1961). Greenhouses also experience internal convective gain from heat exchange through environmental control systems such as artificial lighting, radiators, motors, equipment, etc, present inside the greenhouse. A thermal analysis study indicates that environmental control systems could reduce 13%–56% of the total heating requirements over the year (Ahamed *et al* 2017). This highlights the significant role of environmental control systems in adding heat and regulating the thermal environment of the internal air. The contributions of these systems to the greenhouse's thermal environment are crucial for accurate energy modeling.

2.3.5 Evapotranspiration

Evapotranspiration (ET) is the combined process of evaporation and plant transpiration within a greenhouse. Many studies overlook the inclusion of plants inside the greenhouse or assume a constant evapotranspiration rate due to the complexity of modeling the dynamic nature of plants, leading to huge inaccuracies in predicting the energy requirements (Ahamed *et al* 2019). Plants have a thermal capacity to absorb solar radiation, transferring less heat to the internal air. A major challenge in modeling plant growth from the initial stage, crop development, mid-season, to late season (Gong *et al* 2020), is in altering the crop's leaf area index (LAI). Since it is impossible to simulate the changing LAI throughout the year, an average LAI is often used for energy modeling. LAI is defined as the ratio of total leaf area (m^2) to ground area (m^2). The crops grow to their maximum extent during the mid-season; therefore, most studies assume the LAI and crop coefficient for the mid-season. The reference crop evapotranspiration (ET_o) is estimated using the Stanghellini model suitable for closed greenhouses where the wind speeds are typically less than 1 m s^{-1} (Pamungkas *et al* 2014) and is given by

$$ET_o = 2.\ LAI.\ \frac{1}{\lambda} \frac{s.\left(R_n - G\right) + K_t \frac{VPD.\rho.C_p}{r_a}}{s + \gamma\left(1 + \frac{r_c}{r_a}\right)}. \tag{2.7}$$

The reference crop evapotranspiration (ET_o) represents the evapotranspiration from a standard vegetated surface under specific conditions. To calculate the actual evapotranspiration rate (ET_{actual}) within the greenhouse, the crop coefficient (K_c) is multiplied by the reference evapotranspiration rate. The crop coefficient for mid-season conditions is around 1.10 ± 0.04 (Pamungkas *et al* 2014). Table 2.2 provides

Table 2.2. List of parameters, symbols, and formulas used for determining ET_o using the Stanghellini model.

Parameters	Symbol	Formula	Units	References
Saturation vapor pressure	e_s	$e_s = 0.610\,78.\ \exp^{\left(\frac{17.269\ T}{237.3 + T}\right)}$	kPa	Pamungkas *et al* (2014)
Actual vapor pressure	e_a	$e_a = \frac{e_s.\ RH}{100}$		
Vapor pressure deficit	VPD	$VPD = e_s - e_a$		
Slope of the saturation vapor pressure curve	s	$s = 0.041\,45.\ \exp^{(0.060\,88.T)}$	kPa °C^{-1}	
Atmospheric pressure	P	$P = 101.325$	kPa	—
Latent heat of vaporization	λ	$\lambda = 2.26$	MJ kg^{-1}	—
Specific heat of air	C_p	$C_p = 0.001\,013$	MJ kg^{-1} °C	Pamungkas *et al* (2014)
Psychrometric constant	Υ	$\gamma = \frac{C_p P}{\varepsilon \lambda}$	kPa °C^{-1}	Donatelli *et al* (2005)
Atmospheric density	ρ	$\rho = 1.204$	kg m^{-3}	—
Leaf temperature	T_o	$T_o = 2.52 + (0.84.\ T) + (-0.54.\ VPD)$	°C	Pamungkas *et al* (2014)
Canopy resistance	r_c	$r_c = \frac{100}{0.5.\ LAI}$	s m^{-1}	
Aerodynamics resistance	r_a	$r_a = \frac{665}{1 + 0.54.\ U}$		Donatelli *et al* (2005)
Net radiation	R_n	$R_n = R_{ns} — R_{nl}$	MJ m^{-2} h^{-1}	Pamungkas *et al* (2014)
Net shortwave radiation	R_{ns}	$R_{ns} = (0.07 \times GLSR)/1000/CA$		
Net outgoing longwave radiation	R_{nl}	$R_{nl} = \frac{(0.16)(3600)\ \rho.\ C_p.\left(T - T_o\right)}{r_R}$		
Radiative resistance	r_R	$r_R = \frac{\rho.\ C_p}{4.\ \sigma.\ (T + 273.16)^3}$	s m^{-1}	
Emissivity	ε	$\varepsilon = 0.622$	—	
Specific gas constant	R	$R = 287$	J kg^{-1} K	
Stefan–Boltzmann constant	σ	$\sigma = 5.669 \times 10^{-14}$	MJ K^{-4} m^2 s	

the symbols, relationships, and units for each parameter used in the Stanghellini model.

The soil heat flux (G) is considered negligible for greenhouses due to the absence of any open area (Pamungkas *et al* 2014). Hourly inputs such as temperature (T), relative humidity (RH), ground-level solar radiation (GLSR), and inside air velocity (U), along with leaf area index (LAI) and crop coefficient (k_c) are necessary inputs to model the hourly evapotranspiration rate. Finally, the evapotranspiration convective heat flux $\dot{Q}_{EVT}$ from the plants is given by the equation (Choab *et al* 2021)

$$\dot{Q}_{EVT} = ET_{actual}.\ \lambda.\ CA, \tag{2.8}$$

where

ET_{actual} is the actual evapotranspiration rate in kg m^{-2} h,
λ is the latent heat of vaporization in kJ kg^{-1}, and
CA is the canopy area in m^2.

2.3.6 Infiltration

Infiltration is defined as the unintentional air movement between the interior of the building and the outdoor environment from openings or holes in the greenhouse envelope. This may occur due to the pressure differential between the internal and the external environment. A study identified the infiltration rates of a newly constructed conventional greenhouse and a PSG ranging from 5.63 to 5.92 ACH (Red River College and Proskiw Engineering Ltd 2015). This indicates that infiltration tends to reduce greenhouse efficiency by allowing cooler outside air to enter through openings, thereby increasing the heat loads and decreasing the cooling loads. Depending on factors such as air-tightness, wind speed, and the temperature gradient between the inside and outside air, infiltration accounts for about 20% of total heat loss from a greenhouse (Jolliet *et al* 1991), indicating the necessity to model infiltration. The total infiltration gain $\dot{Q}_{inf}$ can be found using the following equation (Ahamed *et al* 2018):

$$\dot{Q}_{inf} = \dot{V}\rho C_p\left(T_{outside} - T_{air}\right), \tag{2.9}$$

where

$\dot{V}$ is the air exchange rate through infiltration in m^3 h^{-1},
ρ is the air density in kg m^{-3},
C_p is the specific heat capacity of air in kJ kg^{-1} °C,
$T_{outside}$ is the outside ambient air temperature in °C, and
T_{air} is the internal air temperature in °C.

2.3.7 Energy-efficiency features

Energy-efficient parameters are implemented within the greenhouse to create an optimal environment for plant growth. Furthermore, the integration of various energy-saving techniques such as thermal curtains, shade curtains, supplemental heating and cooling, thermal storage, and ventilation are among the most prevalent strategies employed. The implementation of additional energy-efficient strategies depends upon the level of advancements and innovations present within the greenhouse environment. Guidelines for modeling the following energy-efficiency features are given in the following subsections.

2.3.7.1 Thermal and shade curtains

Thermal and shade curtains are often installed on greenhouse glazing to provide resistance to heat flow (Santolini *et al* 2022). Thermal curtains are used at night while shade curtains are used during the day to optimize the energy demand of a greenhouse. Three key parameters are essential for modeling these curtains:

- additional thermal resistance depending on the curtain materials, such as fabric, metals, or plastic;
- location relative to the glazing; and
- a balanced operation schedule.

2.3.7.2 Supplemental heating and cooling

Excessive heat losses at night or heat gained during the day pose a potential threat to the crops, due to which supplemental heating or cooling is required to maintain the greenhouse air temperature. The energy modeling for the supplemental systems involves:

- the source capacity;
- the set-point temperature; and
- the operational schedule.

When the greenhouse air temperature goes above or below the lower set-point, either supplemental cooling or heating is activated to maintain optimal temperature inside the greenhouse. Sometimes, the upper set-point temperature is maintained by natural ventilation, i.e. the opening of vents to allow cooler air to enter the greenhouse. However, ventilation does not provide precise control over temperature limits which may disturb the thermal environment of the greenhouse.

2.3.7.3 Thermal storage

Thermal storage is installed in greenhouses to capture the surplus heat during the day and release this heat to the greenhouse air to maintain optimal temperature at night with minimal or no supplemental heating. Thermal storage walls, phase change material, rock bed storage, water barrels, and more are commonly used thermal storage systems in a greenhouse (Nauta *et al* 2022). To model the thermal storage systems, the following parameters must be included:

- the physical dimension of the storage (height, width, and thickness);
- the thermal properties (conductivity, specific capacitance, absorptance, and emittance); and
- inputs such as greenhouse air temperature, total and beam radiation reaching the surface of the thermal storage, angle of incidence, and inside air velocity are required to calculate heat absorbed by the thermal storage.

This stored energy flows to the greenhouse space acting as a heat gain to the greenhouse air at night.

2.3.7.4 Ventilation

Ventilation classified into natural and forced ventilation is essential to control high temperature, moisture levels, and CO_2 levels inside the greenhouse for good crop production. Natural ventilation supplies fresh air inside the greenhouse without the use of any mechanical systems. It depends on the external wind pressures and the inside and outside temperatures. On the other hand, forced ventilation uses fan assemblies to bring the outside air inside the greenhouse through controlled openings. Cooler air from outside enters the greenhouse to reduce the cooling demand and creates a more uniform distribution of heat, which prevents the accumulation of hot air near the plants. Natural ventilation has a limitation during the summer months when the temperature gradient between the inside and outside air is not significant, and the need for ventilation is the greatest. Ventilation requires the below parameters to be modeled:

- sizing of fan assemblies in the case of forced ventilation;
- the location and size of inlets and outlets;
- the operation schedule for the ventilation system; and
- the opening percentage to the ventilation vents.

The ventilation gain $\dot{Q}_{vent}$ is given by the following equation (TRNSYS-18 2021):

$$\dot{Q}_{vent} = \dot{V}\rho C_p\left(T_{vent} - T_{air}\right), \tag{2.10}$$

where

$\dot{V}$ is the air exchange rate through ventilation in $m^3\ h^{-1}$,
ρ is the air density in kg m^{-3},
C_p is the specific heat capacity of air in kJ kg^{-1} °C,
T_{vent} is the ventilation air temperature in °C, and
T_{air} is the internal air temperature in °C.

Once the modeling of individual components and the connections between them are accomplished, the simulation should be executed. The outputs after the simulation should be compared with the recorded values to observe the accuracy and the performance of the modeling should be quantitatively evaluated using statistical measurements such as root mean square error (RMSE), average predicted or percent error, and mean absolute error (MAE). These parameters are calculated using the following equations:

$$\text{RMSE} = \sqrt{\frac{\sum(y_m - y_s)^2}{n}}, \tag{2.11}$$

$$\text{Average Prediction Error} = \frac{(y_{am} - y_{as})}{y_{am}} \times 100, \tag{2.12}$$

$$\text{MAE} = \frac{\sum|y_m - y_s|}{n}, \tag{2.13}$$

where y_m and y_{am} are the measured and average measured data, y_s and y_{as} are the simulated and average simulated data, and n is the number of data points.

2.4 A case study on a Chinese solar greenhouse

A commercial mono-slope solar greenhouse is located in Elie, Manitoba (49° 55′ N and 97° 28′ W) with a local elevation of 239 m. The measured dataset used for

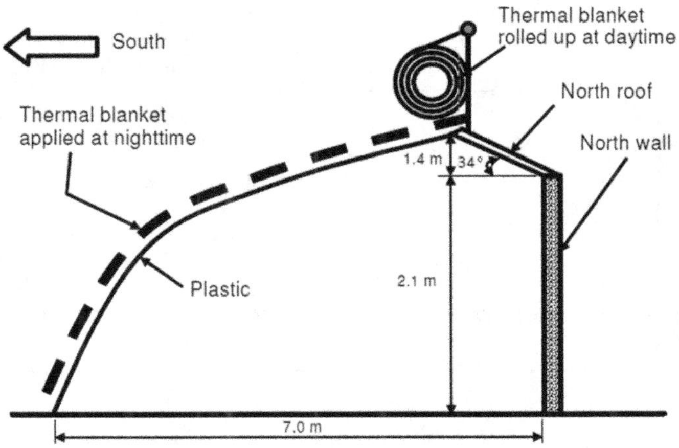

Figure 2.3. Experimental mono-slope Chinese solar greenhouse in Elie, Manitoba. (Reproduced with permission from (Beshada *et al* 2006).)

validation covers the period from 28 March 2017 to 30 March 2017 (Ahamed *et al* 2018). Various sources were consulted to gather information on the greenhouse's specifications. The structure of the experimental CSG in Elie, Manitoba is illustrated in figure 2.3.

2.4.1 Dimensions of the CSG

The CSG was east–west oriented where young tomato plants with a 14 cm height were grown in wet soil. The ground had soil with no cover (Dong 2018). Table 2.3 provides the dimensions of the CSG.

2.4.2 Construction material for CSG

The materials used for constructing this experimental greenhouse are detailed in table 2.4. The construction of the north wall included fiberglass insulation strategically installed to minimize heat transfer, along with sand, chosen for its high specific heat storage capacity (Dong 2018).

2.4.3 Energy-efficiency features for the CSG

The CSG incorporated energy-efficient parameters to regulate the indoor air temperature, ensuring optimal conditions for tomato plant growth, especially in the nighttime. Table 2.5 outlines the energy-efficient parameters integrated into the CSG.

2.4.4 Monitored parameters inside the CSG

The hourly weather data for this location including ambient air temperature, relative humidity, wind speed, and global solar radiation, were recorded every 10 min using a portable weather station positioned near the greenhouse (Ahamed *et al* 2018). Indoor air temperature, soil temperature, and north wall temperature were also

Table 2.3. Dimensions of the CSG in Elie, Manitoba.

Parameters	Values	References
Length	30 m	Ahamed *et al* (2019)
Breadth	7 m	
Footprint area	210 m^2	
North wall height	2.1 m	Ahamed *et al* (2018)
Ridge height	3.5 m	Dong *et al* (2021)
Angle of glazing near the ground (up to 1 m height)	60°	Ahamed *et al* (2019)
Angle of glazing for the rest of the section	26°	
Angle for north roof	34°	

Table 2.4. Materials of construction for the CSG in Elie, Manitoba.

Surfaces	Material of construction	Thickness	References
South glazing	Single-layer polyethylene film	0.152 mm	Dong *et al* (2021)
North wall	Corrugated galvanized sheet steel (external)	2 mm	Dong (2018)
	Fiberglass insulation	152 mm	
	Plywood	13 mm	
	Sand	152 mm	
	Corrugated galvanized sheet steel (internal)	2 mm	
North roof, east and west wall	Corrugated galvanized sheet steel (external)	2 mm	Dong (2018), Dong *et al* (2021)
	Fiberglass insulation	152 mm	
	Plywood	13 mm	
	Plastic film (internal)	2 mm	
Floor/soil	Clay	100 mm	Ahamed *et al* (2019)

recorded every 10 min (Beshada *et al* 2006). Table 2.6 specifies the location of the sensors used to measure these parameters within the CSG.

2.5 Demonstration of the framework

TRNSYS was selected to simulate the CSG due to its wide accessibility and user-friendly simulation platform known for its accuracy and low computational times. It can produce diverse outputs including temperatures, relative humidity, and heating and cooling loads. TRNSYS also offers various extensions, such as SketchUp software, TRNBuild, TRNFlow, and various sub-models to account for factors

Table 2.5. Energy-efficiency features considered fort the CSG in Elie, Manitoba.

Energy-efficient parameters	Material	Values	Schedule	References
Thermal blanket	Cotton	1.2 m^2 K W^{-1}	18:00–9:00	Beshada *et al* (2006)
Electric heater	—	3.6 kW 1.876 kW	18:00–8:30 8:30–9:00	Dong (2018)
Ventilation through ridge roof	—	0.156 m^3 s^{-1} 0.521 m^3 s^{-1} 0.573 m^3 s^{-1} 0.156 m^3 s^{-1}	8:00–11:00 11:00–13:00 13:00–14:30 14:30–17:00	
Thermal storage	Sand	—	—	Beshada *et al* (2006)

Table 2.6. Sensor locations for measuring the monitored parameters inside the CSG in Elie, Manitoba.

Sensors	Location	References
Indoor temperature sensor	122 cm above ground level	Dong (2018)
Relative humidity sensor	122 cm above ground level	
Soil temperature sensor—2 nos.	5 cm under the ground surface 214 cm away from the north wall	
Wall temperature sensor—2 nos.	76 cm above the bottom of the north wall surface	

such as ventilation, infiltration, thermal curtains, evapotranspiration, thermal storage, and more. Based on the energy modeling framework mentioned in section 2.3, the validation process was divided into six categories, which are covered in the respective sections: greenhouse model development (section 2.5.1); weather and initial data for simulation (section 2.5.2); greenhouse construction material (section 2.5.3); solar radiation distribution (section 2.5.4); evapotranspiration (section 2.5.5); and energy-efficiency features including thermal curtain, supplemental heat, thermal storage wall, and natural ventilation (section 2.5.6).

2.5.1 Greenhouse model development

The experimental greenhouse model was developed using Google SketchUp because TRNSYS software can automatically read the geometric information provided by this software. Google SketchUp employs a plug-in known as Trnsys3d, which efficiently generates geometric designs for building models. It is crucial to create Trnsys3d zones within the SketchUp model, as they facilitate the simulation of dynamic energy flow (Hiller and Kendel 2023). Utilizing the tools available within

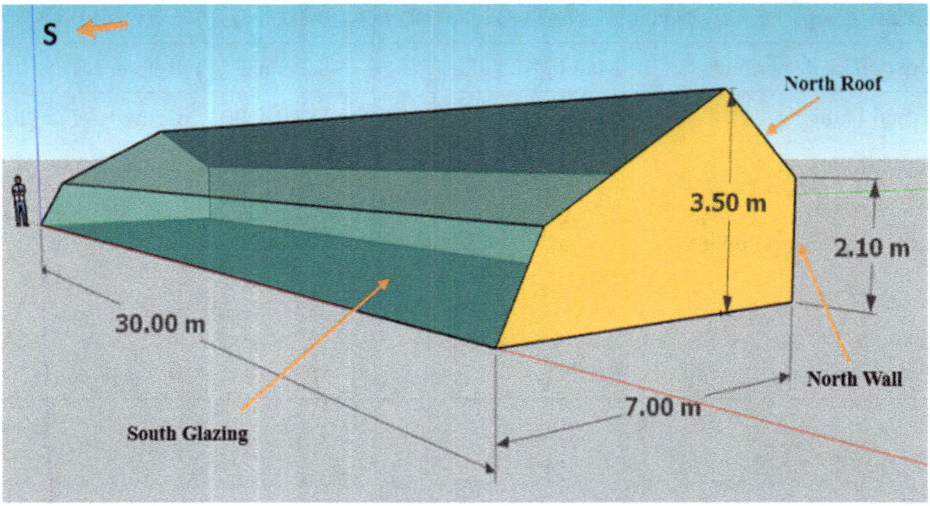

Figure 2.4. Experimental Chinese solar greenhouse model designed using SketchUp software.

the SketchUp module, the experimental greenhouse model was constructed inside the Trnsys3d zone. Trnsys3d enables the creation of fenestration objects directly within the SketchUp software. Consequently, the glazing for this CSG was developed in SketchUp accounting for approximately 98% of the total wall and roof area. Figure 2.4 depicts the greenhouse model created using SketchUp software, highlighting the glazing on the south side.

The greenhouse model from SketchUp was transferred to an integrated tool known as TRNSYS simulation studio, which was used from the project design to the simulation. The interface of the simulation studio consists of many interrelated components that transfer inputs and outputs from one another. A schematic layout, as shown in figure 2.5, was developed for the TRNSYS simulation, encompassing components such as weather data, radiation unit converter, multi-zone building components, evapotranspiration, thermal storage wall, plotters, integrator, and printer.

2.5.2 Weather and initial data for simulation

The weather component in TRNSYS software can read standard weather data formats or user-supplied data files. However, the limitation of this study was that the user-defined weather data for Elie, Manitoba for the year 2017, available from the Government of Canada website (Government of Canada n.d.) had missing/unobserved weather parameters. Therefore, in the absence of user-defined weather inputs a standard weather data format was used. Type-15 weather component in TRNSYS processes the standard files and includes modes for TRNSYS TMY, TMY2, TMY3, EPW, IWEC, CWEC, Meteonorm, and German TRY (TRNSYS-18 2017). These files contain generated values from a data bank for a specific location for a minimum of 12 years (Wikipedia Contributors 2024). The process of identifying an accurate standard weather file for the experimental CSG in Elie, Manitoba for March 2017

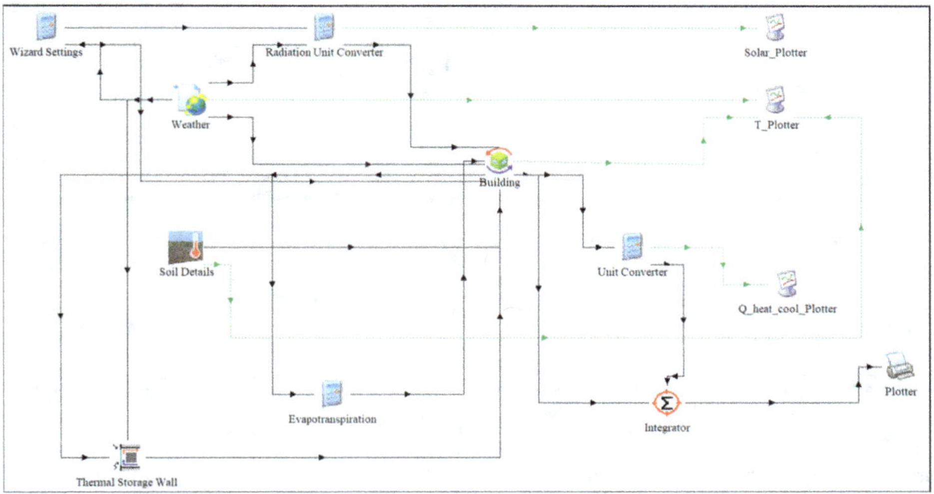

Figure 2.5. Schematic layout of the simulation studio representing the input and output connections for each component.

involved locating the nearest weather station to this greenhouse. The Winnipeg Richardson International Airport station was identified, and all the weather data available for this location were collected (Environment Canada n.d., Climate. OneBuilding n.d.). The datasets under consideration represent the average values observed across multiple years which involved the TMY dataset covering the period from 2007 to 2021, and the CWEC dataset spanning from 2000 to 2017. To ensure the most accurate weather data were used for the simulation, both datasets were compared with the measured values. However, the comparison for only the CWEC dataset was shown because it matched closely for all the weather parameters. The weather parameters such as measured ambient temperature and CWEC ambient temperature, measured relative humidity and the CWEC relative humidity, measured wind speed and the CWEC wind speed, and measured solar irradiance and CWEC solar irradiance were compared, as shown in figure 2.6.

The temperature profile showed some deviations during certain hours within the considered time as indicated in figure 2.6(a). Figures 2.6(b) and 2.6(c) indicate minimal differences in the recorded values. However, significant discrepancies were observed for solar radiation (figure 2.6(d)). This was because these were average values over the years rather than specific values for the year 2017. Table 2.7 illustrates the RMSE, mean value, and total value for ambient temperature, relative humidity, wind speed, and solar irradiance. After analysing these four weather parameters, it was concluded that the CWEC dataset (2000–2017) was the most accurate weather file for the year 2017.

Finally, the selected CWEC dataset was in EnergyPlus format (.epw) which was compatible with the type 15-3 weather component. TRNSYS also requires the initial values of temperature and relative humidity for simulation within the type-56 component. These initial values were assumed to be average values of the measured

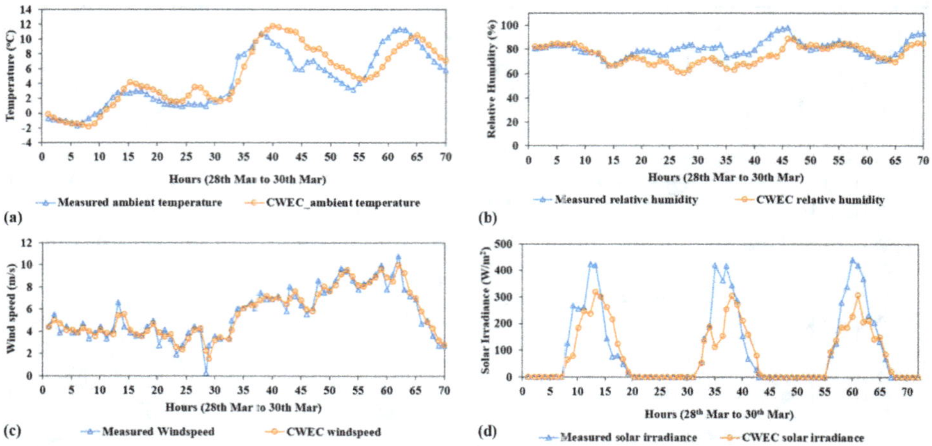

Figure 2.6. Comparison of the measured weather parameters with the CWEC dataset weather parameters from 28 March 2017 to 30 March 2017. (a) Measured ambient temperature versus CWEC ambient temperature. (b) Measured relative humidity versus CWEC relative humidity. (c) Measured wind speed versus CWEC wind speed. (d) Measured solar irradiance versus CWEC solar irradiance.

Table 2.7. Comparison of mean, total, and RMSE values for different CWEC weather parameters.

Weather parameters	RMSE	Mean value	Total value
Measured ambient temperature	—	4.5 °C	—
CWEC ambient temperature	1.3 °C	5 °C	—
Measured relative humidity	—	81.5%	—
CWEC relative humidity	8.7%	76.1%	—
Measured wind speed	—	5.540 m s^{-1}	—
CWEC wind speed	0.6 m s^{-1}	5.548 m s^{-1}	—
Measured solar irradiance	—	—	7606 W m^{-2}
CWEC solar irradiance	77 W m^{-2}	—	6022 W m^{-2}

inside temperature and measured inside relative humidity, which were 19.7 °C and 70%, respectively.

2.5.3 Greenhouse construction material

In TRNSYS software, type-56, or the multi-zone building component was used to model the materials on walls and glazing on the south side. Type-56 consists of a separate pre-possessing program known as TRNBuild. TRNBuild reads and processes the file containing the building description developed by the SketchUp software and generates an information file describing the outputs and required inputs for type-56 (TRNSYS-18 2021). The layers of material were created using the thermal properties of materials (conductivity, density, and specific heat) in TRNBuild, and then combined with respective thicknesses to form opaque walls,

Table 2.8. Thermal properties of different walls, ground, and windows used in TRNSYS software simulations.

Materials	CGSS	Fiberglass	Plywood	Sand	Plastic film	Polyethylene film	Soil
Thickness (mm)	2	152	13	152	2	0.18	100
Density (kg m^{-3})	7830	14	460	2240	900	—	1100
Conductivity (kJ h^{-1} m K)	163.08	0.14	0.334	3.31	19.8	—	1.98
Specific heat (kJ kg^{-1} K)	0.5	0.8	1.88	0.84	1.9	—	1
Solar radiation transmissivity	—	—	—	—	—	0.88	—
U value (W m^{-2} K)	—	—	—	—	—	5.56	2.84
g value	—	—	—	—	—	0.89	—
Visible light transmittance	—	—	—	—	—	0.91	—
References	Dong (2018)					Ahamed *et al* (2019)	

roofs, or floors. The south glazing of the experimental greenhouse was made up of a single polyethylene layer for which the optical and thermal properties were defined using the Lawrence Berkeley National Lab Window program. A DOE-2 file was then created and added to the TRNBuild. Table 2.8 represents the thermal properties of different materials used in the TRNSYS software for thermal simulation.

This component creates greenhouse inputs and defines the desired outputs from the simulation. Inputs such as weather output, soil detail, thermal curtain, ventilation, and others were connected with the greenhouse parameters, providing information every hour. This component calls for regime types, including information regarding infiltration, ventilation, heating, cooling, and gain/loss in the greenhouse, which are elaborated in the subsequent sections. The outputs from the greenhouse were connected to the unit converter to convert them into the desired units, which were then sent to the integrator (type-55) to integrate them over an hour and finally print the output values from the printer (type-25 c).

2.5.4 Solar radiation distribution

This section models the solar distribution on the different surfaces and internal air of the greenhouse. A radiation unit converter was used to convert the total solar radiation and beam radiation from kJ $h^{-1}.m^2$ to W m^{-2}, as well as to calculate the azimuth angle and angle of incidence for different surfaces. This converter receives inputs from the weather file, converts and calculates the respective values, and then provides outputs to the multi-zone building component. In TRNSYS, beam solar radiation was modeled using a detailed approach that uses shading and insolation matrices to distribute the primary direct solar radiation entering the greenhouse. These matrices were based on three-dimensional data of the building and shading surfaces, generated by an auxiliary program known as TRNSHD (Hiller *et al* 2000). Additionally, shortwave diffuse radiation and longwave radiation exchange including multi-reflection were distributed using a view factor matrix, which was generated by another auxiliary program known as TRNVFM (TRNSYS-18 2021). However, for TRNSYS to run these models in a detailed mode, the zone must be a convex and closed volume which means that every surface within the zone must be in the line of sight of all other zone surfaces (no obstructed views) (TRNSYS-18 2021).

TRNSYS also accounts for a factor in calculating the amount of solar heat added to the greenhouse air. Solar-to-air factor (f_{solair}) represents the fraction of solar heat entering the greenhouse air volume through the glazing that is immediately transferred as a convective gain to the internal air (TRNSYS-18 2021). This factor was important for the TRNSYS simulation as it significantly affects energy demand. The fraction may vary from 0 to 1. A study by Imafidon *et al* (2023) indicated that increasing this factor from 0 to 1 reduced the deviation in the daily temperature variation between the simulated and measured data. Many studies use TRNSYS for energy simulation, however, no study focused on the estimation of this factor. Therefore, it was crucial to understand the correct range of f_{solair} so that it could be accurately used for the simulation. Solar-to-air factor depends on the quantity of internal items with very low thermal capacity, such as furniture, because the presence of such materials will lead this fraction to be closer to 1, indicating more heat is added to the internal air. However, the experimental greenhouse consists of plants and thermal storage, both with high thermal capacity to absorb more heat. This suggests that the value of f_{solair} would not be close to 1 for this type of experimental greenhouse. Consequently, to identify the correct range of f_{solair}, it was varied in the interval of 0.2 to determine the values that most closely match the recorded data.

2.5.5 Evapotranspiration

ET was complicated to model because of the dynamic nature of plants. Since the CSG was closed, the hourly evapotranspiration rate was estimated using the Stanghellini model. TRNSYS software modeled evapotranspiration using a sub-component that takes hourly inputs such as inside air temperature, relative humidity, incoming shortwave solar radiation on the ground, and inside air velocity from the type-56 component to calculate the parameters used in equation (2.7) using table 2.2. The values for LAI and wind velocity were considered from the study by Ahamed *et al* (2018). The LAI of tomato crops was considered as 1 $m^2\ m^{-2}$ because the plants were young during the measuring period. Finally, the evapotranspiration heat loss was calculated using equation (2.8) and provided input to the type-56 component. Since evapotranspiration results in heat loss from the greenhouse environment, it was integrated as a loss into the 'Gain/loss type' manager to account for the heat loss. Finally, the evapotranspiration loss was added to the simulation under airnode regime data.

2.5.6 Energy-efficiency features

Energy-efficient parameters relevant to the experimental greenhouse were only considered for the energy modeling. It includes the installation of a thermal curtain, supplemental heating, a thermal storage wall, and natural ventilation.

2.5.6.1 Thermal curtain

The resistance of the thermal curtain, its location, and the operational schedule were modeled within the type-56 component under the 'Window Type' manager. Since the thermal curtain was employed externally in the experimental greenhouse, an additional external thermal resistance of 0.334 h m^2 K kJ^{-1} (as mentioned in table 2.5) was added manually to the south glazing. A daily schedule was defined for the thermal curtain based on the schedule mentioned in table 2.5 using the 'Schedule Type' manager within the TRNBuild. The value was set to '1' when the thermal curtain was completely employed, and '0' when not employed. TRNSYS reads the inputs for this thermal curtain when it was selected in the airnode regime data, and the value for the shade factor was assigned from the schedule.

2.5.6.2 Supplemental heat

The supplemental heating provided by an electric heater of capacity 3.6 kW was modeled using the 'Heating Type' manager under the regime types. The set-point temperature control was set at 18 °C (Dong *et al* 2021). Additionally, an operational schedule was defined for an electric heater in the TRNBuild using the 'Schedule Type' manager. The schedule was based on the literature provided in table 2.5. The value was set to '1' when the electric heater was working at full capacity, '0.5' when working at half capacity, and '0' when the electric heater was off. This schedule came into action when the schedule was integrated into the 'Heating Type' manager for limited sensible heating power, where the schedule was multiplied by the heater's capacity of 12 960 kJ h^{-1}. Finally, the heating feature was set to 'ON' under the airnode regime data.

2.5.6.3 Thermal storage wall

The thermal wall storage was modeled on the north wall by adding the physical dimensions (height, width, and thickness), and the thermal properties (conductivity, specific capacitance, absorptance, and emittance). The thermal storage wall was modeled using a type-36 d component, connected to the weather and type-56 components. The weather data provided inputs such as ambient temperature, wind velocity, and angle of incidence, while the type-56 component supplied the hourly greenhouse air temperature, total, and beam radiation reaching the north wall surface. The type-36 d component calculates the energy flows to the room and provides input to the type-56 component, acting as a heat gain to the greenhouse air. Consequently, a gain was created in the 'Gain/loss type' manager to account for the heat gain from the north wall. Finally, the thermal storage wall gain was added to the simulation under airnode regime data. The heat storage in the sidewall and the north roof of the CSGs was assumed negligible since their construction does not consist of sand for heat storage.

2.5.6.4 *Natural ventilation*

An extension known as TRNFlow was required to model natural ventilation. TRNFlow uses COMIS to model the airflow between the greenhouse and the environment (TRNFlow 2009). Cooling was not considered in the experimental greenhouse because the temperature was controlled through natural ventilation by opening a vent near the ridge (Ahamed *et al* 2018). Dong (2018) made assumptions about the ventilation rates based on indoor temperature fluctuation and CO_2 needs, as indicated in table 2.5. However, the assumed ventilation rates indicated forced ventilation as it had a fixed schedule of operation, whereas natural ventilation varies every hour depending upon the temperature gradient and wind pressure. This posed a major challenge in modeling natural ventilation without any specific details. The first challenge was to identify the size of vents, which were assumed based on the recommendation provided by the American Society of Agricultural and Biological Engineers standards that the combined roof vent area should equal the combined sidewall vent area and each should be at least 15–20 percent of the floor area. For northern climates, 15% may suffice, but warmer climates require greater amounts (Bartok 2015). As a good practice, a screen was installed on the vent to prevent insects from entering the greenhouse. Based on the minimum mesh size to exclude pests and screen availability, the screen size percentage was considered as 41% (The Mesh Company n.d.). This resulted in an actual opening area for natural ventilation of 13 m^2. A large opening of 13 m by 1 m was created in TRNFlow. The vent was located near the ridge on the north roof. In TRNSYS it was necessary to define the airflow link, which was considered from the external environment to the greenhouse air volume (TRNFlow 2009). The opening factor was defined based on the assumption by Dong (2018), with a maximum opening factor of 1 during the peak ventilation rate of 0.573 $m^3\ s^{-1}$, 0.9 during the ventilation rate of 0.521 $m^3\ s^{-1}$, 0.27 during the ventilation rate of 0.156 $m^3\ s^{-1}$, and 0 for all other times. TRNFlow also required inputs such as wind speed, wind direction, and outside temperature from the weather component, as well as internal air temperature from the type-56 component to calculate the hourly ventilation rates.

2.6 Results and discussion

Once the modeling of individual components was done, the TRNSYS simulation was executed to estimate the output values. This section shows the different analyses that were considered to study the variation in the measured and the simulated values. First, it includes the study of the solar-to-air factor (f_{solair}) to identify an appropriate range for greenhouses consisting of plants and thermal storage. Second, it involves the comparison of the simulated internal air temperature of the greenhouse with the measured temperature and the simulated and measured supplemental heat provided to the greenhouse. This is followed by the comparison of the ground temperature and the north wall temperature to observe the deviation offered by the energy modeling process.

2.6.1 Estimation of the solar-to-air factor (f_{solair})

The internal air temperature profile for measured and simulated data with different values of solar-to-air factor (f_{solair}) is shown in figure 2.7. This factor was varied from 0 to 1 in the interval of 0.2 for the analysis. Since no study focused on the estimation of this factor for greenhouses, it was crucial to understand the importance of this factor in the energy simulation using TRNSYS software. As the value of f_{solair} increases from 0 to 1, the spike in temperature also increases during the daytime, whereas the temperature almost remains the same during the nighttime for all the cases as seen in figure 2.7. With $f_{solair} = 0$, the internal air temperature values were underestimated providing a lower temperature graph as compared to the measured internal air temperature, which tends to increase the heating load of the greenhouse. However, a value of f_{solair} greater than 0.4 indicates that the temperature values were overestimated during the daytime, providing a much higher temperature graph as compared to the measured internal air temperature which tends to increase the cooling load. The values of f_{solair} between 0.2 and 0.4 indicate a much closer compliance with the measured internal air temperature. Hence, it was concluded that the f_{solair} value for greenhouses consisting of plants and thermal storage can vary from 0.2 to 0.4. To estimate the temperature profiles of the greenhouse for later analysis, three cases were considered with $f_{solair} = 0.2$, $f_{solair} = 0.3$, and $f_{solair} = 0.4$ because $f_{solair} = 0.2$ had values slightly below the measured values and $f_{solair} = 0.4$ has values above the measured values. The energy modeling results for the indoor air temperature with $f_{solair} = 0.3$ showed much closer compliance to measured values. Therefore, $f_{solair} = 0.3$ was used for the estimation of other output parameters.

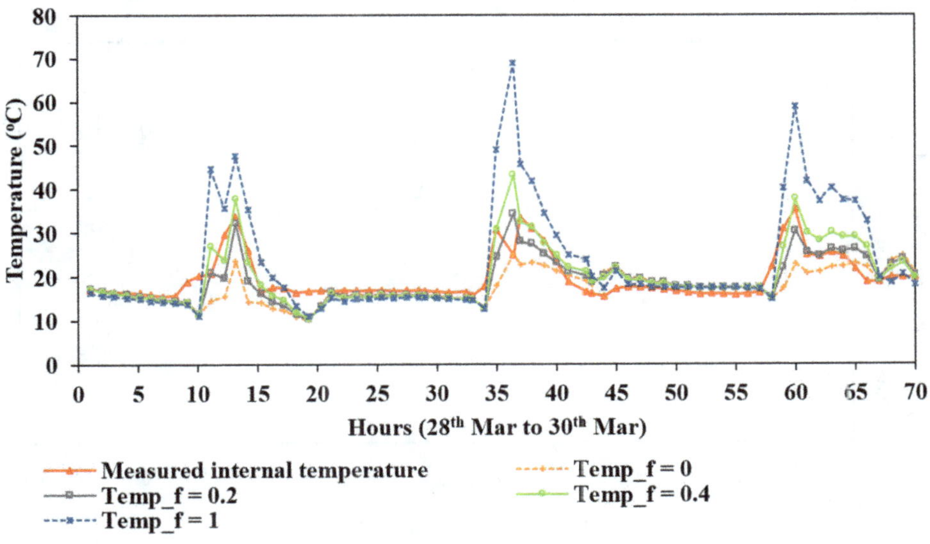

Figure 2.7. Comparison of measured internal air temperature with the simulated internal air temperature profiles for different values of solar-to-air factor (f_{solair}). Note: The temperature profile for $f_{solair} = 0.6$ and $f_{solair} = 0.8$ was also studied and lies between the temperature profile $f_{solair} = 0.4$ and $f_{solair} = 1$, however, it was not included to maintain clarity in the figure.

2.6.2 Comparison of internal air temperature

The comparison between the measured internal air temperature with the simulated internal air temperature is illustrated in figure 2.8. The graph shows that the simulated internal air temperature was in close compliance with the measured internal air temperature with $f_{\mathrm{solair}} = 0.3$. During the early hours of 28 March, the temperature drops below 15 °C because the ambient air temperature falls below 0 °C, leading to more conduction heat loss in the environment. However, during the nighttime, less fluctuation in indoor air temperature can be observed because of the thermal storage wall and thermal curtain. A spike in the indoor air temperature was observed during the daytime over all three days, caused by solar radiation entering the CSG through the south glazing and increasing the internal air temperature. Also, it can be observed that between 15 and 20 h on 28 March, the simulated internal air temperature drops to 10 °C. This could be probably because of the natural ventilation. As already indicated in the literature, growers manually open the vents to allow fresh air to enter the greenhouse, whenever it was required. However, during the simulation, a daily schedule was added to open the vents. Even when the temperature was not that high during that hour on 28 March, the vent was open because of the schedule and cooler air entered the greenhouse due to which the internal air temperature dropped.

The average value of the measured indoor temperature was 19.7 °C, closely matching the simulated average indoor temperature of 19.4 °C. The mean absolute error (MAE) and root mean square error (RMSE) for the indoor air temperature were 2.4 °C and 3.1 °C, respectively, despite the assumption developed for natural ventilation and average weather inputs. Dong (2018) also validated the SOGREEN model with this experimental greenhouse, where the study recorded an average

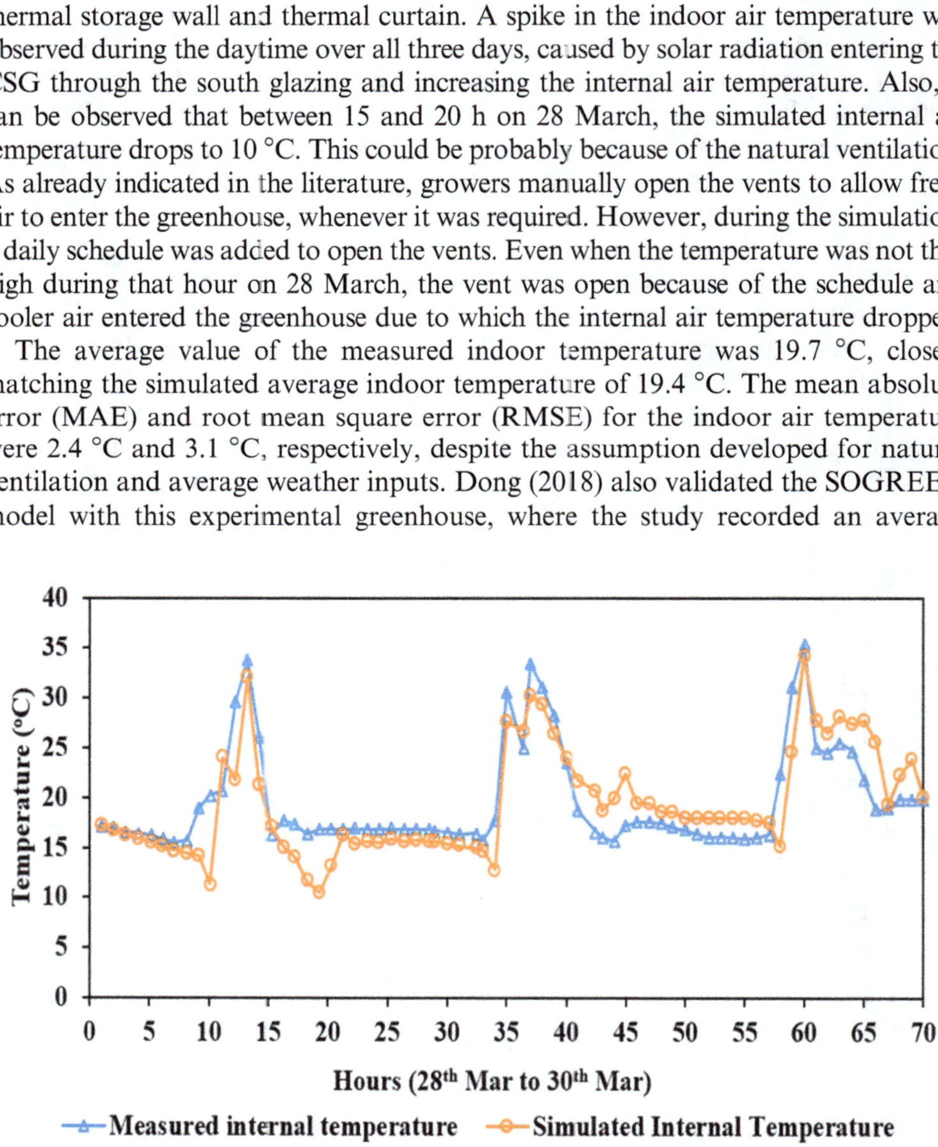

Figure 2.8. Comparison of measured internal air temperature with the simulated internal air temperature from 28 March 2017 to 30 March 2017.

temperature difference of approximately 9.6%. However, the result from this simulation illustrates that the average prediction error in the internal air temperature was just 1.6%. This leads to an important conclusion that this greenhouse energy model accurately predicted the temperature variation in an experimental greenhouse. However, the results were validated for only 72 h, i.e. three days in March, due to the limited availability of measured data during that year. Therefore, validation across different seasons was not conducted. Moreover, similar patterns were observed during both summer and winter, with trends remaining consistent throughout the day and night when the simulations were performed for a complete year.

2.6.3 Comparison of supplemental heating

The graph between the measured and simulated supplemental heat provided by an external electric heater is illustrated in figure 2.9. A 3.6 kW electric heater was used every night to maintain optimal temperature inside CSG for tomato crop growth. The operation of this electric heater was controlled by a thermostat and would turn off when the temperature reached 18 °C (Dong *et al* 2021). The supplemental heating requirements were accurately predicted by the greenhouse energy model from 28 March to 29 March with some deviations observed from 45 to 55 h. This discrepancy in the heating load occurred because of the thermostat setting. As seen in figure 2.8, the simulated internal air temperature of the greenhouse was more than 18 °C and gradually reduced to 16 °C for this period because of which the electric heater was turned off until the internal air temperature was 18 °C and then was gradually turned on when the temperature dropped below 18 °C. The total heat supplied by the electric heater from 28 March to 30 March was approximately 156 kW for the measured supplemental heating and 110 kW for simulated supplemental heating. The RMSE was approximately 1.4 kW, and the MAE was 0.7 kW. Ahamed *et al* (2018) also studied this experimental greenhouse and

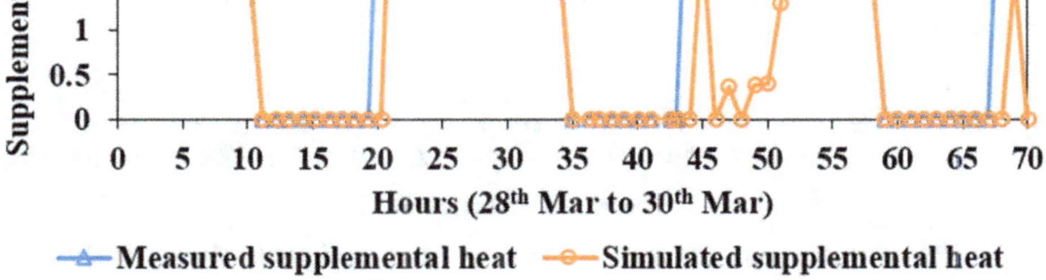

Figure 2.9. Comparison of measured supplemental heat with the simulated supplemental heat from 28 March 2017 to 30 March 2017.

compared it with the CSGHEAT model where discrepancies could be observed during the nighttime of 30 March. The percent error recorded varies from 0.2% to 24.9%, and the average error was around 8.7% when the data for the night of 30 March were excluded from the analysis. However, the average prediction error of the current study was about 29% for the complete period. Although this error was significant, it was mainly observed during some hours on 30 March due to thermostat settings and if the discrepancy due to thermostat setting was not accounted for, the error would reduce to 4.1%.

2.6.4 Comparison of ground temperature

The comparison of the measured ground temperature with the simulated ground temperature from 28 March 2017 to 30 March 2017 is indicated in figure 2.10. It was observed that the simulated ground temperature shows the same trend as the measured ground temperature. However, there was a huge difference in estimating simulated temperature during the nighttime. This deviation in ground temperature may arise because of the differences in the thermal properties of the soil used in the simulation. Additionally, TRNSYS did not simulate the wet behavior of the soil, and it was mentioned that the soil was wet with no cover (Dong 2018). Second, the sensor was located 5 cm under the ground while measuring the temperature (Dong 2018). However, in the greenhouse energy model using TRNSYS simulation, the temperature was calculated on the ground surface. It can be observed that the temperature rises as soon as solar radiation reaches the ground surface and drops because of conduction heat loss between the ground surface and the soil at night. TRNSYS software has a limitation, as it is designed for lumped body analysis, i.e. it does not account for the temperature variation in the *x*- and *y*-axes. Ahamed *et al*

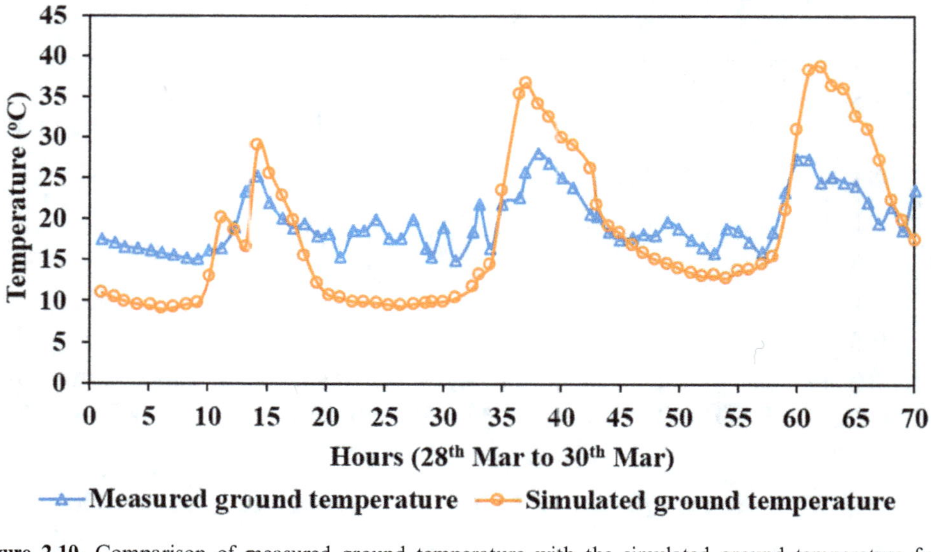

Figure 2.10. Comparison of measured ground temperature with the simulated ground temperature from 28 March 2017 to 30 March 2017.

(2018) studied the variation in the ground temperature with an RMSE of 1.8 °C. The simulation for this study was performed without considering the greenhouse air exchange through the natural ventilation system. In this study with natural ventilation, the average measured value of the ground temperature was 19.8 °C and the simulated average value for this study was 18.3 °C. The RMSE and the MAE were 6.4 °C and 5.6 °C respectively. The predicted error between the measured and simulated ground temperature was approximately 7.3%.

2.6.5 Comparison of north wall temperature

The difference between the measured and simulated north wall temperature is represented in figure 2.11. It was clear from the graph that the simulated north wall temperature accurately predicts the measured values during some hours from 28 March to 30 March. Additionally, the greenhouse energy model results well predicted the peaks and valleys during the simulation period. The north wall temperature rises to above 30 °C during the daytime because it stores more heat during the day. However, the deviation was observed during the evening and nighttime for the 29 and 30 March. In the TRNSYS simulation, the energy gain by the internal greenhouse air from the north wall was gradual. However, the measured data shows a sharp reduction in the temperature. This may be due to the high wind speed and different wind directions in the actual weather conditions for the year 2017. This led to the increased heat loss from the north wall, thereby maintaining slightly lower temperatures as compared to the simulated temperatures. The average measured value of the north wall temperature was 18.6 °C whereas the simulated average value was 21 °C. The RMSE and the mean absolute error were 4.3 °C and 3.5 °C, respectively, with a predicted error of about 13%. Dong (2018) predicted the north wall temperature with an average discrepancy in the wall surface temperature

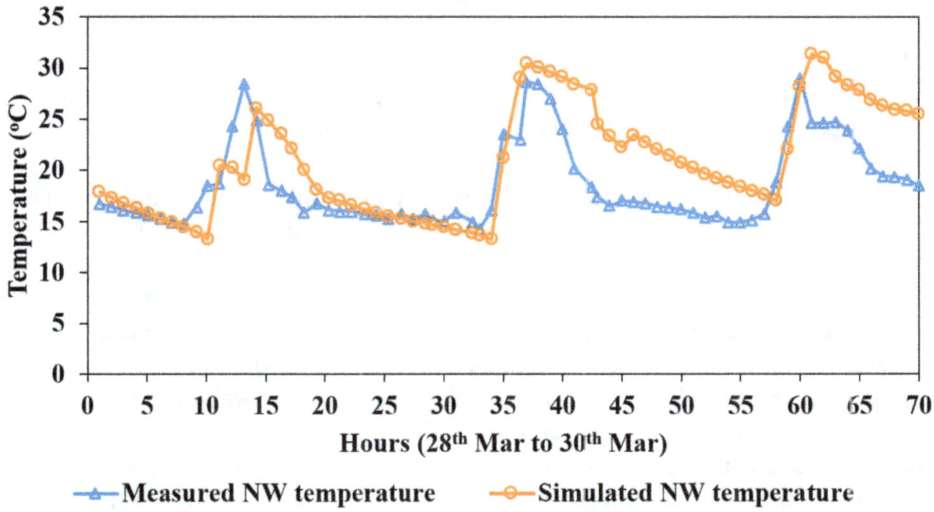

Figure 2.11. Comparison of measured north wall temperature with the simulated north wall temperature from 28 March 2017 to 30 March 2017.

of about 19.4%. In the study by Ahamed *et al* (2018), the RMSE in calculating the north wall temperature was estimated as 2.2 °C. This was because of the different modeling techniques and assumptions considered by each researcher.

2.7 Conclusion

This study demonstrated the importance of energy modeling for the greenhouse industry. Since no comprehensive guidelines or standards were available for greenhouse energy modeling, this study provided an in-depth energy modeling process for greenhouses. This study further used the mentioned energy modeling process to validate the thermal performance and the heating requirements of a greenhouse in Elie, Manitoba. The energy modeling had limitations in modeling the real experimental conditions of the greenhouse because of the absence of crucial data such as precise information on natural ventilation, and accurate weather data for the year 2017. Energy simulations were performed which led to the following conclusions:

1. Many studies used TRNSYS software for energy simulation, however, no researcher studied the solar-to-air factor. This study indicates that the range of solar-to-air factor is from 0.2 to 0.4 for greenhouses consisting of plants and thermal storage. However, this value may vary depending on the different configurations of a greenhouse.
2. Accurate weather inputs are essential for energy modeling, as the greenhouses directly interact with the external environment. However, if accurate weather inputs are unavailable, average weather data can be used.
3. The case study indicated that the natural ventilation was done by opening a vent near the ridge and no precise information about the vent size and opening factor was available. Therefore, assumptions were considered to perform the energy simulations which turned out to be quite accurate in calculating the internal air temperature and the heating requirement inside the greenhouse.
4. Evapotranspiration using the Stanghellini model can be used for a closed greenhouse. An average LAI and crop coefficient can be used to model the evapotranspiration effects without creating much error in the results.

The results obtained after the simulations show close compliance with the measured values. In a greenhouse validation process, the indoor air temperature is the most crucial parameter for verifying the accuracy of the simulation results as it majorly affects heating consumption (Dong 2018). The validation results reveal that the mean predicted error between the measured and the simulated internal air temperature was only 1.6%. This demonstrates that the greenhouse energy model was accurate and can be relied upon for further simulation processes. The validation study was not conducted for a complete year due to the limited availability of measured data during that year. However, discrepancies in the ground and north wall temperatures suggest a more refined modeling tool to be used to estimate other parameters. This study can be extended to check the uniformity and distribution of temperature in the greenhouse space, which was a limitation of TRNSYS software.

However, this energy modeling can be expanded to model the various greenhouse designs under different climatic conditions and estimate the total energy needed to operate a greenhouse. The knowledge from the modeling process can be used to optimize the greenhouse design by maintaining a balance between heat gain and heat loss. This study contributes to the development of reliable modeling guidelines for further simulations to replicate dynamic greenhouse conditions.

Acknowledgments

The authors are very thankful to Dr H Guo for sharing the measured data for the experimental greenhouse as well as to MITACS for providing the research funding.

References

Acosta-Silva Y D J, Torres-Pacheco I, Matsumoto Y, Toledano-Ayala M, Soto-Zarazúa G M, Zelaya-Ángel O and Mendez-Lopez A 2019 Application of solar and wind renewable energy in agriculture: a review *Sci. Prog.* **102** 127–40

Ahamed M S, Guo H and Tanino K 2017 A quasi-steady state model for predicting the heating requirements of conventional greenhouses in cold regions *Inf. Process. Agric.* **5** 33–46

Ahamed M S, Guo H and Tanino K 2018 Development of a thermal model for simulation of supplemental heating requirements in Chinese-style solar greenhouses *Comput. Electron. Agric.* **150** 235–44

Ahamed M S, Guo H and Tanino K 2019 Modeling heating demands in a Chinese-style solar greenhouse using the transient building energy simulation model TRNSYS *J. Build. Eng.* **29** 101114

ASHRAE 2013 *Fundamentals—ASHRAE Handbook* (Peachtree Croners, GA: American Society of Heating, Refrigerating, and Air-conditioning Engineers)

Bartok J W 2015 Natural ventilation guidelines *Greenhouse Management* https://www.greenhousemag.com/article/gm0815-greenhouse-natural-ventilation-guidelines/

Beaulac A, Lalonde T, Haillot D and Monfet D 2024 Energy modeling, calibration, and validation of a small-scale greenhouse using TRNSYS *Appl. Therm. Eng.* **248** 123195

Beshada E, Zhang Q and Boris R 2006 Winter performance of a solar energy greenhouse in southern Manitoba *Can. Biosyst. Eng.* **48** 1–5.8

Chen J, Yang J, Zhao J, Xu F, Shen Z and Zhang L 2016 Energy demand forecasting of the greenhouses using nonlinear models based on model optimized prediction method *Neurocomputing* B **174** 1087–100

Choab N, Allouhi A, Maakoul A E, Kousksou T, Saadeddine S and Jamil A 2021 Effect of greenhouse design parameters on the heating and cooling requirement of greenhouses in Moroccan climatic conditions *IEEE Access* **9** 3047851

Climate.OneBuilding.Org n.d. Data for Canada *Climate.OneBuilding.Org* https://climate.onebuilding.org/WMO_Region_4_North_and_Central_America/CAN_Canada/index.html

Donatelli M, Bellocchi G and Carlini L 2005 Sharing knowledge via software components: models on reference evapotranspiration *Eur. J. Agron.* **24** 186–92

Dong S 2018 Thermal environment modeling of the mono-slope solar greenhouse for cold regions *Thesis for Master of Science* University of Saskatchewan, Saskatoon

Dong S, Ahamed M S, Ma C and Guo H 2021 A time-dependent model for predicting thermal environment of mono-slope solar greenhouses in cold regions *Energies* **14** 5956

Environment Canada n.d. Index of /cmc/climate/Engineer_Climate/CWEC_FMCCE https://collaboration.cmc.ec.gc.ca/cmc/climate/Engineer_Climate/CWEC_FMCCE/

Golzar F, Heeren N, Hellweg S and Roshandel R 2018 A novel integrated framework to evaluate greenhouse energy demand and crop yield production *Renew. Sustain. Energy Rev.* **96** 487–501

Gong X, Qiu R, Sun J, Ge J, Li Y and Wang S 2020 Evapotranspiration and crop coefficient of tomato grown in a solar greenhouse under full and deficit irrigation *Agric. Water Manag.* **235** 106154

Gorjian S, Calise F, Kant K, Ahamed M S, Copertaro B, Najafi G, Zhang X, Aghaei M and Shamshiri R 2021 A review on opportunities for implementation of solar energy technologies in agricultural greenhouses *J. Clean. Prod.* **285** 124807

Government of Canada n.d. Past weather and climate historical data *Government of Canada* https://climate.weather.gc.ca/

Greer F, Raftery P and Horvath A 2024 Considerations for estimating operational greenhouse gas emissions in the whole building life-cycle assessments *Build. Environ.* **254** 111383

Guo H, Li Z, Zhang Z and Cui Y 1994 Simulation of the temperature environment for solar greenhouse: mathematic model and program verification *J. Shenyang Agric. Univ.* **25** 438–43

Hiller M DE, Beckman W A and Mitchell J W 2000 TRNSHD—a program for shading and insolation calculations *Build. Environ.* **35** 633–44

Hiller M and Kendel C 2023 *Trnsys User Manual; Trnsys3d Sketch-Up Plugin* (Stuttgart: Transsolar Energietechnik)

Imafidon O J, Ting D S-K and Carriveau R 2023 Thermal modeling of a passive style net-zero greenhouse in Alberta: The effect of ground parameters and the solar to air fraction *J. Clean. Prod.* **416** 137840

Jolliet O, Danloy L, Gay J-B, Munday G L and Reist A 1991 Horticern: an improved static model for predicting the energy consumption of a greenhouse *Agric. For. Meteorol.* **55** 265–94

Liu B and Jordan R C 1961 Daily insolation on surfaces tilted towards the equator *ASHRAE J.* **10** 53–3

Liu X, Wu X, Xia T, Fan Z, Shi W, Li Y and Li T 2022 New insights of designing thermal insulation and heat storage of Chinese solar greenhouse in high latitudes and cold regions *Energy* **242** 122953

Liu Y H B and Jordan R C 1963 The long-term average performance of flat plate solar collectors: with design data for the US, its outlying possessions and Canada *Sol. Energy* **7** 53–74

Ma C, Han J and Li R 2010 Research and development of software for thermal environment simulation and prediction in the solar greenhouse *North. Hortic.* **15** 69–75

Ma C 2015 Original experimental data (personal communication)

Meng L, Yang Q, Bot G P A and Wang N 2009 Visual simulation model for the thermal environment in Chinese solar greenhouse *Trans. CSAE* **25** 164–70

Mobtaker H G, Ajabshirchi Y, Ranjbar S F and Matloobi M 2016 Solar energy conservation in a greenhouse: thermal analysis and experimental validation *Renew. Energy* **96** 509–19

Nauta A, Lubitz W D, Tasnim S and Han J 2022 Using a greenhouse energy model to examine the potential for maintaining year-round growing conditions in off-grid greenhouses across Canada *The Canadian Society for Bioengineering* Paper No. CSBE22-129

Pamungkas A P, Hatou K and Morimoto T 2014 Evapotranspiration model analysis of crop water use in the plant factory system *Environ. Control Biol.* **52** 183–8

Red River College and Proskiw Engineering Ltd 2015 An investigation of airtightness in Manitoba *Final Report* https://www.rrc.ca/wp-content/uploads/sites/28/2016/03/MB-Hydro-Air-Leakage.pdf

Rorabaugh P A, Jensen M H and Giacomelli G 2002 *Introduction to Controlled Environment Agriculture and Hydroponics* (Tucson, AZ: Controlled Environment Agriculture Center, Campus Agriculture Center, University of Arizona)

Santolini E, Pulvirenti B, Guidorzi P, Bovo M, Torreggiani D and Tassinari P 2022 Analysis of the effects of shading screens on the microclimate of greenhouse and glass façade buildings *Build. Environ.* **211** 108691

Semple L, Carriveau R and Ting D S-K 2017 Assessing heating and cooling demands of closed greenhouse systems in a cold climate *Int. J. Energy Res.* **41** 1903–13

Sethi V P 2009 On the selection of shape and orientation of a greenhouse: thermal modeling and experimental validation *Sol. Energy* **83** 21–38

Sethi V P and Sharma S K 2007 Thermal modeling of a greenhouse integrated to an aquifer coupled cavity flow heat exchanger system *Sol. Energy* **81** 723–41

Sethi V P, Sumathy K, Lee C and Pal D S 2013 Thermal modeling aspects of solar greenhouse microclimate control: a review on heating technologies *Sol. Energy* **96** 56–82

The Mesh Company n.d. 0.42 mm hole stainless steel woven insect mesh midge net—0.22 mm Wire—40 LPI *The Mesh Company* https://themeshcompany.com/shop/insect-mesh/stainless-insect-mesh/0-42mm-hole-stainless-steel-woven-insect-mesh-midge-net-0-22mm-wire-40-lpi/

Tiwari G N 2003 *Greenhouse Technology for the Controlled Environment* (Oxford: Alpha Science International) pp 160–330

Tong G, Christopher D M, Li T and Wang T 2013 Passive solar energy utilization: a review of cross-section building parameter selection for Chinese solar greenhouses *Renew. Sustain. Energy Rev.* **26** 540–8

Tong G, Li B, Christopher D M and Yamaguchi T 2007 Preliminary study on temperature pattern in Chinese solar greenhouse using computational fluid dynamics *Trans. CSAE* **23** 178–85

TRNFlow 2009 A module of an airflow network for coupled simulation with type-56 (multi-zone building of TRNSYS) *TRNFlow Manual Version 1.4* (Stuttgart: TRANSSOLAR Energietechnik GmbH)

TRNSYS-18 2017 *TRaNsient System Simulation Program—Weather Data* **vol 8** (Madison, WI: Solar Energy Laboratory, University of Wisconsin-Madison)

TRNSYS-18 2021 *TRaNsient System Simulation Program—Multizone Building Modeling with Type56 and TRNBuild* **vol 5** (Madison, WI: Solar Energy Laboratory, University of Wisconsin-Madison)

Vadiee A 2011 Energy analysis of the closed greenhouse concept: towards a sustainable energy pathway *PhD Thesis* KTH Royal Institute of Technology, Stockholm

Vadiee A and Martin V 2013 Energy analysis and thermo-economic assessment of the closed greenhouse—the largest commercial solar building *Appl. Energy* **102** 1256–66

Wang T *et al* 2017 Integration of solar technology to a modern greenhouse in China: current status, challenges and prospect *Renew. Sustain. Energy Rev.* **70** 1178–88

Wikipedia Contributors 2024 Typical meteorological year *Wikipedia* https://en.wikipedia.org/wiki/Typical_meteorological_year

Xu H, Zhang Y, Li T and Wang R 2017 Simplified numerical modeling of energy distribution in a Chinese solar greenhouse *Appl. Eng. Agric.* **33** 291–304

Yu H, Chen Y, Hassan S G and Li D 2016 Prediction of the temperature in a Chinese solar greenhouse based on LSSVM optimized by improved PSO *Comput. Electron. Agric.* **122** 94–102

Zhao X T, Xu H, Li T L and Wang R 2019 Establishment of winter energy distribution model for solar greenhouses in Northeast China *J. Shanxi. Univ.* **50** 43–50

IOP Publishing

Renewable Energy Systems
The way forward
David S-K Ting and Jacqueline A Stagner

Chapter 3

Technologies for utilizing solar energy in building

Eyup Koçak and Ece Aylı

Buildings account for a significant portion of global energy consumption, emphasizing the importance of sustainable solutions. This chapter explores the integration of solar energy technologies into buildings, categorized into passive and active methods. Passive techniques, including Trombe walls, solar chimneys, and fenestration, utilize architectural designs to optimize energy efficiency by reducing reliance on conventional energy sources. Active strategies, such as building-integrated photovoltaics (BIPV) and photovoltaic-thermal systems (BIPVT), focus on generating electricity and thermal energy simultaneously, enhancing sustainability. The chapter highlights the evolution of these technologies, addressing their applications, efficiency improvements, and environmental benefits. Advanced materials such as phase change materials and innovative designs such as ventilated facades are discussed as key enablers of efficiency. Recent numerical and experimental studies provide insights into optimizing performance, reducing costs, and promoting large-scale adoption. The findings underscore solar energy's potential to transform building design, achieve energy independence, and reduce carbon footprints. This chapter aims to serve as a comprehensive resource for engineers and architects, bridging the gap between theoretical understanding and practical implementation of solar energy technologies.

3.1 Introduction

Energy consumption in buildings accounts for nearly 30% of total energy consumption (Kalogirou 2015), making the a significant source of consumption. In recent years, serious efforts have been made to reduce this consumption through the effective use of renewable energy sources and energy efficiency. Additionally, minimizing this consumption is crucial to minimize the damage to the environment (Kalogirou 2015). According to EU data, 40% of energy consumption in Europe

doi:10.1088/978-0-7503-6179-8ch3

comes from buildings. When examined on a global scale, current buildings and construction activities account for almost 36% of the world's final energy consumption and about 15% of direct and 39% of process-related carbon emissions (IEA 2020, Islam *et al* 2021). Therefore, the European Parliament published the Energy Performance of Buildings Directive (2002/91/EC) in 2002, which is one of the first significant energy regulations, and it was later revised under the EPBD-Recast in 2010 (European Parliament and Council 2010). The latest EU legislation mandates that new public buildings must be nearly zero-energy buildings from 2018 onwards, and all new buildings from 2020 onwards must also be nearly zero-energy buildings. This strategy is known as the 20/20/20 Strategy, symbolizing three 20% targets. These targets include reducing EU greenhouse gas emissions by 20% compared to 1990 levels, meeting 20% of EU energy consumption from renewable energy sources, and achieving a 20% reduction in energy consumption. Another strategic document is the Energy Roadmap 2050 (IRENA 2019). With this strategy, the EU has committed to reducing greenhouse gas emissions to 80%–95% of 1990 levels by 2050. High energy efficiency and smart energy technologies in buildings and transportation are fundamental steps for this strategy. As of 2024, the EU has progressed beyond these initial mandates. The current regulations focus on achieving net-zero-energy buildings, which are buildings with very high energy performance where the low amount of energy required comes mostly from renewable sources. This effort is part of the broader European Green Deal, which aims to make Europe the first climate-neutral continent by 2050. This initiative builds on previous strategies and includes updated measures and targets for energy efficiency and renewable energy.

During the construction of buildings, fundamental parameters such as building material, thermal properties, shape factor, distance between buildings, building footprint, and building orientation are considered as passive design strategies, which determine the energy requirements. Currently, passive design strategies are taken into account in new constructions to maximize building energy efficiency.

Green buildings are designed to reduce negative impacts on the environment and human health, minimize energy consumption, and maximize the efficient use of renewable energy sources. They consume less fossil-based energy and water compared to traditional buildings. Systems used in heating and cooling, primarily HVAC systems, account for a significant portion of energy consumption. For instance, in the USA, the use of heating, ventilating, air-conditioning (HVAC) systems contributes to 50% of energy use in buildings, which accounts for about 20% of total energy consumption in the USA (Lombard *et al* 2008). In the Middle East, over 70% of building energy is consumed by cooling systems (Vakiloroaya *et al* 2014). Since global electricity production relies heavily on conventional fuels, carbon emissions from HVAC systems are also high. Particularly, consumption and emissions in heating, cooling, and ventilation systems are primary factors driving the trend towards green buildings. The most efficient technique for heating, cooling, and ventilation is passive solar techniques (Bosu *et al* 2023).

To enhance energy efficiency in buildings, two primary techniques are employed: active and passive strategies. Active strategies, which have been in use for over a century, involve systems that require the use of fans and heat pumps/boilers to distribute energy throughout the building, consuming and producing electricity to achieve their goal. On the other hand, passive techniques involve devices/systems/structures/infrastructure/architectural designs that directly utilize natural forces or ambient energy resources. These are more energy-efficient technologies and have been in use for much longer than active strategies, emerging with the onset of humanity's need for shelter. Techniques such as cave dwellings, kang, and hypocaust, which date back thousands of years, were forms of passive utilization of energy even before the concept of passive strategies emerged. Today, the focus is on passive strategies such as sun shading, insulation, interior gardens, water features, atrium spaces, natural ventilation, etc, to reduce thermal loads in indoor environments (Li *et al* 2017, Taherian and Peters 2023, Wanchun *et al* 2020).

The use of solar energy techniques in buildings, both in passive and active methods, is efficient in reducing energy consumption and minimizing greenhouse gas emissions. A substantial body of literature on the application of solar energy techniques in building environments already exists (Erdim and Manioğlu 2014, Hii *et al* 2011, Li *et al* 2017, Lin 1981, Mingfang 2002, Taherian and Peters 2023, Wanchun *et al* 2020). These studies demonstrate that the use of solar energy is an irreplaceably efficient technology that remains open to development. As seen in figure 3.1, according to Web of Science (WOS) data, the number of numerical and experimental studies on this topic is increasing day by day. The keywords used in the creation of figure 3.1 were cyclone separator, cyclone separator efficiency, and DEM. Numerical data matching these keywords were utilized.

The main objective of this review is to examine the current state of solar energy techniques, exploring the impact of different systems on energy and thermal

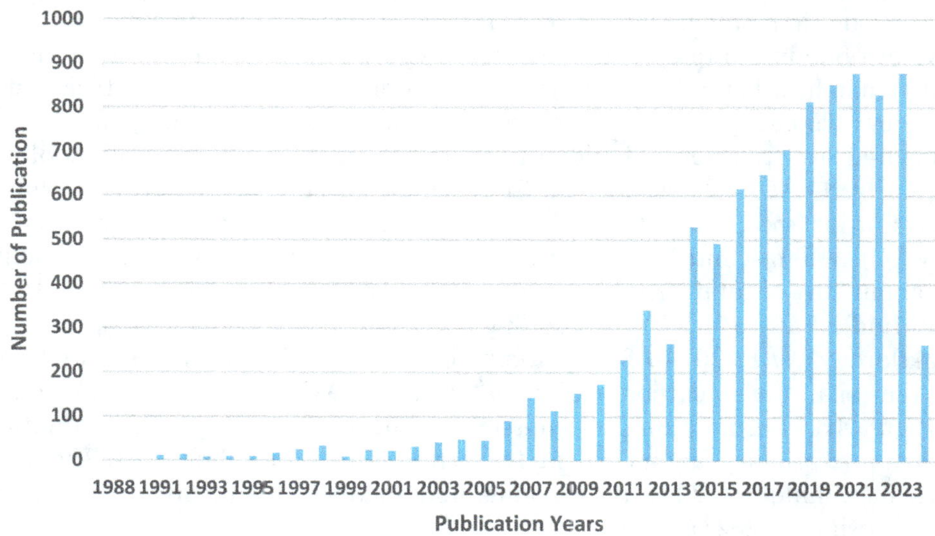

Figure 3.1. Distribution of the number of articles by year (generated using the WOS database).

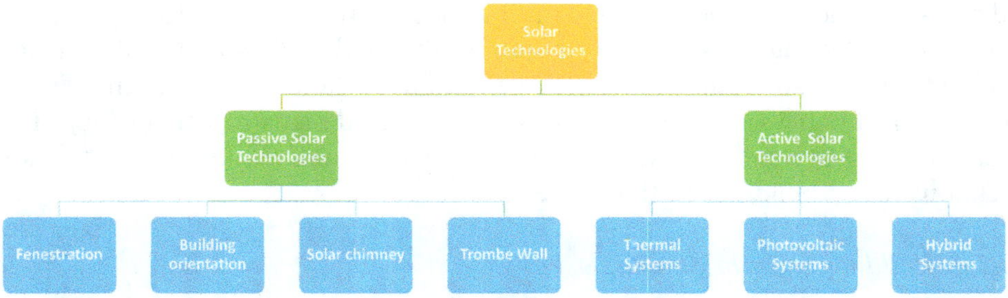

Figure 3.2. Classification of solar technologies (the classification has been made based on Bosu *et al* (2023)).

performance. Particularly, the following sections address studies related to passive architecture, solar chimneys, Trombe walls, photovoltaics, and hybrid methods. The primary goal is review the methods that reduce building energy consumption, investigate various energy-saving techniques, and provide a comprehensive guide for engineers and architects working in this field.

As previously mentioned, solar systems integrated into building design can be classified into two categories: active and passive systems, as illustrated in figure 3.2. In each method, the general principle is to effectively harness solar energy, ensuring its collection, storage, distribution, and electricity generation for thermal comfort, as well as for heating and cooling purposes. Passive solar systems can be fundamentally distinguished by elements such as fenestration, building orientation, solar chimneys, and Trombe walls. A Trombe wall is a thermal mass, usually a dark-colored wall facing the sun, that absorbs heat during the day and releases it into the building at night. A solar chimney is a vertical shaft that enhances natural ventilation by utilizing solar energy to increase the temperature and thus the buoyancy of the air inside the chimney, which drives air circulation. On the other hand, active solar techniques can be divided into solar thermal technology, photovoltaic technology, and hybrid systems.

3.2 Solar technologies applied in buildings

3.2.1 Passive heating techniques

Passive solar heating methods allow the direct use of solar energy in buildings. Fundamentally, these methods involve capturing sunlight and storing it as heat within the building through specific architectural designs, building/exterior facade equipment, and materials suited for this purpose. The initial investment costs for these systems can be substantial, with payback periods ranging from 10 to 20 years, but their operational costs post-installation are significantly lower. Using such systems in buildings also enhances the quality of life by providing more significant potential for natural lighting in interior spaces.

Many architects work without considering external conditions driven by aesthetic concerns. However, lighting, air movement, and heating systems in building designs must be integrated as cohesive factors with external conditions. Various building components directly exposed to the sun, such as walls, doors, windows, ventilators, and roofs, interact with the environment for direct heat gain. Therefore, parameters significantly affecting the solar contribution to the total cooling and heating load in a building design are mainly building orientation, building envelope, and Trombe walls. Solar chimneys can be added to architectural designs to benefit more from passive heating systems. These parameters will be examined separately for solar greenhouses and buildings in this context.

3.2.1.1 Building orientation and design

The success of utilizing solar radiation on a building positioned in a particular location depends on the orientation and size of the windows, the materials used on the exterior, and the shape of the building. Throughout the year, the intensity of solar radiation varies not only with the seasons and weather conditions but also with the orientation of a specific geographical location. Sometimes it peaks, while it receives no sunlight at other times. The sunrise and sunset directions constantly change due to the tilt of the Earth's axis and its annual movement. In the Northern Hemisphere, the sun rises in the east and sets in the west, but the positions of sunrise and sunset change throughout the year. For example, during the winter months, the sun rises in the southeast and sets in the southwest. As the days lengthen, the sunrise and sunset positions shift further north. On 21 June, the longest day of the year, both sunrise and sunset reach their northernmost positions. In the following six months, this movement reverses, and the sun rises a little further southeast and sets a little further southwest each day. In different regions of the Northern Hemisphere, there may be slight differences in sunrise and sunset positions due to variations in latitude, but the general directions and movements are similar. A sun path chart shows the azimuth angles and sun elevation at a given place during the day on different days of the year. The elevation angle, which ranges from 0° to 90° to indicate the horizon and immediately overhead, indicates the height of the sun over the horizon. The azimuth angle, which has 0° as the north, 90° as the east, 180° as the south, and 270° as the west, indicates the sun's position along a horizontal plane. Sun path charts offer a thorough view of the sun's passage across the sky by displaying the time of day for various days of the year. A sun path chart is given in figure 3.3.

The primary objective of building orientation is to minimize solar irradiation on buildings during the summer while maximizing the benefits of solar irradiation during the winter to reduce space heating costs. The heat gain of a building is a function of the surface areas exposed to sunlight. To practically determine the solar irradiation value during the design phase, it is necessary to know the duration of sunshine and the hourly solar intensity for the building region. Despite having the same floor area, exterior surface area, and internal volume, variations in annual energy expenditures can occur in different building forms due to differences in solar radiation gain and loss resulting from facing different directions (Erdim and Manioğlu 2014).

Studies have shown that rectangular-shaped buildings are the least affected by seasonal changes (Lin 1981, Mingfang 2002). Properly oriented rectangular buildings allow for greater utilization of sunlight on larger facades during winter months and less sunlight on shorter facades during summer months across different climatic

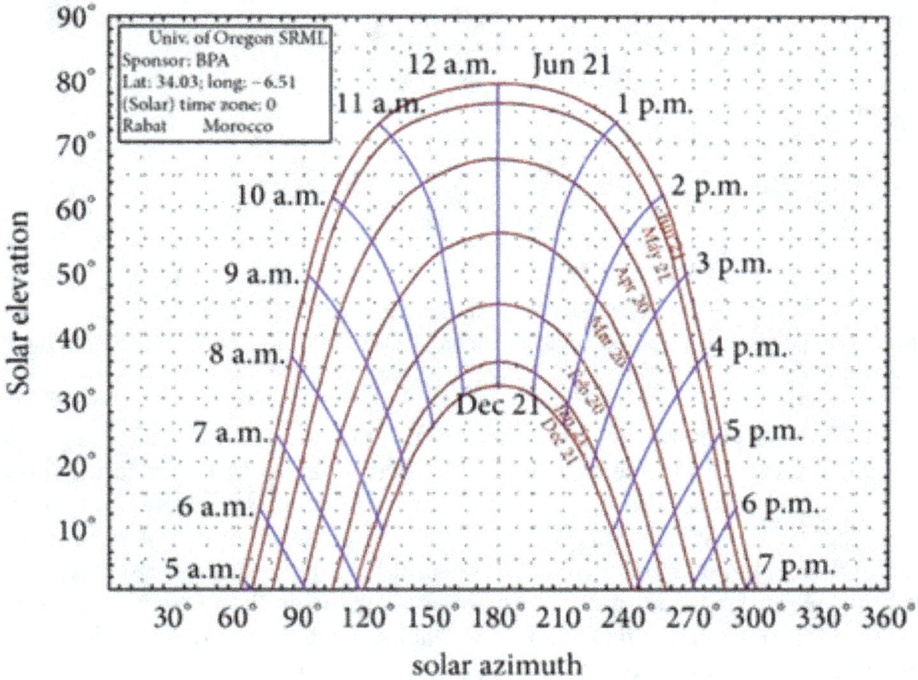

Figure 3.3. Sun path diagram. (Reproduced with permission from (Gharakhani Siraki and Pillay 2012). Copyright 2012 Elsevier.)

conditions. Similarly, minimizing the surface area exposed to solar radiation during summer results in energy savings within the HVAC system. Other building shapes, such as circular or irregular forms, do not optimize solar exposure as effectively as rectangular designs. As mentioned above, solar irradiation for each facade varies in each hemisphere. For example, facades facing south in the Northern Hemisphere and north in the Southern Hemisphere receive higher solar irradiation during summer and lower solar radiation during winter. During this period, north–south facades collect twice as much sunlight during winter, while those facing east–west receive at least four times the sunlight during summer (Hii *et al* 2011). Rectangular buildings, when oriented correctly, are therefore the least affected by seasonal changes compared to other shapes, making them more energy-efficient throughout the year.

The proper building orientation for the total energy gained by solar radiation on the building facades can be calculated using

$$E = A \times \int_{\omega_1}^{\omega_2} (C \times M) \times \left(\frac{\cos i}{\cos \theta_z}\right) d\omega, \tag{3.1}$$

where A is the surface area (m^2), M is the monthly mean daily global radiation on a horizontal surface (J m^{-2}), i is the incidence angle (radians), ω_1 and ω_2 are the hour angles at sunrise and sunset (radians), and θ_z is the zenith angle which is the

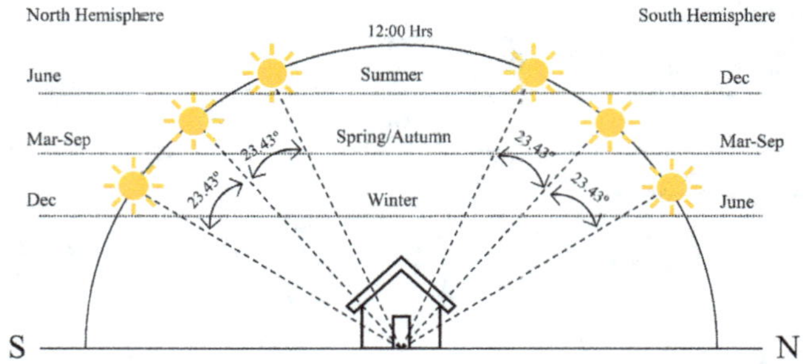

Figure 3.4. A 2D schematic representation of maximum daily solar altitude in both the Northern and Southern Hemispheres. (Adapted with permission from (Valladares-Rendón *et al* 2017). Copyright 2017 Elsevier.)

imaginary line between the observer and the sun (radians) (Gupta and Ralegaonkar 2004). The constant *C* represents an empirical factor that accounts for various effects such as atmospheric conditions, average transmittance of solar radiation through the atmosphere, or specific characteristics of the solar radiation at the location. Gupta and Ralegaonkar optimized the constant *C* value to 0.834 for the months of June and December to use calculations of solar gain for the northern hemisphere (Gupta and Ralegaonkar 2004). This was achieved by employing various shape values and adjusting the orientation angle from 0° to 180°. Therefore, it is possible to identify the best orientation angle to receive the least solar radiation in summer and the most in winter using equation (3.1).

Today, many commercial or open-source software tools are available that can estimate the energy gain on the surface areas exposed to sunlight for any location and specific aspect ratios based on incident beam radiation and hourly sunshine data (Duffie *et al* 1985, Gupta *et al* 2004, Gupta and Ralegaonkar 2004, Lee and Oberdick 1982). Using these programs when determining the design and orientation of the building allows for the analysis of the heat gain effect in buildings, considering factors such as the aspect ratios of the walls, the position of windows, and the number of floors. In the Southern Hemisphere, a similar solar path behavior occurs oppositely. A schematic view illustrating the sun angle and orientation during the summer and winter months is provided in figure 3.4.

3.2.1.2 Fenestration (windows and doors)

Windows and doors are examples of fenestration and shading mechanisms, which are commonly used and crucial architectural elements. In the summer and winter, they regulate ventilation, glare, and solar heat gains passively. These factors are vital and ought to be taken into consideration during the building design stage (Pino *et al* 2012).

Fenestrations affect indoor comfort and energy budgets by considering many factors. They help illuminate buildings throughout the year by letting in daylight and also use solar energy as a passive heating method.

However, these building elements must minimize solar heat gains in summer and maximize them in winter. Therefore, design criteria for fenestration applications in building designs include:

- The orientation of the glazed area.
- Window-to-wall ratio (WWR).
- Self-shading applications.
- Improving the daylight distribution by strategically placing windows and using light shelves or reflective surfaces to ensure even distribution of natural light throughout interior spaces.
- Retaining the view to the outside.

The orientation of glazed areas is one of the important factors for passive solar energy gain. In addition to the orientation of glazed areas, climatic factors, location, humidity, cloudy/clear sky conditions, building/room size, number of occupants, HVAC systems in the building, insulation techniques used in the building, exterior cladding, daylight distribution, and glass materials all have an impact on annual energy savings. There is significant variability in energy demand when large glass areas are used. It is particularly important to pay attention to the type of glass used, especially if it constitutes a significant portion of the facade. In winter conditions, if the glass is single-layered, there will be high heat loss due to conductivity. On the other hand, if the glass is double-layered, heat loss in winter is smaller, but still higher than that of an insulated concrete wall. Therefore, better results are achieved in colder months with a mixed facade. In summer, when cooling is needed, single or double-glazing is detrimental because it allows the entry of the visible spectrum (sunlight) but does not let the infrared spectrum (heat) escape (Pino *et al* 2012).

The solar benefit is lower but still significant if the glass has selectivity. A systematic approach should be taken when determining orientation and the WWR. In the design phase, independent parameters are the building aspect ratio (BAR) and the south window size (SWS) of residential buildings. The dependent parameter is the annual heating, cooling, and total thermal loads of the building in GJ (Inanici and Demirbilek 2000). investigated the effect of the window-to-wall ratio on energy saving in 200 m^2 low-rise residential buildings located in different parts of Türkiye, considering these independent and dependent parameters (Inanici and Demirbilek 2000). The study was conducted in five different locations using computer simulation techniques, and it revealed that south window size and window insulation value are the most significant factors in energy savings. The results of the study on building aspect ratio (BAR) indicated that maximizing elongation in the east–west axis (1:2) could be preferred in hot climates. A cold climate is defined by short, chilly summers and long, hard winters with frequently below-freezing temperatures. The most ideal condition in these types of climates was found to be a compact building design with a BAR of 1:1.2. The effect of windows on heat gain and loss is taken into consideration when calculating the SWS of residential buildings. Because heat gain must be reduced in the summer, buildings with a 25% SWS are generally chosen in hot climates. In cold climates, on the other hand, bigger SWS up to a certain limit are favored to maximize heat gains in the winter. It is highlighted that the results are mostly relevant to apartment buildings without a ground contact or roof. Within the research, when WWR optimization was conducted for a building located in the hot climate region of Antalya, Türkiye, the annual energy savings were observed to be 54.17% (Grynning *et al* 2014).

Self-shading applications are tactics and architectural elements used in building design that allow a structure to throw shadows on itself, minimizing heat gain and direct solar exposure. This can be accomplished in a number of ways, including by using overhangs, balconies, louvers, and other projecting features that prevent sunlight from reaching specific areas of the structure during the strongest hours of the day. Buildings can maintain lower interior temperatures and use less air conditioning by implementing self-shading systems, which improves energy efficiency and occupant comfort.

Pino *et al* (2012) conducted an examination of the WWR parameter while considering the habitability conditions and the necessary energy to achieve these conditions for a mid-rise office building comprising ten stories located in Santiago, Chile (Pino *et al* 2012). In climates such as Santiago's, characterized by low cloud cover during spring and summer months and high temperatures exceeding 28 °C during the day, the occurrence of overheating in office buildings is plausible. Specifically, it is noted that if a building has a WWR of 100%, there is a high likelihood of overheating during summer months when sky conditions are clear and environmental temperatures approach 20 °C, potentially extending to the winter season as well. Therefore, an optimization study on WWR is deemed necessary. The 288 cases examined by Pino *et al* (2012) determined that the size of glazed surfaces is the most influential factor on energy demand (heating and cooling). If the WWR is 20%, any building configuration studied would achieve an annual energy demand below 40 kWh m^{-2}, contingent upon the presence of solar protection and the type of glass selected. Conversely, if the WWR is 50%, the reduction in energy demand ranges would increase by double compared to the 40% WWR case annually. Finally, for buildings with fully glazed facades (100% WWR), the reduction in energy demand would increase threefold each year. Optimal performance is attained through small fenestration (WWR 20%), north-facing overhangs, and east–west oriented horizontal blinds in conjunction with selective double-glazing for Santiago, Chile. Moreover, while slightly increasing heating demand, selective single-glazing decreases cooling demand during summer, providing a viable alternative to the additional costs associated with selective double-glazing. However, it is imperative to consider the significant influence of annual climatic variations on the internal environment of the building (Grynning *et al* 2014, Pino *et al* 2012).

3.2.1.3 Trombe walls

The Trombe wall is a passive system first utilized in the 1960s (Saadatian *et al* 2012). The initial example of a Trombe wall is illustrated in figure 3.5. Its working principle is based on harnessing solar energy to heat a building. For the system to function efficiently, it must be installed on sun-facing facades. Due to its environmentally friendly nature and energy-efficient design, it holds a significant place in architecture.

The Trombe wall is composed of specific layers, each playing a role in collecting, storing, and transferring heat. The first layer is a thermally conductive mass wall, typically with a dark-colored surface that absorbs solar radiation with high

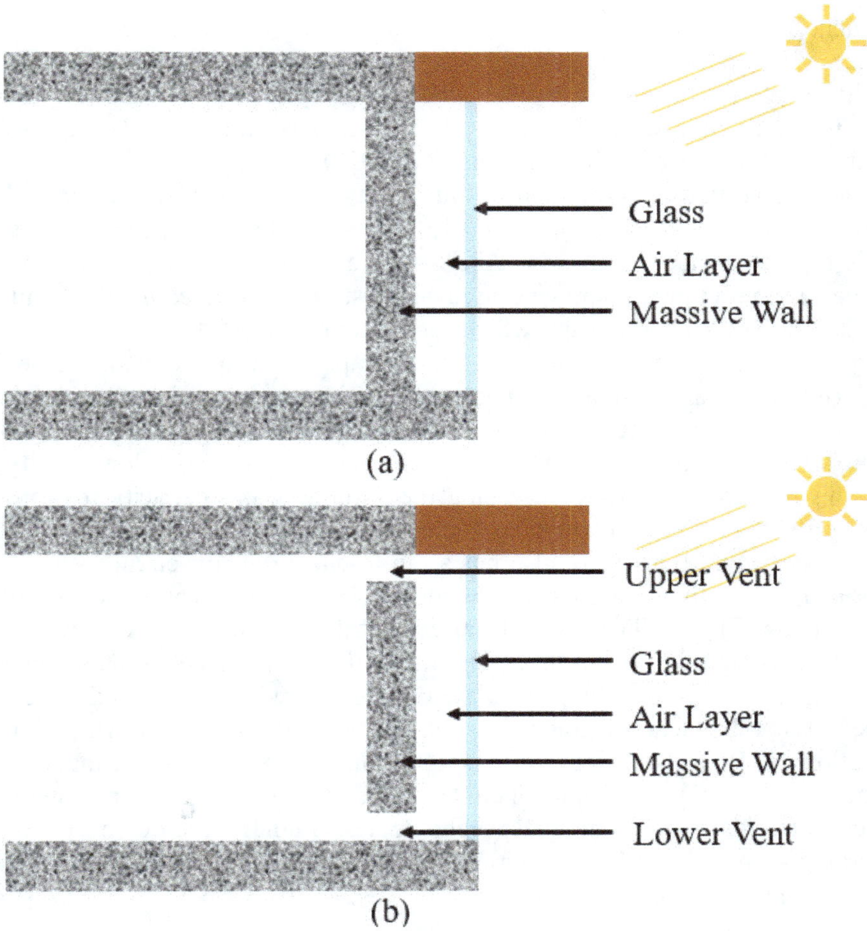

Figure 3.5. Classic Trombe wall: (a) unventilated and (b) ventilated. (Adapted with permission from (Xiao *et al* 2024). Copyright 2024 Elsevier.)

emissivity capability. This allows the system to achieve maximum heat collection capacity. On the exterior of the wall, a double-layered glass or polycarbonate panel is used. The primary purpose of these panels is to create a greenhouse effect, enabling solar radiation to reach the wall.

Between the wall and the panel, there is an air gap that enhances heat transfer and provides natural ventilation. To maintain controlled airflow within the air gap, vents are placed at the top and bottom, facilitating natural convection. Additionally, an insulation layer on the interior side of the wall minimizes heat loss.

These systems, with low energy costs, lose their efficiency if not designed to receive direct sunlight. On the other hand, without adequate ventilation and shading measures, they may lead to overheating during summer. As they do not involve any mechanical systems, Trombe walls require low maintenance and are cost-effective.

One modification in Trombe wall design is the introduction of water-based walls. The development of solar transwalls and solar hybrid walls has emerged from this approach. In water-based variants, water is used as a heat storage medium to enhance system performance and optimize energy efficiency. The high specific heat capacity of water ensures more efficient heat absorption. In water-based Trombe walls, the traditional design is preserved, but water is utilized as the heat storage medium (Prozuments *et al* 2023, Zhang *et al* 2020).

Some notable studies in this field include research conducted by Xiao *et al* (2024), Charqui *et al* (2022) and Rabani *et al* (2015). Although the system is simple, research on this topic continues, and more efficient systems are still being developed. For instance, another improvement was discussed in studies by Ma *et al* (2018) and Xiao *et al* (2024). The classic design, which lacks thermal insulation, causes heat loss during nighttime. With the new design, composite Trombe walls have been introduced, significantly enhancing the heat retention capacity of the original design.

As an alternative to the traditional Trombe wall, the fluidized Trombe wall was also introduced (see figure 3.6). This design fills the air layer with a low-density fluid that is highly heat-absorbent, increasing the Trombe wall's capacity to store heat (Xiao *et al* 2024). The fluid must be filtered to prevent leakage into the building's interior spaces, and the system also has a fan to encourage air circulation.

Another innovative approach to enhancing system efficiency is the use of low-emission (Low-E) glass Trombe walls and aerogel Trombe walls (Xiao *et al* 2024). These systems significantly improve the thermal insulation performance of glass, minimizing energy losses. In particular, double-layered Trombe walls effectively reduce heat loss by incorporating insulation and ventilation layers. Furthermore, the integration of PID-controlled smart systems into Trombe walls enhances their operational efficiency. Examples include temperature-controlled fan systems and electrochromic/thermochromic glass technologies. Studies have demonstrated that temperature-controlled fans can significantly boost the energy efficiency of Trombe walls (Xiao *et al* 2023).

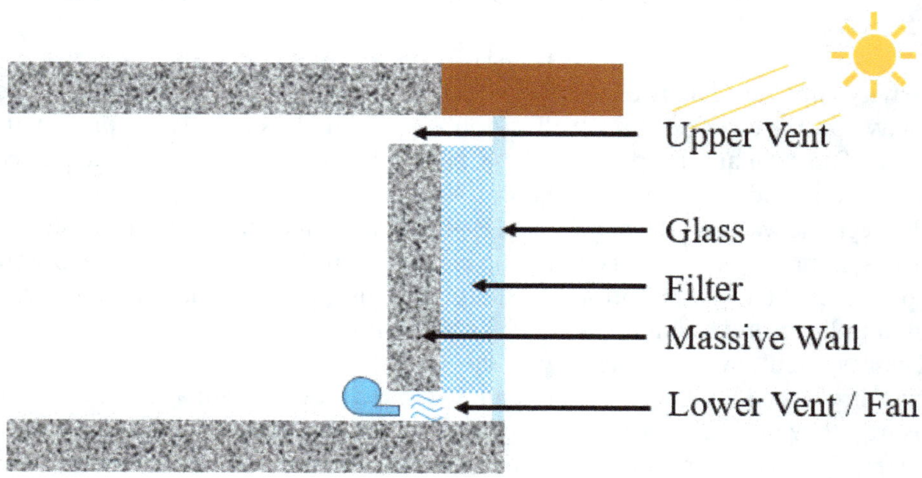

Figure 3.6. Fluidized Trombe wall. (Adapted with permission from (Xiao *et al* 2024). Copyright 2024 Elsevier.)

Electrochromic and thermochromic glass dynamically adjusts transparency in response to electrical voltage or temperature changes, optimizing solar heat gain. This feature enhances the efficiency of Trombe walls under varying seasonal and climatic conditions (Ma *et al* 2019). Additionally, advanced materials and technologies have been integrated into innovative Trombe wall systems (TWSs) to further improve their performance. One such innovation is the phase change material (PCM) Trombe wall. PCMs can store substantial amounts of thermal energy by transitioning between phases (solid to liquid or vice versa) at specific temperatures. Compared to traditional materials, such as concrete, PCMs offer higher heat storage capacity within a smaller volume, allowing for more compact Trombe wall designs and increased usable space within buildings. Furthermore, the gradual release of heat by PCMs helps maintain stable and comfortable indoor temperatures.

Another significant advancement is the development of air-purifying Trombe walls, designed to enhance indoor air quality. These systems incorporate materials such as photocatalytic coatings or air filters (Xiao *et al* 2024). Photocatalytic coatings, activated by sunlight, degrade harmful chemicals and microorganisms in the air, while air filters capture dust, pollen, and other particulates, improving air quality. This feature provides a notable advantage for buildings in regions with high air pollution levels.

Additionally, photovoltaic (PV) Trombe walls represent a major innovation by integrating PV panels into Trombe wall designs. These systems offer dual benefits by simultaneously generating electricity and storing solar heat. PV panels convert sunlight directly into electrical energy, while the Trombe wall system stores the energy as heat. This integrated design enhances building energy efficiency, promotes the use of renewable energy, and reduces energy costs, making it an ideal solution for achieving energy independence (Xiao *et al* 2024). The role of PV panels in active systems will be discussed in detail in the next section, active techniques.

3.2.1.4 Solar chimney

Solar chimney technology is a type of power plant that can recover waste heat at low temperatures and convert solar energy into electricity. Additionally, it is a method that can be used to improve the thermal performance of buildings for natural ventilation and to minimize the use of equipment used for heating and cooling purposes. One of the options for natural ventilation of buildings is the solar chimney. Solar chimneys are natural ventilation systems that use solar radiation to produce convective airflows. Convective flows extract air from inside a building or room, distributing excess heat to improve thermal comfort and remove contaminants, thereby enhancing indoor air quality. When designed appropriately according to the building's orientation, whether on the facade or the roof, solar chimneys can capture a significant amount of solar energy and create a structure capable of generating high mass flow rates. These chimneys are more commonly used in tropical climates and lightweight structures that store small amounts of thermal energy. They are characterized by high thermal inertia, accumulating heat during periods of high solar irradiance and dissipating it at night to create nocturnal

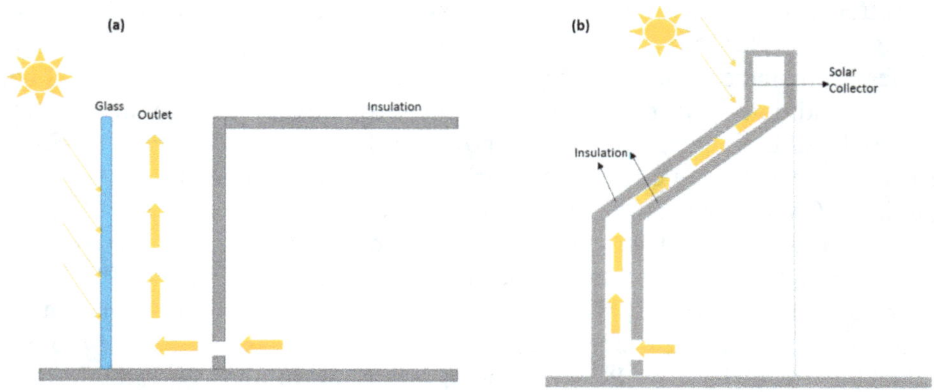

Figure 3.7. Solar chimney: (a) vertical type and (b) roof type.

ventilation. The glazed surfaces of solar chimneys are positioned to receive maximum solar irradiation, allowing the absorber plate to absorb radiant energy.

The temperature difference between the chimney channel and the interior of the house creates a pressure difference, causing indoor air to flow outward, thereby providing ventilation in the room. Solar chimneys must be connected to the interior through a ventilation hole. There are two common types of solar chimneys: vertical and roof solar chimneys. Vertical chimneys use a vertical glass surface to capture solar heat, while roof chimneys use solar collectors for the same purpose. However, roof chimneys encounter additional airflow resistance, making their use more challenging (Jiménez-Xamán *et al* 2019, Vargas-López *et al* 2019). In figure 3.7, vertical and roof solar chimneys are depicted.

3.2.2 Active techniques

As shown in figure 3.2, active techniques are essentially divided into two main categories and thirdly emerge as a hybrid method by combining these two energy conversion techniques. Systems are fundamentally based on the conversion of solar energy into electricity or heat. Photovoltaic panels are used to convert solar light into electrical energy, while solar collectors are used to convert solar light into heat energy. These active techniques are methods that increase energy efficiency and maximize the utilization of solar energy. The use of hybrid systems, which can simultaneously obtain electricity and heat energy, is also crucial.

3.2.2.1 Photovoltaic systems

Reducing energy consumption, which is one of the significant issues contributing to ecological degradation, can be addressed through innovative approaches in buildings, one of which is designing energy harvesting levels with photovoltaic panels. Photovoltaic (PV) systems consist of panels containing semiconductor materials, which do not require complex technology. This simple system can be widely used in the horizontal or vertical envelope of buildings. However, the high temperatures reached by solar cells due to the absorbed solar radiation reduce their efficiency, and insufficient airflow can exacerbate heating, especially in summer conditions, posing a more serious problem. Therefore, placing a naturally ventilated open channel on the back of the PV panel emerges as a general solution method. This allows for effective heat dissipation and reduces heat gain on the building's exterior facade. The airflow in the channels behind the PV modules is actively circulated using fans, pumps, or compressors to address the issue of high temperatures (Han *et al* 2013, Khedari *et al* 1996, 2000). The flow rate of the fluid, air gap, tilt angles, and the positioning of PV panels are fundamental parameters that affect the efficiency in photovoltaic panels.

According to International Energy Agency 2024 data (Zhang *et al* 2018), PV capacity additions saw an increase of over 80% from 2022 to 2024, reaching a record-breaking 420 gigabytes. In 2023, the European Union reached a PV usage increase of up to 52 gigabytes. India commissioned 12 gigabytes of solar PV usage in 2023.

In Brazil, with energy incentives, PV additions increased by over 20% as illustrated in figure 3.8.

In the early 1800s, French scientist Edmond Becquerel stumbled upon a fascinating phenomenon during his experiments with electrochemistry—light could influence electricity. Decades later, Adams and Day took a major step by crafting the first rudimentary solar cell, albeit with low efficiency. Charles Fritts then improved upon this design by doubling the efficiency with a selenium component. A turning point arrived in 1905 with Albert Einstein's revolutionary theory explaining the photoelectric effect. In the 1940s, Bell Labs created the first silicon solar cell, but its low efficiency limited its use to sensors at that time. The 1950s witnessed the first practical application of solar cells, showcased at the National Academy of Sciences in the US. This paved the way for widespread adoption in space exploration, spearheaded by NASA. The space race significantly fueled advancements in photovoltaic technology. Progress in related fields such as electronics and quantum mechanics further boosted the development of solar cells. By the 1980s, companies were actively producing prototypes, and by 1986, systems for harvesting solar energy were introduced. The following years saw increasing recognition and programs dedicated to photovoltaic technology around the world, including China's entry in 2009. This historical journey underlines the remarkable

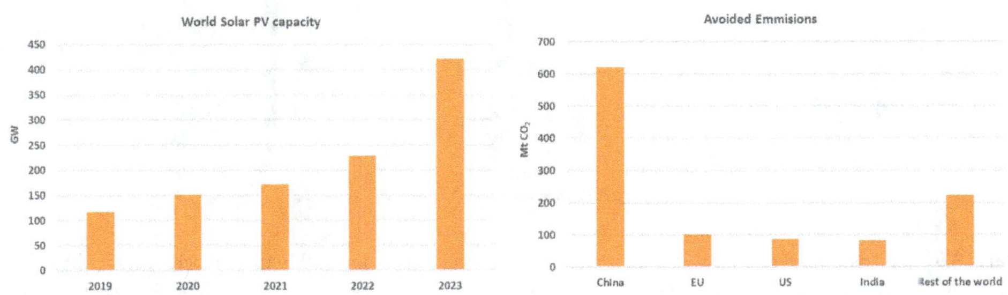

Figure 3.8. Solar PV capacity additions and avoided emissions in worldwide. (Adapted from Zhang *et al* (2018). CC BY 4.0.)

strides made in photovoltaic technology, highlighting its growing significance across diverse applications (Biyik *et al* 2017).

Three main methods can be mentioned for integrating PV panels into buildings. The first method is the one where there is no ventilation gap and the panel is directly attached to the building. The second situation involves the presence of an air gap, while the third situation is a ventilated facade with indoor airflow. In the literature, numerous numerical, theoretical, and experimental studies have been conducted on the fluid flow and heat transfer in the air channel behind PV modules. Figure 3.8 schematically illustrates the integration of photovoltaic panels into buildings (Biyik *et al* 2017, Kim and Lee 2007). Creating an air channel between PV modules and the external envelopes of a building, benefiting from the air circulation in the channel, can reduce the operating temperature of PV modules comparing to a BIPV system directly on a building wall without any ventilation space (figure 3.9(a)), thus providing an effective method to increase the energy efficiency of PV modules.

As shown in figure 3.9(b), PV systems are integrated into the facade of the building, and the open-air system enters from the bottom and exits from the top in systems where heat absorption leads to decreased temperatures, resulting in increased efficiency and extended lifespans of PV modules.

3.2.2.1.1 Structure of solar cells and the photovoltaic cell model

Photovoltaic cells are produced using semiconductor technology and are formed by combining p-type conductive material with n-type material that consists of charge-carrying electrons. When sunlight strikes the junction of these two conductors, it generates a current through the circuit. At the n–p semiconductor junction, electrons move into the p-type region, filling the electron deficiency near the junction surface and forming negative ions (−). Simultaneously, a positive ion (+) wall forms in the n-type region. When sunlight hits this area, the current in the external circuit flows from the p-type to the n-type region. The structure of the photovoltaic cell, as seen in figure 3.10, involves charge carriers being driven towards the regions where they are predominant when photon light energy is absorbed to a certain extent. In

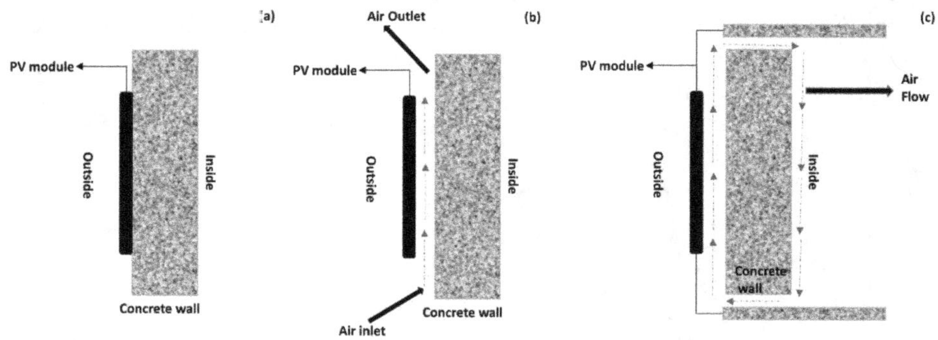

Figure 3.9. (a) Installing a BIPV system directly onto the building wall without any ventilation space, (b) installing a BIPV system as a facade with an air gap, and (c) affixing a BIPVT system as a ventilated facade with internal airflow utilized for PV panel cooling. (Adapted with permission from (Baljit *et al* 2016). Copyright 2016 Elsevier.)

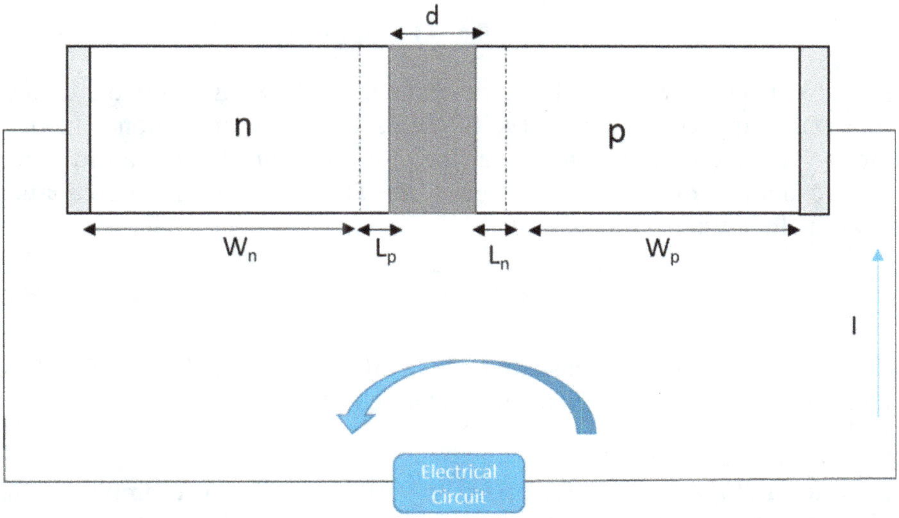

Figure 3.10. The p–n junction surface provides conductivity through the energy of photons. (Adapted from Biyik *et al* (2017).)

this situation, a current I_s flows through the junction, causing the n region to become negatively charged (−) and the p region to become positively charged (+) (Biyik *et al* 2017).

In a nutshell, PV cells turn sunlight into electricity using light particles (photons). When light hits the cell, it energizes electrons in a special material, making them flow and creating a voltage. This is called the photovoltaic effect. The Sun's energy is the source of this electricity. Basically, sunlight bumps electrons loose in the cell, it frees electrons, producing an electric current. In summary, PV cells convert solar energy directly into electrical energy, and this conversion occurs when the photons in the light activate the charge carriers (electrons and holes) in the semiconductor material (Kim and Lee 2007). To obtain energy in a PV cell, the material that absorbs sunlight must be a semiconductor compatible with the solar spectrum within the energy difference between the valence band and the conduction band (band gap) in a semiconductor material and capable of producing energy at the desired level. These semiconductors can include silicon (Si), gallium arsenide (GaAs), cadmium telluride (CdTe), among others. In a PV cell, p-type and n-type regions are created in the semiconductor. At the junction between these p- and n-type regions, an electric field spontaneously forms. For a PV cell to generate electrical energy, the photovoltaic event must occur at the n–p junction in the semiconductor material used in production (Marques Lameirinhas *et al* 2022).

In equation (3.2), the relationship between current and voltage is given for the p–n junction. The I_{ph} term (amperes) is related to the semiconductor behavior under illumination and the other term is related to the observed current when a voltage is applied. In the equation, I_{is} (amperes) is the reverse saturation current, in other words the dark current, and n is the junction non-ideality factor:

$$I = I_{is}\left(e^{\frac{V}{nu_T}} - 1\right) - I_{ph}. \tag{3.2}$$

Equation (3.2) is only valid for the arrow directions shown in figure 3.3, assuming that the entire applied voltage V (volts) drops across the transition region. This assumption is based on the premise that the resistance outside the transition region and at the ohmic contacts can be neglected. The u_T (volts) in the equation is expressed as follows:

$$u_T = \frac{K}{q}T, \tag{3.3}$$

where K is the Boltzmann constant (1.38×10^{-23} J K^{-1}), T^{-1} is the absolute temperature (K), and q is the elementary charge (1.6×10^{-19} C).

3.2.2.1.2 Classification of PV

Solar PV technology can be categorized in various ways. For example, it can be classified based on the technology as crystalline, thin-film and emerging, or according to the application method as roof or facade. Another classification method is based on design approach (Biyik *et al* 2017). In figure 3.11, a classification of PV technology is depicted (Pillai *et al* 2022).

- *PV technologies*: The classification of PV technology is divided into three main groups: first, crystalline silicon structures; second, thin-film structures; and third, a combination of first and second-generation technologies. Second-generation technologies have lower efficiency compared to first-generation technologies, but they are preferred due to their lower costs. Similarly, third-generation technologies are highly efficient but expensive (Pillai *et al* 2022). Crystalline silicon solar cells initially evolved from single-crystal silicon rods and later transitioned to polycrystalline silicon structures. Single-crystal silicon rods are sliced into thin wafers and utilized. In the production of polycrystalline silicon, the process starts with melting silicon, then casting the liquid silicon into squares, solidifying them, and transforming them into crystals with various orientations. Afterward, thin slices are cut, resulting in polycrystalline silicon. The efficiency of monocrystalline cells is mostly around 22%. Polycrystalline silicon is produced with a lower-cost process compared to monocrystalline silicon. While the size of polycrystalline cells is similar to monocrystalline cells, their efficiency ranges from 14% to 18%. While monocrystalline cells are homogeneous, polycrystalline cells have many reflective points. Thin-film solar cells consist of thin semiconductor material layers applied to a solid backing material, resulting in lower costs due to significantly reduced semiconductor material compared to silicon wafers. Depending on the type of material used, thin-film solar cells are further categorized into different types. Particularly preferred for solar energy production in areas such as forests, solar fields, and traffic and street lights, these cells feature thin and flexible arrangements of various layers. Their ease of use, flexibility compared to traditional cells, and lower costs are

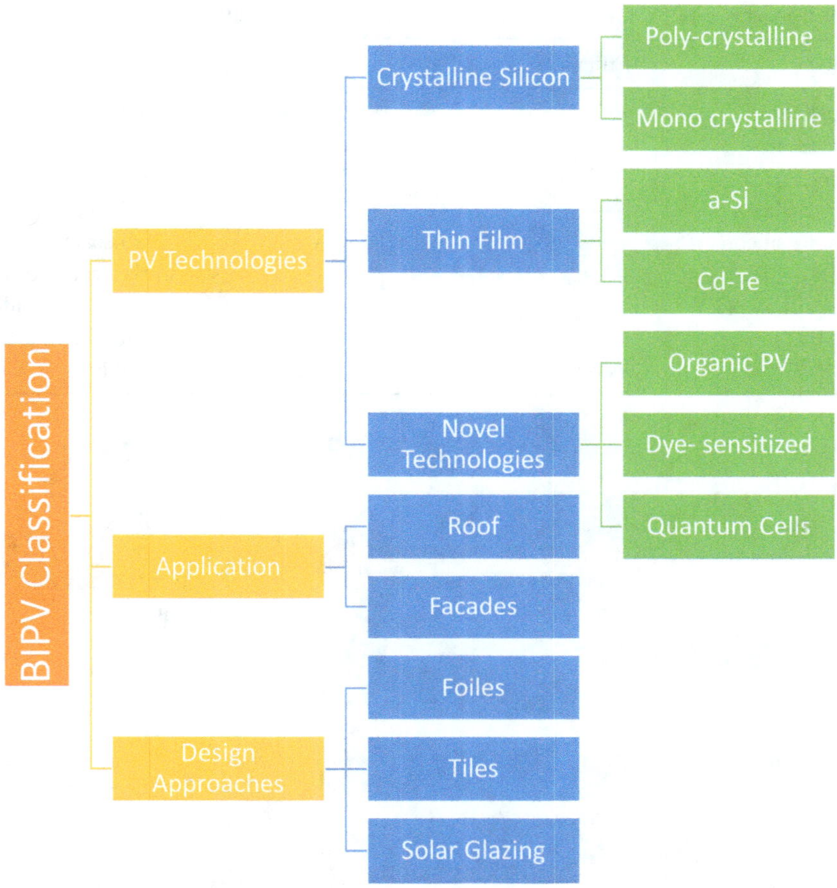

Figure 3.11. PV classification.

advantages, while their lower efficiency and complex structures are disadvantages. To increase cell efficiency, one technique involves creating a layered structure of several cells due to their low efficiency (Tripathy *et al* 2016). The comparison of PV technologies due to the efficiency and cost is given in table 3.1 (Khan and Arsalan 2016). Figure 3.12 illustrates the overall structure of these two types of PV modules.

- *Application*: Another method of BIPV classification is based on the type of application. These are generally divided into roofs and facades. Facades are improvement processes aimed at enhancing the energy performance, aesthetics, and overall sustainability of building exteriors. With this application, both the exteriors are updated, and energy efficiency is increased, thereby extending the building's lifespan. It has been shown in the literature that this is an effective strategy. Developing facade renovation solutions based on the existing building envelope properties, climate conditions, and technical and economic constraints has been demonstrated to be an effective solution for

Table 3.1. Comparison of photovoltaic technologies (Khan and Arsalan 2016).

Technology	Efficiency	Cost	Lifespan	Application
Monocrystalline	+++	+++	+++	Residential, commercial
Polycrystalline	++	++	+++	Residential, commercial
Thin film	+++	+	+	Large-scale, commercial
Crystalline silicon	+	++	+++	Residential, commercial
Amorphous silicon	+++	+	+	Small-scale, portable
Concentrated PV	+++	+++	+++	Large-scale, commercial
Non-concentrated PV	++	+	+++	Residential, small-scale
Organic PV	+	+	+	Small-scale, portable
Hybrid PV	Variable	Variable	Variable	Variable

+++: high, ++: moderate, +: low.

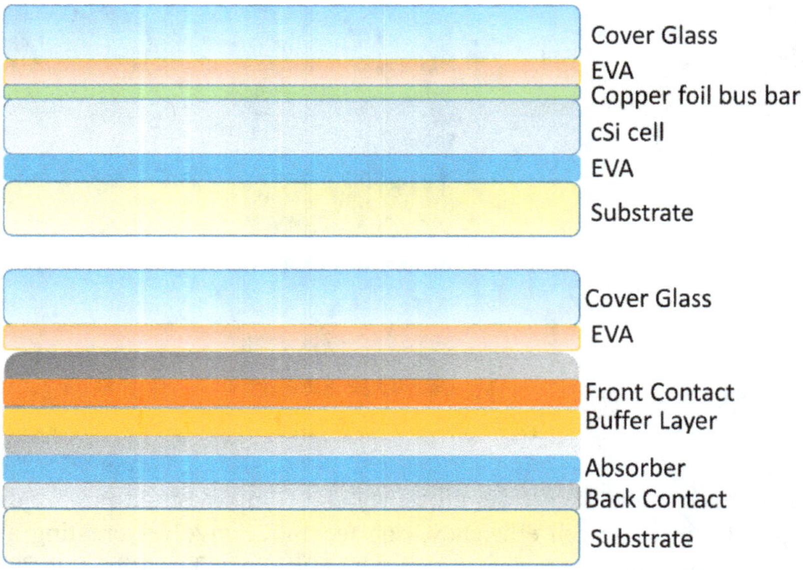

Figure 3.12. Composition of (a) c-Si wafer-based and (b) thin-film PV modules. (Adapted with permission from (Tao and Yu 2015). Copyright 2015 Elsevier.)

improving the energy performance of the building envelope and indoor thermal comfort.

The use of BIPV elements during facade renovation is a good design option for achieving an efficient improvement solution, as it also enables electricity generation from solar energy. Although it is a more challenging process compared to solutions placed on roofs, due to mutual shading of buildings, urban obstacles, and architectural elements that significantly affect efficiency potential, this application is a strategy that provides important benefits in terms of energy efficiency, aesthetics, indoor comfort, and sustainability. It also contributes to reducing emissions and allows for the effective and sustainable renovation of the building envelope (Lazaroiu *et al* 2023, Shukla *et al* 2017, Vulkan *et al* 2018). In rooftop PV panel applications, the angle of solar incidence, shading, the level of dustiness in the environment, and the temperature of the environment where the panels are located are among the most important parameters affecting efficiency.

Another important factor that can increase efficiency is the variability of panel angles according to the months. From an architectural design perspective, reducing the visibility of roof surfaces can further enhance their usability. Especially in rooftop applications, which can potentially damage the structural integrity of buildings, their usage from the outset can reduce their impact on the building's lifespan. Therefore, one of the most important aspects to consider in rooftop applications is the structural capacity of the building. Rooftop PV placement is typically oriented towards the equator, with the tilt angle dependent on the latitude. Solar radiation is one of the most influential factors affecting efficiency in rooftop applications.

Additionally, whether the PV module is flat or inclined also affects performance. For instance, Atmaja's study demonstrated that a curved PV module produced 15% more energy than a flat PV module (Atmaja 2013).

- *Design approaches*: There are several basic categories of products that can be made according to the design approach, including foils, tiles, modules, solar energy cells, and glazing products. These modules can generally be used with various roof materials and can be mounted on both facades and roofs, providing integrated energy systems and offering aesthetic solutions.

  Building-integrated photovoltaic (BIPV) foils are flexible, lightweight, easy to produce, and easy to install. Their use is quite common, but there may be restrictions depending on the configuration of the roof. Foil applications have the advantage of being easy to use and having fewer errors than other products.

  BIPV tile products can replace a certain number of traditional tiles and shingles, achieving efficiency by covering all or some of the roofs. Tile products have great potential due to their easy retrofit conditions. The use of curved ceramics in tile products is more aesthetically effective than flat tiles. All surfaces are resistant to the sun, and protection from exposure is the most important parameter affecting efficiency in tile use.

  Solar cell glazing products can be used on facades and roofs with suitable glass or tile coating. These products provide water and sun protection while transmitting daylight. They transmit the diffusion ray between the cells, typically between 3 and 50 mm. Glazing products can offer different colors and transparency levels, exhibiting aesthetically superior properties. These products are generally designed to provide both sun protection and waterproofing and are long-lasting.

  Sandberg and Moshfegh (1998) have made a comparison of these three types of products. Other products can be used with traditional roofing materials or cover the roof completely

3.2.2.1.3 Numerical and experimental studies related to PV performance

In the study of Sandberg and Moshfegh (1998), the experimental approach involved using heating foils to simulate the heat transferred from the solar cells to the air. They measured characteristics such as airflow velocity and local temperatures in their system designed with air gaps. According to their results, shifting the module's position to the bottom side of the air gap affects the module temperature but does not have a dominant effect on the airflow velocity. Positioning the air gap on the bottom side becomes more important for convective heat transfer.

In another study, Fossa *et al* (2008) experimentally investigated the effect of geometric configuration on heat transfer in double-layer photovoltaic facades. The results showed that increasing the channel spacing reduces average operating temperatures. Another study demonstrated that using Monier concrete tiles as the material for the roof solar collector, and gypsum board on the inner side, increases efficiency. A similar study examined the thermal behavior of a double-skin photovoltaic facade system experimentally, indicating the significant impact of cooling system design on efficiency, as well as highlighting lower solar exposure values with vertically placed PV panels (Zogou and Stapountzis 2011).

In Gan's study (2009), the effect of air gap size on the performance of air gap PV panels and the effects of different roof types and panel lengths were numerically investigated. The study suggests that minimizing heating issues requires different air gap sizes for different types of modules, posing an optimization problem. It is claimed that air gap optimization can be achieved using computational fluid dynamics (CFD) techniques. In the study by Han *et al* (2013), it was stated that when PV is integrated into the building's exterior facade, it is exposed to overheating issues, hence requiring cooling.

Improving the thermal distribution and cooling of solar panels is a topic that researchers frequently work on. Mankani *et al* (2022) focused on the effect of material thickness, fin height, and fin thickness on heat dissipation rate (Mankani *et al* 2022). Moreover, the fundamental characteristics of the heat receiver were examined by converting the material to aluminum.

The base model, with a cooler, was able to reduce the PV cell temperature from 42 °C to 27 °C at an ambient temperature of 42 °C. With optimized fin spacing of 7, base plate thickness of 0.0025 m, fin height of 0.12 m, and fin thickness of 0.002 m, they were able to reduce average panel temperatures by 3.5%, 4%, and 9%, respectively. Mesgarpour *et al* (2021) investigated a porous medium designed to obtain the optimal design of a two-combination PV panel cooler (Mesgarpour *et al* 2021). After testing discrete vertical and double horizontal layered structures with solid fins under the same conditions, they evaluated the thermal performance of these two structures with different porosities. The two models used a hybrid configuration of sensitivity analysis, CFD, and evolutionary differential optimization algorithm to optimize the porosity of the heat absorber. As a result, the use of a porous medium increased the surface area by 18% compared to solid fins of the same dimensions, while reducing the volume by 14%. Implementing this porous medium resulted in a 20.3% increase in heat transfer under the same conditions and dimensions compared to solid fins of the same size.

Yang *et al* (2004) conducted an experimental study on the performance of the initial BIPV system installation (Yang *et al* 2004). An air gap was incorporated between the PV panels and the wall to facilitate cooling and enhance efficiency. Their findings indicated that the highest power output was observed on the roof. The system's total annual energy output was approximated at 6878 kWh^{-1}. The estimated power price of the BIPV system ranged between HK$1.5 kWh^{-1} and HK$2 kWh^{-1}, surpassing the electricity rates offered by two local utility companies. El-Nagar *et al* (2024) have investigated the performance change resulting from the integration of paraffin carbon foam composite structure into PV panels. In their study, they compared three different scenarios: the first being a standalone PV panel naturally cooled, the second being a building-integrated photovoltaic system, and the third being a PV panel utilizing paraffin wax as a phase change material. According to their findings, the PV system utilizing paraffin reduces module temperature by 18.1 °C. The panel containing PCM and carbon foam exhibited better performance compared to the other systems.

The evaluation of BIPV retrofit potentials is not an easy task at the urban scale, as it involves determining both the potential locations for energy reduction through energy retrofit and the areas where BIPV is feasible and necessary. Saretta *et al* (2019) conducted a study to assess BIPV retrofit potential, suggesting that retrofit potential studies constitute a separate research topic. They emphasized the existence of some fundamental energies, such as a comprehensive database containing the geometric and energy-related characteristics of buildings, which is seen as an effective support for defining policies and measures for BIPV integration into building envelopes. Martin-Chivelet *et al* (2018) highlight that BIPV is an attractive solution for effectively and sustainably retrofitting building envelopes, resulting in both material savings and savings in traditional electricity consumption. Moreover, an increase in efficiency is another advantage. In their study, they integrated standard PV modules into ventilated facades, demonstrating that this is a good solution for the rehabilitation of the building envelope. They predicted that by using PV modules integrated into the facade instead of concrete panels, and by renewing lighting and windows, the percentage of a building's energy needs that can be met by its integrated photovoltaic system (PV self-sufficiency index) would reach 6.6%. The self-sufficiency index is a measure of how self-sufficient a building is in terms of the energy it produces on its own. Fu *et al* (2024) compared the energy-saving potentials of BIPV facades for office buildings in different climatic regions. They found that for slab-type buildings, the energy contribution rate ranged from 5% to 15%, while for barrel-type buildings, the energy-saving rate ranged from 8% to 12%. Useful energy contribution rates varied from 5% to 16%.

Tiagarajan and Go (2024) conducted the photovoltaic system analysis and energy analysis of BIPV systems, reducing the annual energy consumption of the selected building by 250 tons. Sun *et al* (2024), conducted an experimental study in China, investigating heat dissipation issues in building-integrated photovoltaic roofs. By employing numerical methods, they optimized three key parameters: air gap thickness, PV panel spacing, and heating power, finding that the air gap thickness had the most significant impact on PV panel temperature. El Samanoudy *et al* (2024) numerically examined the efficiency of BIPV application in office use, finding that energy consumption in high-rise office buildings decreased by 13%–32% when alternative window replacement types and exterior wall finishing types were applied.

One of the biggest problems in PV applications is the potential for fires. The fire behavior of BIPV modules is triggered by the electrical activity of PV modules. Especially in roof applications, fire problems arise due to ventilation restrictions and limited walking areas, which are among the biggest risks in fires. Sometimes fires that are not initiated by PV arrays can become even more dangerous when the fire reaches the PV system. BIPV fires have been categorized into four categories by Yang *et al*. The first is ignition due to hot spots, arcing, and installation errors, the second is potential fire risks to building occupants, and the third is potential risks to fire and rescue personnel. The hot-spot phenomenon arises from partial shading, material defects, manufacturing errors, and degradation over time. Pandian *et al*'s study (2016) indicates that PV modules reaching surface temperatures above 85 °C are at risk of fire. Arcing is a fire cause resulting from ionization of air across an air gap. Incorrect module installations can generate high heat energy and start a fire. Even improper wiring can cause arcing and ignite fires. The study of Yang *et al* (2023) provides detailed information on potential fire hazards.

3.2.2.3 Building-integrated photovoltaic-thermal (BIPVT) systems

3.2.2.3.1 Working principle of the BIPVT systems

BIPV is used solely for generating photovoltaic energy, whereas BIPVT is designed to produce both photovoltaic and thermal energy. In these systems, the heat used to reduce the temperature of PV modules is reused with recovery systems, leading to thermal energy production. While BIPV systems are typically integrated into the roofs or facades of buildings, this hybrid system is often installed in places used for hot water or heating. Due to its production of both thermal and electrical energy, it generally exhibits higher performance in terms of total energy efficiency. Building-integrated photovoltaic-thermal configurations offer a highly valid means of reducing energy consumption and meeting low-energy building requirements. They have various applications such as air-cooled systems, water-cooled systems, concentrator-based systems, and PCM-based systems. This hybrid system holds significant potential in the residential cooling and heating sector. It is known that high temperatures reduce photovoltaic efficiency; a PVT collector can minimize this undesirable effect and provide high energy savings. PVT collectors are classified as water-based or air-based, with water-based ones being more advantageous in terms of thermal efficiency. Air-based ones are usually limited to small-scale independent collectors. BIPVT systems constitute a subset of PVT systems through integration with the PVT concept itself. The operation and design of air-based collectors and full-scale systems are the same, with the main difference being the system size. BIPVT covers a large portion of the building's facade or roof and has an added

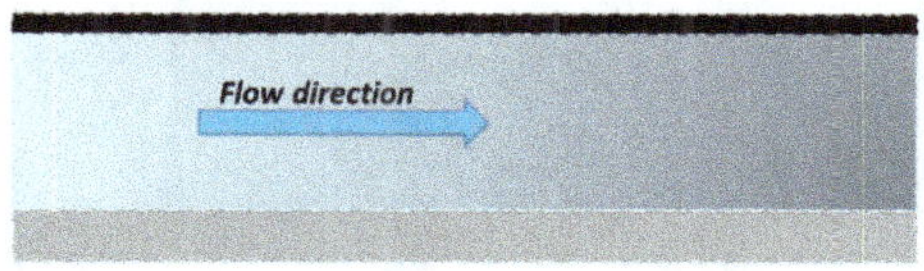

Figure 3.13. Structure of air-based PVT and BIPVT design. (Adapted with permission from (Rounis *et al* 2021). Copyright 2021 Elsevier.)

building function. Figure 3.13 illustrates a PVT design (Abdelrazik *et al* 2022, Debbarma *et al* 2017, Rounis *et al* 2021).

Photovoltaic-thermal technology began to be used in the 1990s in America, and later its usage started in Switzerland and Europe, spreading to the whole world. The usefulness of BIPVT design for generating both electricity and useful heat was demonstrated during these years, and its exergy efficiency was examined by Fujisawa and Tani (1997), observing that the hybrid design achieved higher efficiency. In general, the BIPVT system converts structural elements of a building such as its roof or facade into energy production, providing both electricity and heat energy. This increases the energy efficiency of buildings and ensures environmental sustainability. When examining the operating principle of the system, a portion of the incoming solar energy is first converted into electricity by the PV module and then progresses along the facade. When a cooling medium is used, a portion of the absorbed solar energy is removed as heat from the system. The heat transfer coefficient of the building envelope is altered by BIPVT components, facilitating heat flow between the environment and the interior space. These systems gather sunlight using flat surfaces or reflective concentrating devices, and cooling fluid mediums can be utilized to cool the PV.

A BIPVT system can be either fully integrated into a building or simply mounted onto it. These two installation methods include the following: roof elements that directly replace traditional roofing tiles or are mounted on flat or low-sloped roofs, ventilated facade systems used as external cladding, window elements, rain protection coatings and curtain walls. The best integration results are achieved when the BIPVT system design is considered during the building design phase.

The performance of the system is described by the electrical and thermal efficiency of the system as follows (Evans and Florschuetz 1975):

$$\eta_{PV} = \eta_{STC}(1 - \beta_{PV}(T_{PV} - T_{STC})), \tag{3.4}$$

where $\beta_{PV} = 0.0046$ (PV module temperature coefficient), $\eta_{STC} = 0.125$ (PV module efficiency at standard test conditions (STC)), and $T_{STC} = 25$ C (PV module cell temperature at STC).

The thermal output, Q (watt) is given by

$$Q = \dot{m}C\Delta T, \tag{3.5}$$

where $\dot{m}$ is the mass flow rate of the airflow (kg s^{-1}), C is the specific heat capacity of the airflow (J kg^{-1} K^{-1}) and ΔT denotes the temperature difference between the inlet and outlet temperature of the airflow.

The thermal efficiency is given as

$$\eta_{\text{th}} = \frac{Q}{A_{\text{c}}G}, \tag{3.6}$$

where A_{c} is the collector area (m^2), and G is the solar irradiance on the collector (W m^{-2}). Thermal efficiency, serves as a valuable indicator for comparing the thermal performance of various design configurations in systems operating under identical conditions (Bambara *et al* 2011, Evans and Florschuetz 1975).

3.2.2.3.2 Numerical and experimental studies related to BIPVT performance
Abdalgadir *et al* (2023) have conducted a numerical study to investigate the performance of a BIPVT system. The study, carried out in MATLAB, aimed to reduce the energy required for heating during winter and prevent high photovoltaic cell temperatures during summer. The results they obtained indicate that cooling the BIPVT system with exhaust air in summer can reduce cell temperatures by 9 °C–19 °C. It has been claimed that this system could provide significant energy savings.

Ge *et al* (2022) have implemented a BIPTV design with a water-cooled wall. The water-cooled wall does not directly contact the heat exchanger PV module but distributes the heat from the module to the air, allowing the heat to be used to heat water in the house. This system, which heats domestic water, has been shown to provide hot water above 35 °C for more than 7 h a day.

In their study, Yang *et al* (2021) examined the effect of BIPVT geometry and materials on performance. Sensitivity analysis results show that in climates with high humidity in summer and mild winters, the thermal transmittance of the internal window and the solar heat gain coefficient of the external window are the design parameters that most affect indoor thermal comfort and energy consumption.

Rounis *et al* (2021) conducted a study on modeling the performance of air-based PVT and BIPVT, as well as air duct and wind-induced convection phenomena. The results suggest that shading issues and excessive heating problems are generally avoided by avoiding lively configurations (complex building shapes, wind funneling effects, vegetation or obstacles) and that direct exposure of the absorber to the wind is a more advantageous design.

Hailu *et al* (2015) theoretically examined the performance of a two-stage variable-capacity air-source heat pump combined with a photovoltaic-thermal system. The results indicate that the COP value improves when the average ambient temperature is above −3 °C.

Tardif *et al* (2017) numerically compared the performance of three different heat management strategies. The results show that preheating fresh air does not provide significant energy savings, but combining the collector with an ASHP achieves a success rate of 62% in mechanical energy savings.

Ghazali Ali *et al* (2024) aimed to minimize panel temperature increase by using passive evaporative cooling and investigating the effect of different thermal conductive materials. The results show that the composite iron oxide clay layer provides a 10% decrease in panel temperature, and the use of a hollowed clay layer is the material that provides the greatest decrease.

Wang *et al* (2016) argued that a BIPVT system created by filling the gap between finned heat pipes and insulation with a mixture of metal wires and phase change material can be reduced with mass production and waste material recycling.

3.3 SWOT analysis

In order to analyse the strengths, weaknesses, opportunities, and threats of implementing solar energy in buildings, a SWOT analysis examined separately in four distinct columns and presented in figure 3.14.

The major strengths include the potential for long-term reduction in operating costs, assistance in reducing emissions, being a renewable system, and ease of integration into buildings. Weaknesses include issues with storage, space requirements, the possibility of low efficiency depending on specific criteria, and initial costs. The recent major opportunities lie in the rapid technological advancements and ongoing development of new systems that increase efficiency and reduce costs. Increasing demands will lead to increased demand for solar energy systems, bringing along numerous opportunities

3.4 Conclusion

The purpose of this chapter is to explore solar application technologies in buildings. The study is primarily divided into two main application forms of solar techniques, and examined in detail under two headings. These headings are categorized as passive and active control techniques. Passive control techniques encompass three fundamental techniques, namely fenestration, Trombe walls, building orientation, and design. Active methods are divided into BIPV and BIPVT systems, and detailed classifications are discussed under each heading. The most fundamental conclusion drawn from the results is that solar energy applications in the building sector are

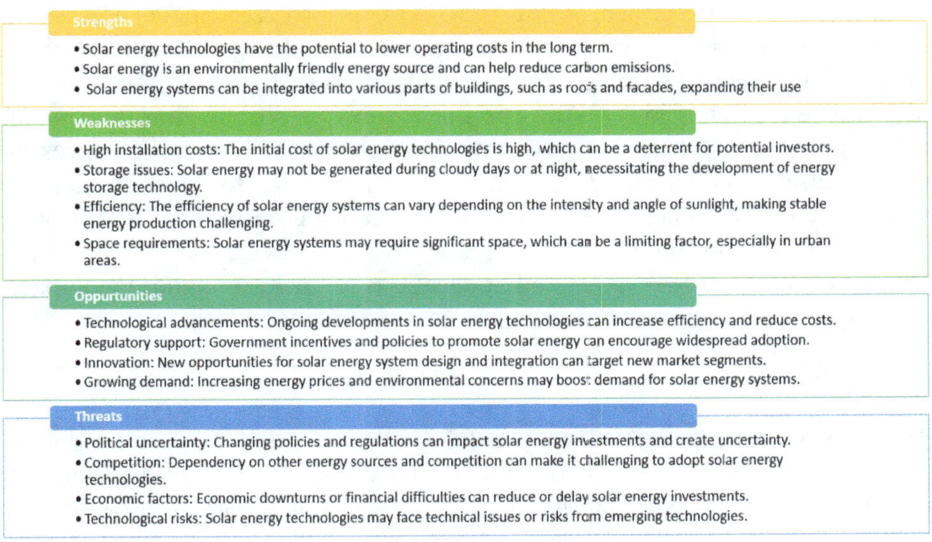

Figure 3.14. SWOT analysis for solar applications in buildings.

increasing every day, and research on this subject is becoming more intense by the day.

In particular, solar thermal collector systems have the ability to replace conventional fossil fuels for heating and cooling, as heating accounts for more than one-third of the world's total energy consumption. Therefore, adopting an intelligent investment method that leads to energy savings by harnessing solar energy is a research topic that attracts the attention of the whole world. Technological advancements contribute to sustainable solutions for thermal systems, and the potential to cope with increasing energy demands and environmental concerns is very high. It has been emphasized in many studies that paying attention to design and local climatic conditions is necessary to make the most of solar technologies. BIPVT, one of the most effective systems, has started to be developed and has come a long way to date. With this system, it is possible to establish buildings that produce both heat and electricity with zero-energy consumption.

Building-integrated photovoltaic-thermal systems have not be widely used due to their high cost. On the other hand, it emerges as a promising method for the future due to its high efficiency in both heat and electricity generation. Solar technology applications in both facade and roof applications are frequently encountered in the literature. Many factors such as shading effect, ambient temperature, building orientation, and PV tilt have been mentioned in the literature to achieve a highly efficient system.

Literature review also shows that especially in recent years, numerical studies have increased significantly and have become a preferred tool in optimization studies for the high-efficiency use of solar energy. Simulation and numerical studies enable the understanding of performance, calculation of power generation capacity, facade energy potential, the effect of shading factors, and electricity usage amounts. The greatest advantage of numerical studies is that they allow easy modification of the system for different configurations. The compatibility of numerical studies with experimental studies also demonstrates the high predictive ability of numerical tools. The examined solar energy technologies address the process of collecting, storing, and utilizing solar energy from the roofs or facades of buildings. These technologies, when applied, increase the energy independence of buildings, meet their electricity needs, and reduce emissions. With appropriate design, localization, and maintenance requirements taken into account, these technologies can be successfully applied. Recent research and development efforts continue to focus on increasing the efficiency of solar energy technologies and reducing costs. Finally, this article emphasizes how important a role solar energy technologies can play in the building sector and expects future research and applications to further enhance the impact of solar energy technologies on the sustainability of buildings.

References

Abdalgadir Y, Qian H, Zhao D, Adam A and Liang W 2023 Daily and annual performance analyses of the BIPV/T system in typical cities of Sudan *Energy Built Environ.* **4** 516–29

Abdelrazik A S, Shboul B, Elwardany M, Zohny R N and Osama A 2022 The recent advancements in the building integrated photovoltaic/thermal (BIPV/T) systems: an updated review *Renew. Sustain. Energy Rev.* **170** 112988

Atmaja T D 2013 Façade and rooftop PV installation strategy for building integrated photo voltaic application *Energy Procedia* **32** 105–14

Baljit S S S, Chan H Y and Sopian K 2016 Review of building integrated applications of photovoltaic and solar thermal systems *J. Clean. Prod.* **137** 677–89

Bambara J, Athienitis A K and O'Neill B 2011 Design and performance of a photovoltaic/thermal system integrated with transpired collector *ASHRAE Trans.* **117** 403–10

Biyik E *et al* 2017 A key review of building integrated photovoltaic (BIPV) systems *Eng. Sci. Technol. Int. J.* **20** 833–58

Bosu I, Mahmoud H, Ookawara S and Hassan H 2023 Applied single and hybrid solar energy techniques for building energy consumption and thermal comfort: a comprehensive review *Sol. Energy* **259** 188–228

Charqui Z, El Moutaouakil L, Boukendil M and Hidki R 2022 Numerical study of heat transfer in a tall, partitioned cavity confining two different fluids: application to the water Trombe wall *Int. J. Therm. Sci.* **171** 107266

Debbarma M, Sudhakar K and Baredar P 2017 Comparison of BIPV and BIPVT: a review *Resour Eff. Tech.* **3** 263–71

Duffie J A, Beckman W A and McGowan J 1985 Solar engineering of thermal processes *Am. J. Phys.* **53** 382

El-Nagar D H, Emam M, El-Betar A A and Nada S A 2024 Performance improvement of building-integrated photovoltaic panels using a composite phase change material-carbon foam heat sink: an experimental study *J. Build. Eng.* **91** 109623

El Samanoudy G, Abdelaziz Mahmoud N S and Jung C 2024 Analyzing the effectiveness of building integrated photovoltaics (BIPV) to reduce the energy consumption in Dubai *Ain Shams Eng. J.* **15** 102682

Erdim B and Manioğlu G 2014 Building form effects on energy efficient heat pump application for different climatic zones *A/Z ITU J. Fac. Archit.* **11** 335–49

European Parliament and Council 2010 *Directive of the European Parliament and of the Council of 16 July 2002 on the Energy Performance of Buildings* Official Journal of the European Union

Evans D L and Florschuetz L W 1975 Cost studies on terrestrial photovoltaic power systems with sunlight concentration *Int. Solar Energy Congress and Expo, Extended Abstract, Sol Use Now – A Resource for People* pp 114–5

Fossa M, Ménézo C and Leonardi E 2008 Experimental natural convection on vertical surfaces for building integrated photovoltaic (BIPV) applications *Exp. Therm Fluid. Sci.* **32** 980–90

Fu Y, Xu W, Wang Z, Zhang S, Chen X and Du X 2024 Numerical study on comprehensive energy-saving potential of BIPV façade under useful energy utilization for high-rise office buildings in various climatic zones of China *Sol. Energy* **270** 112387

Fujisawa T and Tani T 1997 Annual exergy evaluation on photovoltaic-thermal hybrid collector *Sol. Energy Mater. Sol. Cells* **47** 135–48

Gan G 2009 Numerical determination of adequate air gaps for building-integrated photovoltaics *Sol. Energy* **83** 1253–73

Ge M, Zhao Y, Xuan Z, Zhao Y and Wang S 2022 Experimental research on the performance of BIPV/T system with water-cooled wall *Energy Rep.* **8** 454–9

Gharakhani Siraki A and Pillay P 2012 Study of optimum tilt angles for solar panels in different latitudes for urban applications *Sol. Energy* **86** 1920–8

Ghazali Ali M, Hassan H, Ookawara S and Nada S A 2024 Performance enhancement of porous clay cooler for BIPV applications using hollowed clay structures and metallic extended fins as evaporation and thermal conductivity enhancers *Energy Build.* **314** 114298

Grynning S, Time B and Matusiak B 2014 Solar shading control strategies in cold climates—heating, cooling demand and daylight availability in office spaces *Sol. Energy* **107** 182–94

Gupta R, Ralegaonkar R and Asati A 2004 A simulation technique using LGT for temperature prediction of passive solar heated building model *Int. Conf. Energy Environ.* pp 181–5

Gupta R and Ralegaonkar R V 2004 Estimation of beam radiation for optimal orientation and shape decision of buildings in India *J. Inst. Eng. (India): Arch. Eng. Div.* **85** 27–32

Hailu G, Dash P and Fung A S 2015 Performance evaluation of an air source heat pump coupled with a building-integrated photovoltaic/thermal (BIPV/T) system under cold climatic conditions *Energy Procedia* **78** 1913–8

Han J, Lu L, Peng J and Yang H 2013 Performance of ventilated double-sided PV façade compared with conventional clear glass façade *Energy Build.* **56** 204–9

Hii D J C, Heng C K, Malone-Lee L C, Zhang J, Ibrahim N, Huang Y C and Janssen P 2011 Solar radiation performance evaluation for high density urban forms in the tropical context *Proc. Building Simulation 2011: 12th Conf. Int. Building Performance Simulation Association* pp 2595–602

IEA 2020 Energy Efficiency 2020, IEA, Paris https://www.iea.org/reports/energy-efficiency-2020, Licence: CC BY 4.0

Inanici M N and Demirbilek F N 2000 Thermal performance optimization of building aspect ratio and south window size in five cities having different climatic characteristics of Turkey *Build. Environ.* **35** 41–52

IRENA 2019 *Global Energy Transformation: A Roadmap to 2050* (Abu Dhabi: International Renewable Energy Agency)

Islam N, Irshad K, Zahir M H and Islam S 2021 Numerical and experimental study on the performance of a photovoltaic trombe wall system with Venetian blinds *Energy* **218** 119542

Jiménez-Xamán C, Xamán J, Moraga N O, Hernández-Pérez I, Zavala-Guillén I, Arce J and Jiménez M J 2019 Solar chimneys with a phase change material for buildings: an overview using CFD and global energy balance *Energy Build.* **186** 384–404

Kalogirou S A 2015 Building integration of solar renewable energy systems towards zero or nearly zero energy buildings *Int. J. Low-Carbon Technol.* **10** 379–85

Khan J and Arsalan M H 2016 Solar power technologies for sustainable electricity generation—a review *Renew. Sustain. Energy Rev.* **55** 414–25

Khedari J, Hirunlabh J and Bunnag T 1996 Experimental study of a roof solar collector towards the natural ventilation of new habitations *Renew. Energy* **8** 335–8

Khedari J, Mansirisub W, Chaima S, Pratinthong N and Hirunlabh J 2000 Field measurements of performance of roof solar collector *Energy Build.* **31** 171–8

Kim D-E and Lee D-C 2007 Feedback linearization control of three-phase AC/DC PWM converters with LCL input filters *7th Int. Conf. Power Electronics (Daegu, South Korea)* pp 766–71

Lazaroiu A C, Gmal Osman M, Strejoiu C-V and Lazaroiu G 2023 A comprehensive overview of photovoltaic technologies and their efficiency for climate neutrality *Sustainability* **15** 16297

Lee K S and Oberdick W A 1982 Development of a passive solar simulation technique using small-scale models *UMR-DNR Conf. on Energy* **8**

Li X, Shen C and Yu C W F 2017 Building energy efficiency: passive technology or active technology? *Indoor Built. Environ.* **26** 729–32

Lin H 1981 Building plane, form and orientation for energy saving *J. Archit.* **4** 37–41

Lombard P, José Ortiz L and Christine P 2008 A review on buildings energy consumption information *Energy Build.* **40** 394–8

Ma Q, Fukuda H, Lee M, Kobatake T, Kuma Y and Ozaki A 2018 Study on the utilization of heat in the mechanically ventilated Trombe wall in a house with a central air conditioning and air circulation system *Appl. Energy* **222** 861–71

Ma Q, Fukuda H, Wei X and Hariyadi A 2019 Optimizing energy performance of a ventilated composite Trombe wall in an office building *Renew. Energy* **134** 1285–94

Mankani K, Nasarullah Chaudhry H and Kaiser Calautit J 2022 Optimization of an air-cooled heat sink for cooling of a solar photovoltaic panel: a computational study *Energy Build.* **270** 112274

Marques Lameirinhas R A, Torres J P N and de Melo Cunha J P 2022 A photovoltaic technology review: history, fundamentals and applications *Energies* **15** 1823

Martín-Chivelet N, Gutiérrez J, Alonso-Abella M, Chenlo F and Cuenca J 2018 Building retrofit with photovoltaics: construction and performance of a BIPV ventilated façade *Energies* **11** 1719

Mesgarpour M, Heydari A, Wongwises S and Reza Gharib M 2021 Numerical optimization of a new concept in porous medium considering thermal radiation: photovoltaic panel cooling application *Sol. Energy* **216** 452–67

Mingfang T 2002 Solar control for buildings *Build. Environ.* **37** 659–64

Pandian A, Bansal K, Thiruvadigal D J and Sakthivel S 2016 Fire hazards and overheating caused by shading faults on photo voltaic solar panel *Fire Technol.* **52** 349–64

Pillai D S, Shabunko V and Krishna A 2022 A comprehensive review on building integrated photovoltaic systems: emphasis to technological advancements, outdoor testing, and predictive maintenance *Renew. Sustain. Energy Rev.* **156** 111946

Pino A, Bustamante W, Escobar R and Pino F E 2012 Thermal and lighting behavior of office buildings in Santiago of Chile *Energy Build.* **47** 441–9

Prozuments A, Borodinecs A, Bebre G and Bajare D 2023 A review on Trombe wall technology feasibility and applications *Sustainability* **15** 3914

Rabani M, Kalantar V, Dehghan A A and Faghih A K 2015 Empirical investigation of the cooling performance of a new designed Trombe wall in combination with solar chimney and water spraying system *Energy Build.* **102** 45–57

Rounis E D, Athienitis A and Stathopoulos T 2021 Review of air-based PV/T and BIPV/T systems–performance and modelling *Renew. Energy* **163** 1729–53

Saadatian O, Sopian K, Lim C H, Asim N and Sulaiman M Y 2012 Trombe walls: a review of opportunities and challenges in research and development *Renew. Sustain. Energy Rev.* **16** 6340–51

Sandberg M and Moshfegh B 1998 Ventilated-solar roof air flow and heat transfer investigation *Renew. Energy* **15** 287–92

Saretta E, Caputo P and Frontini F 2019 A review study about energy renovation of building facades with BIPV in urban environment *Sustain. Cities Soc.* **44** 343–55

Shukla A K, Sudhakar K and Baredar P 2017 Recent advancement in BIPV product technologies: a review *Energy Build.* **140** 188–95

Sun C, Lu Y and Ju X 2024 Experimental and numerical study to optimize building integrated photovoltaic (BIPV) roof structure *Energy Build.* **309** 114070

Taherian H and Peters R W 2023 Advanced active and passive methods in residential energy efficiency *Energies* **16** 3905

Tao J and Yu S 2015 Review on feasible recycling pathways and technologies of solar photovoltaic modules *Sol. Energy Mater. Sol. Cells* **141** 108–24

Tardif J M, Tamasauskas J, Delisle V and Kegel M 2017 Performance of air based BIPV/T heat management strategies in a Canadian home *Procedia Environ. Sci.* **38** 140–7

Tiagarajan T and Go Y I 2024 Integration of BIPV design and energy efficient technologies for low energy building in meeting net zero target *e-Prime.* **8** 100554

Tripathy M, Sadhu P K and Panda S K 2016 A critical review on building integrated photovoltaic products and their applications *Renew. Sustain. Energy Rev.* **61** 451–65

Vakiloroaya V, Samali B, Fakhar A and Pishghadam K 2014 A review of different strategies for HVAC energy saving *Energy Convers. Manage.* **77** 738–54

Valladares-Rendón L G, Schmid G and Lo S-L 2017 Review on energy savings by solar control techniques and optimal building orientation for the strategic placement of façade shading systems *Energy Build.* **140** 458–79

Vargas-López R, Xamán J, Hernández-Pérez I, Arce J, Zavala-Guillén I, Jiménez M J and Heras M R 2019 Mathematical models of solar chimneys with a phase change material for ventilation of buildings: a review using global energy balance *Energy* **170** 683–708

Vulkan A, Kloog I, Dorman M and Erell E 2018 Modeling the potential for PV installation in residential buildings in dense urban areas *Energy Build.* **169** 97–109

Wanchun S, Jinxin F, Zhengguo Z and Xiaoming F 2020 Research progress of phase change heat storage technology for passive energy conservation *Chem. Ind. Eng. Prog.* **39** 1824–34

Wang Z, Zhang J, Wang Z, Yang W and Zhao X 2016 Experimental investigation of the performance of the novel HP-BIPV/T system for use in residential buildings *Energy Build.* **130** 295–308

Xiao Y, Yang Q, Fei F, Li K, Jiang Y, Zhang Y, Fukuda H and Ma Q 2024 Review of Trombe wall technology: trends in optimization *Renew. Sustain. Energy Rev.* **200** 114503

Xiao Y, Zhang T, Liu Z and Fukuda H 2023 Thermal performance study of low-e glass Trombe wall assisted with the temperature-controlled ventilation system in hot-summer/cold-winter zone of China *Case Stud. Therm. Eng.* **45** 102882

Yang H, Zheng G, Lou C, An D and Burnett J 2004 Grid-connected building-integrated photovoltaics: a Hong Kong case study *Sol. Energy* **76** 55–9

Yang R *et al* 2023 Fire safety requirements for building integrated photovoltaics (BIPV): a cross-country comparison *Renew. Sustain. Energy Rev.* **173** 113112

Yang S, Fiorito F, Prasad D, Sproul A and Cannavale A 2021 A sensitivity analysis of design parameters of BIPV/T-DSF in relation to building energy and thermal comfort performances *J. Build. Eng.* **41** 102426

Zhang L, Hou Y, Liu Z, Du J, Xu L, Zhang G and Shi L 2020 Trombe wall for a residential building in Sichuan-Tibet Alpine valley—a case study *Renew. Energy* **156** 31–46

Zhang T, Wang M and Yang H 2018 A review of the energy performance and life-cycle assessment of building-integrated photovoltaic (BIPV) systems *Energies* **11** 3157

Zogou O and Stapountzis H 2011 Experimental validation of an improved concept of building integrated photovoltaic panels *Renew. Energy* **36** 3488–98

Chapter 4

Measuring heat flux at solar photovoltaic plants

Daniel Trevor Cannon and Ahmad Vasel-Be-Hagh

Total heat flux at a photovoltaic (PV) panel's surface, including radiation and convection, is one critical parameter in assessing solar farms' environmental impact (e.g. the heat island effect). This article compares two approaches for measuring the total heat flux. One approach is to employ a direct-contact surface heat flux made of series-connected thermocouples known as thermopiles. Attaching such a sensor to a PV panel's surface under solar irradiance while subject to wind-induced convection cooling can create confusion about what precisely this sensor measures. Do such measurements include both incoming and outgoing radiation? Do they factor in convection? Does radiative heating from the Sun impact the sensor's perception of surrounding temperature, introducing errors in heat flux measurements? A more sophisticated approach uses a net radiometer, comprising two pyranometers and two pyrgeometers, to measure radiation alongside a shielded surface heat flux sensor to gauge convection. The summation of the two measurements yields the total heat flux. This more expensive technique produces more accurate measurements. This chapter explores whether a single surface heat flux sensor directly attached to the panel's surface generates acceptable measurements or necessitates coupling with a net radiometer and a shielded direct-contact sensor to accurately measure heat fluxes.

4.1 Introduction

The prominence of solar energy in the global electricity generation landscape is on the rise, with a nearly 26% increase in solar photovoltaic (PV) generation in 2022 alone, reaching almost 1300 TWh [1]. Solar PV is expected to exceed coal's contribution to the global electricity market by 2027, exemplified by the construction of massive solar farms, such as the 2.245 GW Bhadla Solar Park in India, which spans 14 000 acres with a total of 10 million PV panels. Such extensive arrays of dark surfaces can potentially alter net surface radiation fluxes. One might argue that PV panels convert energy to electricity and ship it away from the site, so they must have cooling effects, if any. However, it must be noted that they convert light, not heat;

doi:10.1088/978-0-7503-6179-8ch4

thus, long-term radiation (i.e. heat) is not converted to electricity. Also, it is essential to note that PV panels do not reflect more than bare soil or a vegetated canopy as they are darker and, thus, their absorptivity is larger. So, they absorb more heat via radiation. Similarly, compared to the soil or vegetated canopies, they have a larger emissivity due to their darker color, and their surface temperature is generally greater, too; hence, they emit more heat into the surroundings. They cover the ground, preventing a nighttime heat discharge from the underlying canopy into the sky via radiation. They extend the surface area, leading to more radiation and convection exchanges. They add to the ground's surface roughness, impacting its convection heat coefficient. Also, they are tilted versus the horizontal background canopy, which, again, impacts the convection heat transfer. These alterations are believed to result in a change in the total surface heat flux.

These alterations in total heat flux, particularly across vast areas such as the 14 000 acre Bhadla Solar Park, can impact the dynamics of the viscous sublayer of the atmospheric boundary layer. Understanding such changes, if any, is essential, as the viscous sublayer significantly impacts the mechanics of the boundary layer and can influence properties such as the near-surface temperature [2] and vertical temperature profiles, atmospheric stability, and the rate of turbulence generation, destruction, and transportation. It might also lead to other environmental consequences, such as the heat island effect [3]. Given the consequential nature of these transformations, a thorough assessment of surface heat fluxes becomes imperative. This assessment also requires evaluating surface heat fluxes at the neighboring natural canopies to generate a baseline for isolating the solar farm's influence. This provides insight into how and to what extent mega-scale solar canopies, whether PV or concentrated, influence the environmental dynamics of their surroundings.

In addition to such environmental studies, measuring solar panels' surface temperature and heat flux has engineering applications as they directly impact the plant's power performance. It is a known fact that an increase in the surface temperature of the panels decreases their efficiency [4, 5]. Hence, assessing solar panels' surface heat flux and temperature within utility-scale solar plants' supervisory control and data acquisition systems is common [6].

The canopy's surface heat flux can be measured via three approaches. Approach one includes a net radiometer, made of an upward-facing pyranometer to measure incoming short-wave radiation, an upward-facing pyrgeometer measuring the incoming long-wave radiation, a downward-facing pyranometer recording the outgoing short-wave radiation, and a downward-facing pyrgeometer to measure outgoing long-wave radiation. The summation of these four components yields total net radiation at the canopy's surface. The apparatus also required a shielded surface heat flux sensor to measure convection. This sensor needs to be shielded to ensure it is not affected by the radiation, which is already measured by the net radiometer. The convective heat flux measured by this surface heat flux sensor is less than the actual convection and must be corrected. This is because the spot where the heat flux sensor attaches to the panel's surface has a smaller temperature due to the shield used to cover the sensor from radiation. This correction requires a knowledge of the air temperature, requiring an extra thermocouple. Therefore, this set-up requires a total of six sensors.

A second approach to measuring the heat flux at the panel's surface is to simply employ one surface heat flux sensor. One can attach this sensor to the panel's surface without shielding the sensor, as doing so would block the radiation, a significant heat flux component. These thin sensors attach to the panel as they are exposed to the air on the opposite side. Consequently, one side reflects the panel's temperature, while the other is intended to correspond to the temperature of the surroundings. The performance of this technique for solar facilities with extensive incoming and outgoing radiation accompanied by outgoing convection is unknown. It is yet to be determined if such sensors can capture the outgoing radiation as they directly attach to the surface that emits heat (panel's surface) and whether the incoming radiation manipulates their measurement by heating the sensor's surface facing the air, pushing the temperature of the top side of the thermocouples inside the sensor beyond the surrounding temperature. This would increase the temperature difference across the sensor, resulting in positive bias. On the other hand, the sensor attaching to the surface shades the panel locally; hence, it is not unlikely for the temperature of the panel's surface right under the sensor to become less than the actual panel temperature. This also increases the temperature difference between the sensor's two sides, increasing the measured heat flux. Conversely, the heat flux sensor may not absorb all of the radiative heat flux, reflecting a portion of it and consequently introducing a negative bias. This negative bias is anticipated to partly counterbalance the positive bias previously described for the sensor. However, the dominant bias remains uncertain, and it remains unclear whether the absolute error will exceed tolerable limits, necessitating the abandonment of this approach. The uncertainty of the difference between values measured by approaches one and two defines these tolerable limits. Thus, if the variance between the measurements obtained through the initial and secondary approaches falls within the uncertainty margin, the viability of the secondary approach can be deemed acceptable.

A third method for quantifying total heat flux involves employing a high-frequency anemometer alongside an air temperature sensor to capture temperature and fluctuating velocity components. By applying the eddy covariance relationship, one can calculate the vertical heat flux as $\rho c_{\mathrm{p}} \overline{w'T'}$, with ρ, c_{p}, w', and T' represent the air density, specific heat capacity, the fluctuating component of the vertical velocity, and the fluctuating component of temperature.

The third approach is not within the scope of this study as we aim to focus on measuring the heat flux right at the surface ($z = 0$), while one must install the anemometer at a distance from the surface. The eddy covariance cannot measure heat flux right at the surface as velocity, including its vertical component, is zero at the surface. This study aims to assess the performance of the second approach by comparing it against the first one. Clearly, the first approach, utilizing more sensors, is more sophisticated and accurate. This study considers the heat flux measured by approach one as the true value and evaluates the second approach accordingly. Uncertainties of both approaches are estimated and considered.

The article unravels by describing the experimental set-up, followed by a presentation of the methodology. After presenting and discussing the measurements obtained from approaches one and two, the article concludes by summarizing the critical observations, including whether or not approach two is reliable and why we care about assessing the quality of this approach if the reliability of approach one is already established.

4.2 The experiment and methodology

The experimental set-up illustrated in figure 4.1 allows estimating the errors associated with simply using an unshielded surface heat flux sensor to measure the total heat flux at the panel's surface. This experiment was set-up at a latitude and

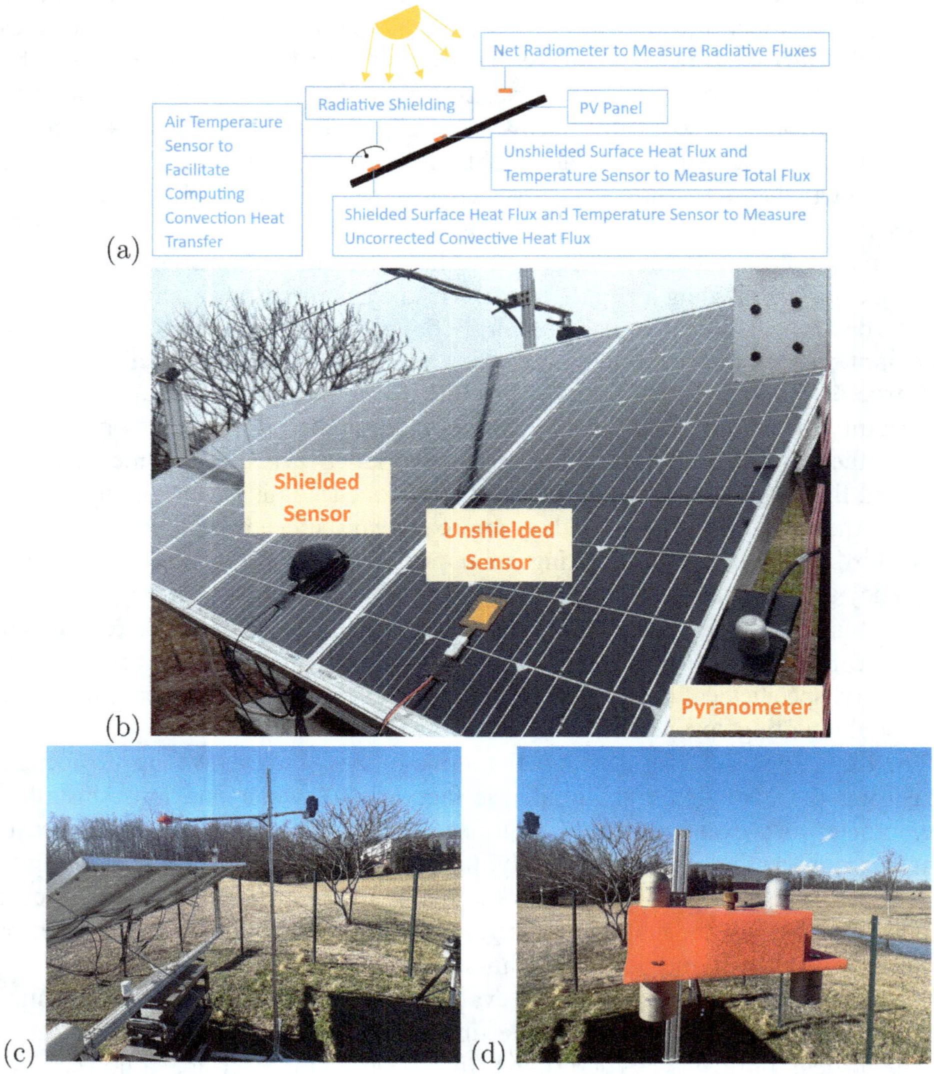

Figure 4.1. The experimental set-up: (a) overview, (b) the shaded and unshaded heat flux sensors, and (c) and (d) the net radiometer including two pyranometers and two pyrgeometers.

longitude of 36 and −85. Panels were facing south at a fixed tilt angle of ~30° with respect to the horizon, which is the optimal fixed tilt angle for the experiment's location. Hereafter, q_{us} represents the value recorded by this sensor, and q_{err} indicates this sensor's error. The goal is to evaluate $q_{err} = q_{true} - q_{us}$. This requires measuring q_{true}. To measure q_{true}, the configuration incorporates two pyranometers, two pyrgeometers, a shaded surface heat flux and temperature sensor, and an air temperature sensor. The four pyranometers and pyrgeometers are labeled 'net radiometer' in figure 4.1. The net radiometer measures the total radiative heat flux, and the shielded surface heat flux sensor enables the measurement of convective heat flux. The summation of the two results in true heat flux.

After acquiring q_{true} data, we investigate the correlation between the error ($q_{err} = q_{true} - q_{us}$) and downward and upward long-wave and short-wave radiations, along with net radiation, to discern which of these components predominantly influences the error. Subsequently, we plot q_{err} against the most influential parameter, illustrating error bars to indicate whether q_{err} falls within the measurement uncertainty. If so, q_{err} is considered insignificant. Nevertheless, we will employ linear regression to establish a linear relationship between q_{err} and the primary influencing parameter. This equation can then be utilized to estimate q_{err}.

It is important to emphasize that direct sunlight will be absorbed by the surface of the heat flux sensor and the embedded thermocouples, leading to errors in their measurement of the surrounding air temperature. This prompted the study to quantify the error in terms of total radiation. Since we have no reason to believe that other factors, such as wind speed and ambient temperature, affect the accuracy of a heat flux or temperature sensor, we did not quantify them in this study.

4.2.1 Radiation

As shown in figures 4.1(c) and (d), the net radiometer employed for measuring short-wave and long-wave components of radiation included upward and downward pyranometers and pyrgeometers. The upward-looking pyranometer and pyrgeometer were Apogee SP-510-SS and SL-510-SS to measure the incoming shot-wave and long-wave radiation components, respectively [7]. Combining these two measurements gives us the total solar irradiance received by the solar panel. The downward-looking pyranometer and pyrgeometers were Apogee SP-610-SS and SL-610-SS models, measuring outgoing radiation. Combining these two measurements yields the total irradiance emitted from the canopy, including the solar panel. The utilized pyranometers could measure wavelengths ranging from 370 nm to 2240 nm, which accounts for nearly 97% of the total radiative energy [8]. The pyrgeometers measured wavelengths from 5 to 30 μm. This set-up allowed us to measure individual radiation components, i.e. upward and downward short-wave and long-wave radiation, which was essential as we did not know which of these components the radiative error (q_{err}) was most correlated with. Thus, we evaluated the correlation between q_{err} and all four radiation components and their net effect to understand which component the q_{err} varies with the most.

4.2.2 Convection

This study used Hukseflux FHF05 sensors to measure surface temperature and heat flux. Shielding a surface heat flux sensor blocks radiative heat flux. Thus, the shielded sensor will only measure the heat flux from convection heat transfer. The readings from the shielded sensor provide information about the convection heat transfer at its location. However, this value is lower than the convection heat transfer measured by an unshielded sensor, as the shield reduces the local surface temperature by blocking direct exposure. The employed surface heat flux sensors, both shielded and unshielded, are also capable of measuring the surface temperature. Data analysis indicated a linear increase in temperature difference with solar irradiance, reaching up to 10 °C on very sunny days. Figure 4.2 shows the surface temperature variations on two different days. Figures 4.2(a) and (b) correspond to 27 October 2023 and 8 October 2023, with an average solar radiation flux of 192.3

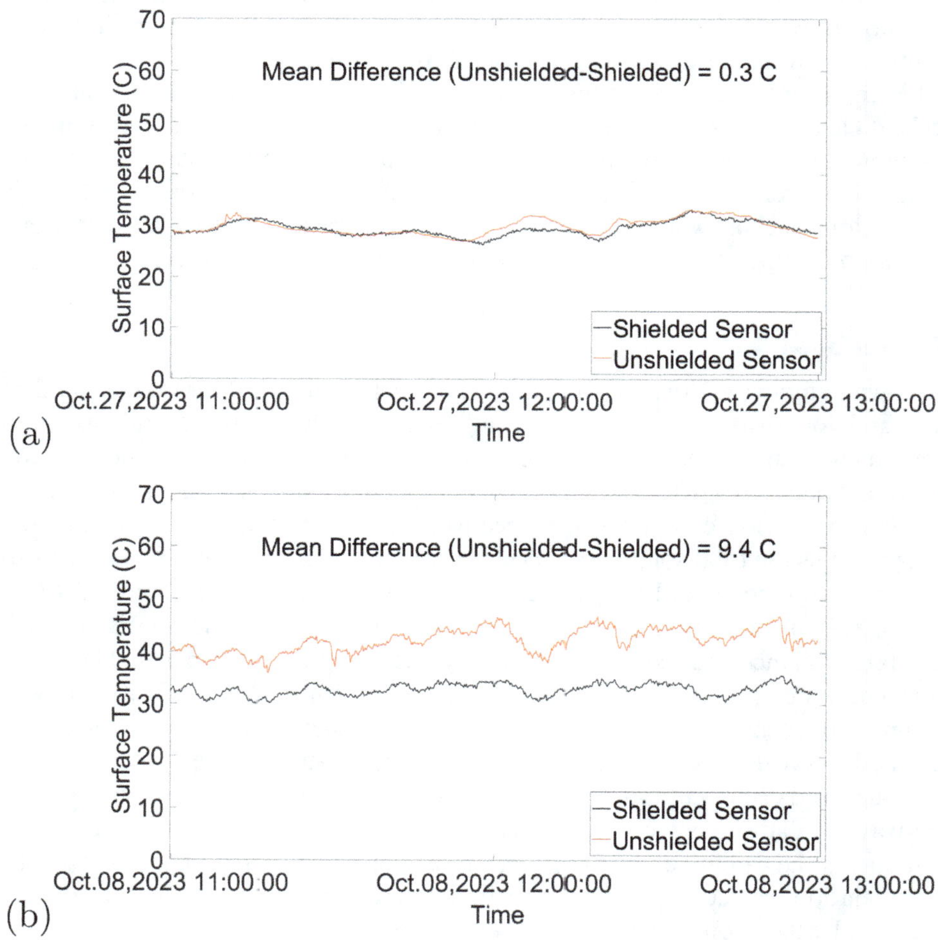

Figure 4.2. Surface temperature: (a) an overcast day and (b) a sunny day.

and 612.3 W m^{-2}, respectively. The figure clearly illustrates how the temperature difference caused by the shield's shade varies from 0.3 °C to 9.4 °C between these two days as the solar radiation flux increases from 192.3 to 612.3 W m^{-2}. Therefore, the actual convection at the panel's surface and that obtained by the shielded sensor are not equal, as larger surface temperatures increase the rate of convective cooling. However, the convection measured by the shielded sensor can be adjusted to accurately represent the rest of the panel's surface. The convective heat flux at the shielded sensor ($q_{c,s}$) is

$$q_{c,s} = h \times \left(T_{p,s} - T_{\infty}\right) \tag{4.1}$$

in which h stands for the convection heat transfer coefficient, $T_{p,s}$ is the surface temperature measured by the shielded sensor, and T_{∞} is measured by an air temperature sensor above the panel. Similarly, one can compute the convective heat flux at the location of the unshielded sensor as $q_{c,us} = h \times (T_{p,us} - T_{\infty})$. Therefore, the difference between the two terms ($q_{c,us}$ and $q_{c,s}$) is given by

$$\Delta q_c = h \times \left(T_{p,us} - T_{p,s}\right). \tag{4.2}$$

Solving equation (4.1) for h and substituting it in equation (4.2) yields

$$\Delta q_c = \frac{q_{c,s}}{\left(T_{p,s} - T_{\infty}\right)} \times (T_{p,us} - T_{p,s}) \tag{4.3}$$

in which Δq_c is what one must add to the shielded sensor's readings to correct them to the convection at the unshielded locations across the panel; hence, $q_{c,us}$ can be computed as what the shielded sensor reads plus the corrective value estimated via equation (4.3). In equation (4.3), $q_{c,s}$ and $T_{p,s}$ are from the shielded heat flux sensor, ($T_{p,us} - T_{p,s}$) is provided in figure 4.2, and T_{∞} is measured by an air temperature sensor protected by radiative shielding.

4.2.3 Error

Accordingly, the error in measurements of the unshielded heat flux sensor (approach two) can be deduced using

$$q_{err} = q_{r,net} + q_{p,s} + \Delta q_{c,s} - q_{p,us}, \tag{4.4}$$

which can also be expressed by $q_{err} = q_{true} - q_{p,us}$, where $q_{true} = q_{r,net} + q_{p,s} + \Delta q_{c,s}$ is the true value of heat flux, $q_{r,net}$ is the net radiation heat flux measured by the pyranometers and pyrgeometers, $q_{p,s}$ is the heat flux measured by the shaded panel and is negative as it measures convection heat flux, $\Delta q_{c,s}$ is the corrective convection term computed via equation (4.3), and $q_{p,us}$ is the heat flux measured by the unshielded sensor. q_{err} is the estimated error in using a single surface heat flux sensor to measure a panel's surface heat flux.

4.2.4 Data quality control

An essential aspect of this field campaign involved the rigorous quality control of data. This investigation thoroughly examined the raw data to identify and document various anomalies, including missing data, gaps in time sequences, data points falling outside acceptable ranges, deviations from expected trends, unreasonable sudden fluctuations, and stagnant data. Subsequently, time-based filters were employed to eliminate outlier data from the study. The data quality control protocol outlined above was executed using specialized software known as Pecos [9–11], which was developed by researchers at Sandia National Laboratories.

4.2.5 Uncertainty

Data measurement campaigns inherently include various sources of error, including calibration discrepancies, repeatability issues, non-stability (long-term drift), and digital rounding (resolution). This study assessed the uncertainties linked to each error source and combined them into a combined uncertainty, u_c, using $u_c = \sqrt{\sum_{i=1}^{N} u_i^2}$, where N is the number of uncertainties being combined and u_i is the corresponding uncertainty with a source of error. In most cases, sensor manufacturers did not provide specified uncertainties for each individual error source. Instead, they consolidated multiple error sources into a single calibration uncertainty. To account for potential excluded uncertainty, the ultimate uncertainty, often called the expanded uncertainty, was estimated by multiplying the combined uncertainty u_c by a confidence factor of $k = 1.5$, hence, uncertainties presented in this study are $U = 1.5 \times u_c$, corresponding to a confidence level of 87%. If we had used a coverage factor of $k = 2$, it would have resulted in a slightly higher confidence level of 90%. However, all the measuring instruments we used in this work were new and had been recently calibrated by the manufacturers. As a result, we were confident in the accuracy of the calibration errors provided. Additionally, the measurement campaign was planned for only six months, so there were no concerns regarding long-term drift uncertainties. Therefore, we decided to use a moderate coverage factor of 1.5.

Table 4.1 presents the uncertainties associated with the sensors utilized in this investigation [12]. These uncertainties were transformed into error bars, which are visually represented on graphs within section 4.3. All sensors used in this study are analog; hence, their resolution was determined by the employed 18 bit data logger and the voltage conversion coefficients provided by the sensor manufacturers. To achieve greater precision, we adjusted the voltage range of the data logger for each sensor, minimizing it as much as possible. This range could vary from as narrow as 60 mV to as wide as 100 V.

4.3 Results and discussion

The study utilized data collected over three weeks, specifically the final seven days of September and the initial 14 days of October 2023. Data recorded between 11:00 am and 1:00 pm, when the Sun's radiation was perpendicular to the set-up, was selected to ensure that the shielded sensor did not capture any radiation flux. The correlation

Table 4.1. Instrument uncertainties [12]. Repeatability and long-term drift uncertainties are included in the calibration uncertainties, except for the Apogee pyronometer, for which the repeatability is ~1%, and long-term drift is ~2% per year.

Sensor	Measured parameter	Calibration	Resolution	Combined	Total
Hukseflux FHF04 (50 x 50 mm^2)	Panel's front and back surface temperature	5%	1 °C	Measurement dependent	Measurement dependent
	Panel's front and back surface heat flux	5%	0.025 W m^{-2}	Measurement dependent	Measurement dependent
Omega EWSA-PT1000	Air temperature above the panels	0.15 + 0.002T °C	0.1 °C	Measurement dependent	Measurement dependent
Apogee SL-510-SS and SL-610-SS pyrgeometer	Long-wave radiation coming off or going into the sky	5%	0.008 W m^{-2}	Measurement dependent	Measurement dependent
Apogee SP-510-SS and SP-610-SS pyranometer	Short-wave radiation coming off or going into the sky	5%	0.006 W m^{-2}	Measurement dependent	Measurement dependent

Table 4.2. The correlation between q_{err} and radiation components.

Long-wave radiation downward	Long-wave radiation upward	Short-wave radiation downward	Short-wave radiation upward	Net radiation	
Correlation coefficient	0.29	0.23	–0.13	–0.28	0.93

between q_{err}, as computed using equation (4.4), and various radiation components, as well as the total net radiation, was analysed. As indicated in table 4.2, the measurement error associated with using a single surface heat flux sensor to gauge the total heat flux at the panel's surface shows the strongest correlation with total radiation, with a correlation coefficient of 0.93. Interestingly, this error (q_{err}) correlates negatively with upward and downward short-wave radiation. However, its correlation with upward and downward long-wave radiations is positive, although small.

Figure 4.3(a) shows sample data from 25 September 2023. Figure 4.3(b) shows the same data using the error bars. In figure 4.3(a), positive values indicate a heat flux

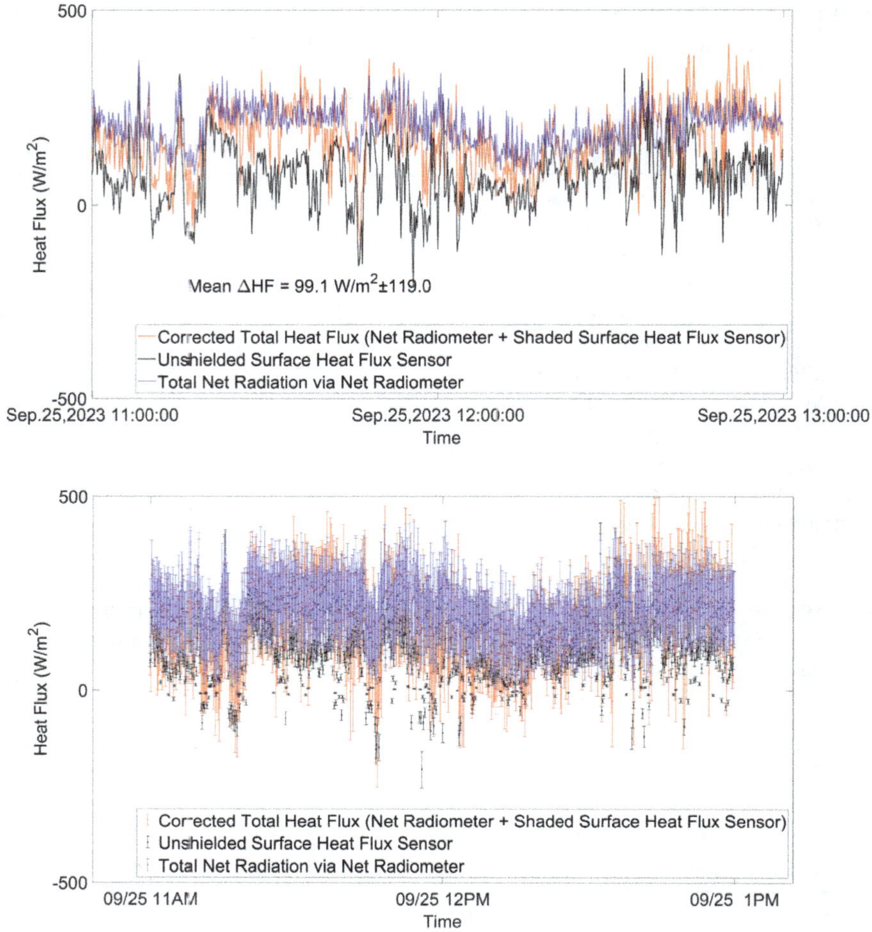

Figure 4.3. Sample recorded data illustrating the time variation of heat flux: (a) average values and (b) error margins.

going into the panel's surface. The blue line represents the net radiation measured by the pyranometer and pyrgeometer sensors. The red line shows the true heat flux obtained via combining the net radiation with the corrected convection based on the measurements of the shielded sensor. Since convection takes heat away from the surface, it is negative. Hence, the true value of the total heat flux shown in red is generally less than the total net radiation shown in blue (the red line is below the blue line, in general). The black line illustrates the heat flux measured by the unshielded sensor. The difference between this measurement (shown in black) and the true heat flux (shown in red) indicates the measurement error, shown by q_{err}, which was 99.1 $\mathrm{W\,m^{-2}}$ on average on 25 September 2023. This value was calculated on 14 separate days and plotted against net radiation in figure 4.4. Two main observations could be made.

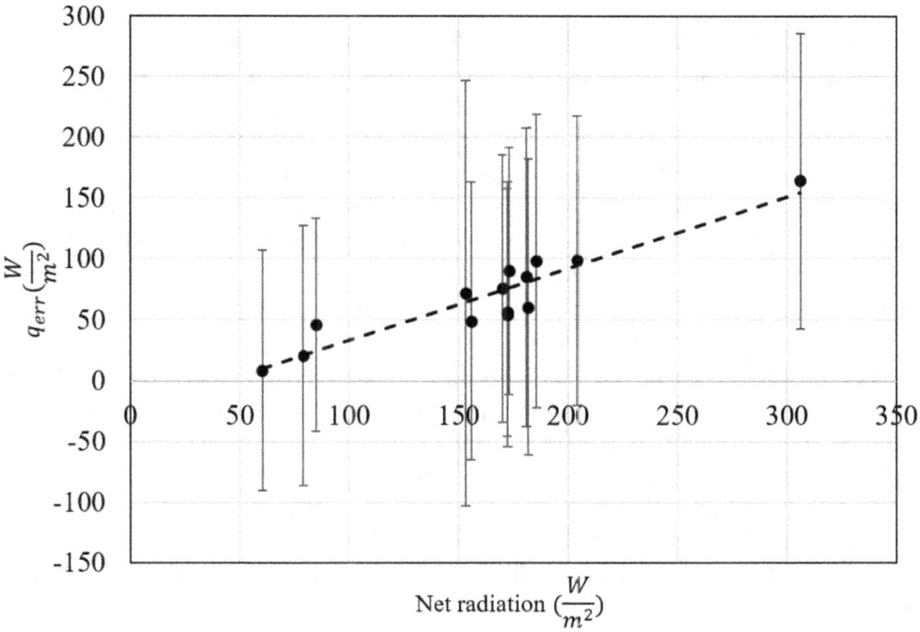

q_{err} (W/m²)	8.5	20.7	46.0	48.9	54.5	56.1	60.3
Net Radiation (W/m²)	60.2	79.0	84.9	155.8	172.3	172.2	181.8
Uncertainty (+/-)	98.5	106.9	87.6	114.2	108.9	101.6	121.6

q_{err} (W/m²)	71.6	75.8	85.1	90.1	98.4	99.1	164.6
Net Radiation (W/m²)	153.4	170.0	181	173.0	185.5	204.0	306.1
Uncertainty (+/-)	174.9	109.8	122.4	101.5	120.4	119.0	121.7

Figure 4.4. The variation of q_{err} with net radiation. The table provides the raw data used in the figure.

First, it is evident that the value of q_{err} consistently falls within the margin of error, rendering this error insignificant. This observation is quite meaningful as it permits utilizing a single direct-contact surface heat flux sensor for measuring total surface heat flux at solar facilities and similar outdoor settings when exposed to direct solar radiation. This simplification is pivotal, eliminating the necessity for employing two pyranometers, two pyrgeometers, an air temperature sensor, and a shielded heat flux sensor. Consequently, there is also a reduction in the requirement for a multi-channel data acquisition unit. These adjustments not only streamline such measurements but also mitigate complexity and lower costs while minimizing the size of the experimental set-up.

Another crucial observation is the linear correlation between the estimated q_{err} and the net radiation, facilitating the development of a straightforward linear equation to rectify measurements. In the case examined in this study, this equation manifested as

$$q_{\text{err}} = 0.5871 \times q_{\text{r,net}} - 25.6 \tag{4.5}$$

with a coefficient of determination of beyond 0.86, meaning that the line explains more than 86% of the data. While q_{err} proves to be insignificant, we recommend the utilization of this equation to rectify total surface heat flux data obtained from direct-contact sensors positioned on PV panel surfaces. A potential query may arise regarding the adaptability of this equation across different sensor models or brands, prompting consideration for why one would choose to correct an unshielded sensor rather than employing a set-up equipped with four pyranometers and pyrgeometers, a shielded sensor, and an air temperature sensor to obtain the true value. While this approach may be practical for applications necessitating only one measuring station, the requirement for multiple stations is common. Constructing several replicas of the described six-sensor set-up and necessitating a more sophisticated data acquisition unit would incur substantial costs and demand more extensive maintenance compared to employing a single surface heat flux sensor. With that, one must note that the purpose of this study was to show that the errors associated with using a single direct-contact heat flux sensor are insignificant and that one can correct for those errors as they appear to have a linear relationship with net radiation; that relationship, however, requires recalibration for different sensor models or brands.

4.4 Conclusion

This study investigated the feasibility of utilizing a single, direct-contact surface heat flux sensor to evaluate the total heat flux at the surface of solar panels. By employing a refined set-up consisting of six measuring probes, the study established a precise baseline for heat flux measurement, subsequently comparing it with data acquired from the direct-contact surface heat flux sensor. The discrepancy between these two measurements, denoted as q_{err}, was examined to assess the accuracy of the direct-contact sensor. Key observations from this investigation include:

- The degree of underestimation exhibits a linear relationship with the net radiation, given by $q_{err} = 0.5871 \times q_{r,net} - 25.6$.
- q_{err} did not correlate to any of the radiation components (i.e. short- or long-wave radiation, either upward or downward) but to the net radiation (i.e. the summation of the four components). The correlation between q_{err} and the net radiation was strong, with a correlation coefficient 0.93.
- q_{err} appeared to fall within the margin of error on all 14 days of measurements; hence, it is insignificant.

These observations are significant as they authorize the utilization of a direct-contact surface heat flux sensor for measuring the total heat flux at the panel's surface. The advantages of employing such sensors are manifold, including:

- A set-up incorporating a single sensor is more cost-effective, easier to maintain, and demands smaller and simpler data acquisition requirements. These attributes hold particular significance for field studies exposed to varying weather conditions and uncontrollable external factors, such as interference from birds disrupting the sensor's operations.

- Utilizing a single sensor instead of a set-up comprising multiple sensors simplifies data quality control and analysis processes, thereby minimizing the potential for human-induced errors.
- Pyranometers and pyrgeometers displayed sensitivity to clouds and their edge effects, amplifying radiative fluxes and necessitating the filtration of specific data points during the data quality control process. Contrarily, the surface heat flux sensor did not encounter such issues.

Acknowledgements

This work was made possible due to a National Science Foundation CAREER grant (award #2144299).

Data availability

Both the raw and processed data used in this research can become available upon request. Please contact the corresponding author.

References

[1] International Energy Agency n.d. Solar PV *IEA* https://www.iea.org/energy-system/renewables/solar-pv (Accessed: 12 September 2023)

[2] Broadbent A M, Krayenhoff E S, Georgescu M and Sailor D J 2019 The observed effects of utility-scale photovoltaics on near-surface air temperature and energy balance *J. Appl. Meteorol. Climatol.* **58** 989–1006

[3] Barron-Gafford G A, Minor R L, Allen N A, Cronin A D, Brooks A E and Pavao-Zuckerman M A 2016 The photovoltaic heat island effect: larger solar power plants increase local temperatures *Sci. Rep.* **6** 35070

[4] Bilen K and Erdoğan I 2023 Effects of cooling on performance of photovoltaic/thermal (PV/T) solar panels: a comprehensive review *Sol. Energy* **262** 111829

[5] Kaiprath J and Kishor Kumar V A 2023 Review on solar photovoltaic-powered thermoelectric coolers, performance enhancements, and recent advances *Int. J. Air-Cond. Refrig.* **31** 6

[6] Vasel A and Iakovidis F 2017 The effect of wind direction on the performance of solar pv plants *Energy Convers. Manage.* **153** 455–61

[7] Cannon T and Vasel-Be-Hagh 2024 A improving heat flux measurement accuracy in solar farm environmental studies: a corrective equation approach *9th Thermal and Fluids Engineering Conf. (TFEC) (Corvallis, OR, 21–24 April)* pp 1661–4

[8] Kambezidis H 2012 3.02—The solar resource *Comprehensive Renewable Energy* ed A Sayigh (Oxford: Elsevier) pp 27–84

[9] Lindig S, Louwen A, Moser D and Topic M 2020 Outdoor PV system monitoring—input data quality, data imputation and filtering approaches *Energies* **13** 5099

[10] Holmgren W F, Hansen C W, Stein J S and Mikofski M A 2018 Review of open source tools for PV modeling *IEEE 7th World Conf. on Photovoltaic Energy Conversion (A Joint Conf. of 45th IEEE PVSC, 28th PVSEC and 34th EU PVSEC) (Waikoloa Village, HI)* (Piscataway, NJ: IEEE) pp 2557–60

[11] King B H, Robinson C D, Carmignani C, Riley D and Jones C B 2018 Application of the sandia array performance model to assess multiyear performance of fielded CIGS PV arrays *IEEE 7th World Conf. on Photovoltaic Energy Conversion (WCPEC) (A Joint Conf. of 45th IEEE PVSC, 28th PVSEC and 34th EU PVSEC) (Waikoloa Village, HI)* (Piscataway, NJ: IEEE) pp 3607–12

[12] Cannon D T and Vasel-Be-Hagh A 2024 Daytime thermal effects of solar photovoltaic systems: field measurements *J. Renew Sustain Energy* **16** 056501

IOP Publishing

Renewable Energy Systems
The way forward
David S-K Ting and Jacqueline A Stagner

Chapter 5

Production of clean fuels by thermochemical conversion and photocatalytic splitting of water

Lokeshwar Puri, Nasim Mia, Yulin Hu and Greg F Naterer

Reducing greenhouse gas (GHG) emissions from the transportation sector plays a critical role in decarbonization. A promising solution is liquid and gaseous fuels from clean and renewable sources. Among them, bio-oil and H_2 are potential solutions. Instead of conventional production methods, this chapter covers the fundamentals and recent studies regarding (i) bio-oil production by hydrothermal liquefaction (HTL) and pyrolysis and (ii) H_2 production by solar based water splitting over a photocatalyst are discussed. This chapter provides insights and fundamental knowledge on fabricating semiconductor photocatalysts and water splitting for H_2 production and bio-oil production using HTL and pyrolysis of biomass.

Abbreviations

AQE	apparent quantum efficiency
AQY	apparent quantum yield
E_g	bandgap energy
Bi_2O_3	bismuth oxide
$CdWO_4$	cadmium tungstate
CO_2	carbon dioxide
CeO_2	cerium(IV) oxide
CB	conduction band
DRS	diffuse reflectance spectra
Fe_2O_3	ferric oxide
FTIR	Fourier transform infrared
GC–MS	gas chromatograph–mass spectrometer
g-C_3N_4	graphitic carbon nitride
GHG	greenhouse gas
HTL	hydrothermal liquefaction

doi:10.1088/978-0-7503-6179-8ch5

H_2	hydrogen
IRi	nfrared radiation
CH_3OH	methanol
MoS_2	molybdenum disulfide
NHE	normal hydrogen electrode
O_2	oxygen
QY	quantum yield
GR	reduced graphene oxide
RuO_2	ruthenium(IV) oxide
SMR	steam methane reforming
$SrTiO_3$	strontium titanate
TiO_2	titanium dioxide
VB	valence band
XRD	x-ray diffraction
ZnO	zinc oxide
ZnSe	zinc selenide
ZnS	zinc sulfide
ZrO_2	zirconium dioxide

5.1 Introduction

In recent decades, climate change, depleting fossil reserves, high greenhouse gas (GHG) emissions and other environmental issues have led scientists to increasingly examine the production of fuel from cleaner and sustainable sources. A fuel can be generally categorized as a solid, liquid, or gas. Among them, liquid fuel, such as bio-oil, has received significant attention due to its ability to potentially replace, fully or partially, fossil fuel-derived liquid hydrocarbon fuels upon downstream upgrading.

Owing to this growing interest in bio-oil, the usage of biomass as an alternative energy source that is more sustainable, cost effective and environmentally friendly, has increased dramatically. However, the selection of an appropriate biomass conversion approach remains a challenge. Past literature suggests a range of thermochemical conversion methods have been utilized such as hydrothermal processing (Lachos-Perez *et al* 2022), pyrolysis (Grafmüller *et al* 2022), torrefaction (Abdulyekeen *et al* 2021), and gasification (Ayub *et al* 2024). Hydrothermal liquefaction (HTL) and pyrolysis are the two most commonly used bio-oil production methods. HTL involves thermochemical conversion and break-down of a biopolymeric structure of biomass using hot and pressurized water at 250 °C–375 °C and 4–22 MPa pressure (Gollakota *et al* 2018). Unlike HTL, pyrolysis is a 'dry' thermochemical conversion process. It is known as the only carbon negative technology for energy conversion. Pyrolysis results in three main products including bio-oil, biochar, and gases, and can be categorized into flash pyrolysis, fast pyrolysis, and slow pyrolysis in accordance to heating rate and residence time.

Other methods of H_2 production routes have been developed to replace steam methane reforming (SMR) and coal gasification. Among them, photocatalytic water splitting harnesses solar energy as the source of energy to drive the water splitting reaction that produces H_2. Solar energy is a free, clean, abundant, and enduring renewable resource (Molaei 2024). One of the key elements ensuring a high and stable H_2 production rate from water splitting reaction is the design and fabrication of a semiconductor photocatalyst. A semiconductor is a material that has an electrical conductivity falling between a conductor (e.g. generally metals) and insulator (e.g. such as most ceramics and glass).

In general, this chapter consists of two main sections: (i) bio-oil production by pyrolysis and HTL and (ii) H_2 production by photocatalytic water splitting. Fundamental knowledge about the reaction process, and primary process conditions about the production routes are discussed. Specifically, for photocatalytic water splitting, the fundamental factors of semiconductor photocatalysts that predominantly affect the water splitting efficiency are discussed, followed by a description of the main reaction conditions of the photocatalytic water splitting. Furthermore, bio-oil production by pyrolysis and HTL, their basic reaction processes and recent studies will be presented.

5.2 Photocatalytic water splitting for H_2 production

The basic principle of an overall water splitting reaction is illustrated in figure 5.1 (A). Upon solar irradiation, electrons (e^-) in the valence band (VB) are excited to move to a conduction band (CB) and leave holes (h^+) in the VB. These photo-generated electrons and holes result in reduction and oxidation for producing H_2 and O_2, respectively. To achieve the overall water splitting reaction, the bottom of the CB must be more negative than the redox potentials of $H^+ + 2e^- \rightarrow H_2$ (i.e. 0 V versus NHE at pH of 0). On the other hand, the top of the VB must be more positive than the redox potential of $2H_2O \rightarrow O_2 + 4H^+ + 4e^-$ (i.e. 1.23 V versus NHE at pH of 0). As a result, the minimal photo energy thermodynamically required to drive the overall water splitting reaction is 1.23 eV. This corresponds to a wavelength of approximately 1000 nm within the near-infrared (IR) region (Maeda 2011).

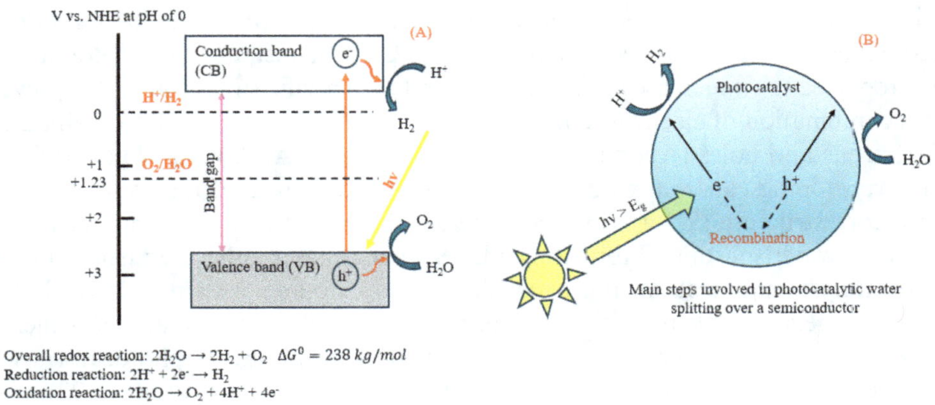

Figure 5.1. (A) Basic principle of a photocatalytic water splitting reaction and (B) the main steps involved in photocatalytic water splitting over a semiconductor. (NHE: normal hydrogen electrode for measuring the electrode potential for the thermodynamic scale of redox potentials.) (Adapted with permission from (Maeda and Domen 2007). Copyright 2007 American Chemical Society.)

As shown in figure 5.1(B), the overall water splitting reaction over a semiconductor photocatalyst is achieved in three steps. Initially, the photon energy excites the semiconductor, and the absorbed photon energy ($h\nu$) is larger than the bandgap energy (E_g). This releases the electron–hole pairs in bulk. Then the photogenerated electrons and holes are separated and migrated to the surface of the semiconductor to avoid recombination. Finally, both the reduction and oxidation reaction take place using the electron–hole pairs to produce H_2 and O_2, respectively (Maeda and Domen 2007).

The TiO_2 semiconductor has been used as a photocatalyst since the 1970s owing to its non-toxicity, photochemical stability, and relatively low cost. However, the wide bandgap of TiO_2 of 3.2 eV makes it only harvest a small portion of the light spectrum, typically in the ultraviolet range. To broaden the use of the spectrum, TiO_2 has been coupled with other semiconductors (e.g. graphitic carbon nitride, g-C_3N_4 with a bandgap of 2.7 eV) to narrow the bandgap that is active within the visible light range and promote the separation of photogenerated charge carriers (Wang and Wang 2022). Another disadvantage of TiO_2 is that the recombination rate of photogenerated electrons and holes is high which lowers the photocatalytic performance. The following equation (5.1) is used to determine the solar-hydrogen conversion efficiency (η_{STH}). According to this equation, it can be concluded that efficient photocatalysts are required to exhibit high performance in light absorption, excellent separation of photogenerated electrons and holes, and a lower recombination of charge carriers (Eidsvåg *et al* 2021):

$$\eta_{STH} = \eta_A \times \eta_{CS} \times \eta_{CT} \times \eta_{CR}. \tag{5.1}$$

Here η_A, η_{CS}, η_{CT}, and η_{CR} represent the efficiencies of light absorption, charge separation, charge transport, and charge reaction, respectively.

To date, different photocatalyst modification techniques have been employed to enhance properties such as defect engineering, bandgap engineering, heterojunction system, metal doping, non-metal doping, nanocomposites, and co-catalysts. Recent review articles have examined photocatalyst modification strategies (Aziz Aljar *et al* 2020, Chen *et al* 2022, Grushevskaya *et al* 2022). For instance, metal doping has been reported to (i) increase surface defects and the specific surface area; (ii) prevent the recombination of electrons and holes; and (iii) enhance the separation efficiency of electrons and holes (Abdullah *et al* 2022). However, Al Abri *et al* (2019) found that over-doping causes the aggregation of nanoparticles, and the formation of a dense core structure and a larger cluster size, thereby lowering the photocatalytic activity in water splitting. Fabricating a heterojunction is another frequently utilized photocatalyst modification approach. Aside from doping with metals (including transition metals, noble metals, and alkaline earth metals), non-metal atoms such as C, N, O, S, and F have also been doped with semiconductor photocatalysts to enhance the water splitting efficiency (Han *et al* 2022, Lu *et al* 2022). The heterojunction system is another modification method that has been widely adopted for enhancing the H_2 evolution rate. The heterojunction photocatalyst system can be fabricated by using two or more semiconductors with matched energy bands to form a Type-I, Type-II, *Z*-scheme, or Schottky junction. In this system, the photo-generated charge carriers can be separately spatially at the interface to avoid the recombination of electrons and holes (Liu *et al* 2022).

5.2.1 Key factors affecting photochemical reactions

5.2.1.1 Band gap

The bandgap represents the distance between the VB and CB. It is the minimal energy needed to excite an electron up to a state in the conduction band where it can participate in conduction. It is a critical factor to differentiate between a conductor, semiconductor, and insulator. As indicated in figure 5.2(A), the bandgap between the VB and CB is essentially small enough that it can be bridged by excitation such as photon energy. Generally, a lower bandgap ensures a greater photocatalytic activity (Khatun *et al* 2023). In bandgap engineering, making composites with metal oxides, hybridization with nanomaterials, metal and non-metal doping, and hydrogenation, are techniques to narrow down the bandgap of a photocatalyst to enable larger visible light absorption.

Samsudin and Abd Hamid (2017) modified TiO_2 by both doping with nitrogen (triethylamine as the precursor) or fluorine (ammonium fluoride as the precursor) or co-doping with nitrogen or fluorine. Co-doping led to variations of the electronic structure by forming a mid-band state to lower the bandgap from 3.2 eV of TiO_2 to 2.24 eV of N,F–TiO_2. Metal doping was reported by Anandan *et al* (2010) to prepare Cd or Cu doped ZnO photocatalyst. It was found that a mid-band was formed below CB and extended the absorption within visible range.

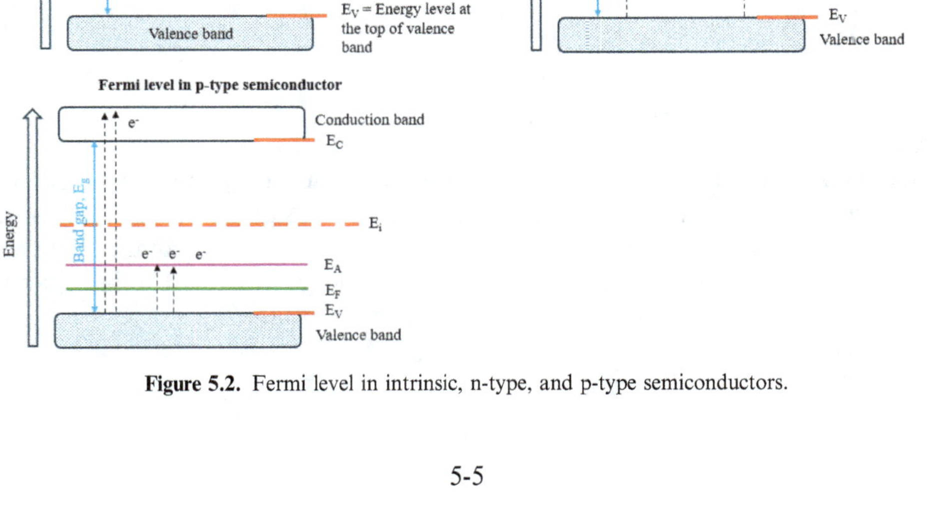

Figure 5.2. Fermi level in intrinsic, n-type, and p-type semiconductors.

Zhang *et al* (2012) reported that the formation of chemical bonding between ZnS and reduced graphene oxide (GR) made GR/ZnS nanocomposite exhibit visible light responsiveness. Xu *et al* (2012) prepared nanocomposites by coupling Bi_2WO_6 with TiO_2 and observed that an internal field was generated and the migration of photogenerated electrons and holes was promoted to different directions.

Other methods have been developed for determining the bandgap including the photoluminescence method, density functional theory, Cody plot, and Tauc plot. Among them, due to its simplicity and high accuracy, the most common method is a Tauc plot (Mursyalaat *et al* 2023). The Tauc method was developed by Jan Tauc in 1966 and based on the assumption that the energy-dependent absorption coefficient (α) can be expressed by (Makuła *et al* 2018)

$$(\alpha \cdot h\nu)^{1/\gamma} = B(h\nu - E_g), \tag{5.2}$$

where h is the Planck constant, ν is the photon's frequency, E_g is the bandgap energy, and B is a constant. Here γ is equal to ½ or 2 for the direct and indirect transition bandgap, respectively.

Photon energy, $h\nu$, can be calculated using (Bo *et al* 2017)

$$h\nu = \frac{1240}{\lambda}, \tag{5.3}$$

where λ is the wavelength.

Data obtained from UV–vis diffuse reflectance spectra (DRS) is typically used to determine the bandgap energy. Initially, DRS results are required to be converted to the corresponding absorption spectra using a Kubelka–Munk function ($F(R_\infty)$). It is expressed by the following equation:

$$F(R_\infty) = \frac{K}{S} = \frac{(1 - R_\infty)^2}{2R_\infty}, \tag{5.4}$$

where K and S are the absorption and scattering coefficients, respectively. R_∞ is the reflectance of an infinitely thin specimen. The following equation can be used to determine R_∞:

$$R_\infty = \frac{R_{\text{sample}}}{R_{\text{standard}}}. \tag{5.5}$$

Using $F(R_\infty)$ to replace α in equation (5.1) to yield

$$(F(R_\infty) \cdot h\nu)^{1/\gamma} = B(h\nu - E_g). \tag{5.6}$$

In making a plot of $(F(R_\infty) \cdot h\nu)^{1/\gamma}$ against $h\nu$ (photon energy), the region shows a steep, linear increase of light absorption with increasing photo energy. This is the characteristic of a semiconductor, and the x-axis, the intersection point of the linear fit of the Tauc plot, provides an estimate of the bandgap energy. The bandgap for some commonly used semiconductors is summarized in table 5.1.

Table 5.1. Summary of bandgap of photocatalysts used in water splitting.

Photocatalysts	Bandgap (eV)	References
TiO_2	3.24	Kunthakudee *et al* (2022)
Au/TiO_2 (synthesized at pH of 3.2)	2.87	
Au/TiO_2 (synthesized at pH of 10.0)	2.71	
Bi_2O_3	2.81	Khalid *et al* (2020)
MoS_2	1.80	
$3\%Bi_2O_3/MoS_2$	1.79	
$7\%Bi_2O_3/MoS_2$	1.76	
$11\%Bi_2O_3/MoS_2$	1.69	
$15\%Bi_2O_3/MoS_2$	1.53	
$TiO_2/RuO_2/CuO$	2.50	Juliya *et al* (2021)
TiO_2/RuO_2	2.90	
$SrTiO_3$	3.28	Deng *et al* (2023)
$SrTiO_3/SrCO_3$	3.34	
$g\text{-}C_3N_4$	2.60	Hu *et al* (2022)

5.2.1.2 Crystallinity and crystallite size

Crystallinity and crystallite size have been widely considered as two critical factors affecting the photocatalytic activity of semiconductor nanoparticles. Both parameters play a role in controlling the mobility and concentration of charges in the photocatalysts (Abdulkadir *et al* 2023).

Wang *et al* (2016) observed the crystallite size of CuO/SiO_2 affected the light absorptivity of the photocatalyst. Wang *et al* (2014) reported that a larger crystallite size of TiO_2 produced a higher photocatalytic performance; however, the crystallinity played no role in the photocatalytic performance. Crystallite size is correlated to the photo-reactivity by determining the surface area, surface energy, light absorptivity, and lattice distortion. The detailed relationship between the crystallite size and photocatalyst activity of semiconductors is under debate. Some literature has suggested that the optimal size of a crystallite is between 7 and 15 nm for most of the photocatalytic reactions (Liu *et al* 2008).

The Debye–Scherrer equation is used to calculate the crystallite size of the nanoparticle photocatalyst and is provided in equation (5.7). X-ray diffraction (XRD) results are required to complete this calculation:

$$D = \frac{K\lambda}{\beta \cos \theta}. \tag{5.7}$$

Here D is the crystallite site of nanoparticles. K is a dimensionless shape factor and has a typical value of 0.9. λ is the wavelength of the x-ray. β denotes the full width at half maximum (FWHM) (Sivagami and Asharani 2022). θ is the diffraction angle.

Table 5.2. Summary of crystallite size of photocatalysts used for water splitting.

Photocatalyst	Crystallite size (nm)	References
ZnO	47.4	Ullah *et al* (2023)
Ag-ZnO	45.2	
γ-ZnO	43.6	
Ag-γ-ZnO	40.2	
Fe_2O_3	39.0	Reli *et al* (2021)
CeO_2	21.9	
ZnSe	5	Sanchez-Martinez *et al* (2021)
Se/ZnSe	6	
ZrO_2	8	Jawhari (2023)
$CdWO_4$	26	

Table 5.2 shows a summary of the crystallite size of some photocatalysts used for the water splitting reaction.

Crystallinity is another parameter affecting photocatalytic performance. In general, it has been reported that a higher crystallinity leads to a higher photoreactivity, which could be due to an enhanced electron transport from the conduction band to the surface adsorbed molecules (Benkö *et al* 2003). Moreover, past literature has suggested that a higher crystallinity lowers the number of defects in the crystal structure and could reduce the number of sites for electron–hole recombination, thereby resulting in better catalytic activity (Eidsvåg *et al* 2021). XRD graphs can also be applied to determine the crystallinity by the ratio of the area under the crystalline peaks and the area under all peaks, as shown in the following equation (University of Utah 2024):

$$\%\,\text{Crystallinity} = \frac{\text{Area under the crystalline peaks}}{\text{Area under all peaks}} \times 100\%. \tag{5.8}$$

5.2.1.3 Fermi level

The Fermi level in semiconductors indicates the probability of the occupancy of electrons at different energy levels. As frequently mentioned about the conduction band and valence band, they represent the groups of energy states of the electrons. According to the Pauli exclusion principle, each allowed energy level can only be occupied by no more than two electrons of opposite spin. This means that all available states up to a certain energy level will be occupied by two electrons at low temperatures. This certain energy level is called the Fermi level. The Fermi level is helpful for calculating the (i) density of electrons; (ii) number of holes within the material; and (iii) the relative proportion of holes and density of electrons with respect to the temperature. The Fermi level in semiconductors can also be defined as the energy level that corresponds to the center of gravity of conduction band electrons and valence band holes weighted according to their energies. The Fermi–Dirac distribution function (equation (5.9)) gives the probability that electrons will occupy a certain energy level at a certain temperature:

$$f(E) = \frac{1}{1 + e^{(E-E_F)/kT}}. \tag{5.9}$$

Here E is the allowed energy state. E_F is the Fermi energy level of a given material. k is the Boltzmann's constant and T is temperature.

Figure 5.2 shows the Fermi level in an intrinsic semiconductor (or called pure or undoped semiconductor). Based on the Fermi level, an n-type and p-type semiconductor (introducing dopant) can then be defined. For the intrinsic semiconductor, the number of electrons in the conduction band is equal to the number of holes in the valence band and hence the Fermi level is always to be the middle point of bandgap. For the position of the Fermi level in the n-type semiconductor, the impurities are added to the semiconductor, and this introduces a donor energy level (E_D in the) just below the conduction band. This donor energy level provides a number of electrons to the conduction band, which leads to a larger number of electrons in the conduction band than the number of holes in the valence band. The center of gravity will shift towards the conduction band and therefore the Fermi level in the n-type semiconductor close to the conduction band. In p-type semiconductors, a new energy level (E_A) is introduced just above the valence band and is called the acceptor level. This acceptor level can accept the electrons moving from a valence band so that the number of electrons moving from a valance band to a conduction band will be lower. As a result, the number of holes in the valence band will be larger than the number of electrons in the conduction band. The center of gravity will shift towards the valence band and therefore the Fermi level in a p-type semiconductor close to the valence band.

5.2.1.4 Dimensionality

In accordance with the dimensionality, nanomaterials can be divided into zero-dimensional (0D), one-dimensional (1D), two-dimensional (2D), and three-dimensional (3D) nanostructures. 0D nanomaterials are characterized by those materials having dimensions that are all under the nanometer range. Quantum dots are the most common 0D nanomaterials used to decorate the photocatalyst in photocatalytic water splitting to absorb more visible light and prevent the recombination of the electron and hole. 1D nanostructures are those materials that have a dimension within the range of 1–100 nm, and cover different morphologies, compositions, and structures such as wires, rods, tubes, and ribbons. Nanotube photocatalysts may provide a higher surface area for the overall redox reaction. Nanowire and nanorod photocatalysts could contribute to an effective charge carrier transport and collection. 2D nanomaterials are atomically thick and consist of a single to a few layers of atoms (Baig 2023). The use of 2D nanomaterials in photocatalytic water splitting is capable of reducing the travel distance for the photogenerated electrons and holes, thereby enhancing the light absorption efficiency. The beneficial impact of 2D nanomaterials may be related to their high surface area and small thickness. Furthermore, 3D nanostructures are characterized by the combination of different nanoscale features in one single structure (Byakodi *et al* 2022). Examples of 3D nanostructures utilized in photocatalytic water splitting are $SrTiO_3$ (Moniruddin *et al* 2017), α-Fe_2O_3/TiO_2 (Han *et al* 2017), and g-C_3N_4 (Li *et al* 2019).

5.2.1.5 *Apparent quantum efficiency*

For photocatalysis, quantum yield (QY) is defined as the ratio of the number of electrons participating in the reaction to the total number of photons absorbed by the system under a specific wavelength. However, for the heterogeneous photocatalytic reaction system, it is challenging to determine the actual number of photons absorbed by the system owing to the light reflection and/or scattering. QY can be calculated by replacing the determination of the total number of photons absorbed by the system with the incident photons count. This new parameter refers to the apparent quantum yield (AQY), or apparent quantum efficiency (AQE), and it can be calculated by the following equation:

$$\mathrm{AQY} = \frac{N_e}{N_P} \times 100\% = \frac{(r \times N_A \times 2)}{(I \times A/E_\lambda)} \times 100\%, \tag{5.10}$$

where N_e and N_P represent the total number of transferred electrons and incident photons, respectively. r is the H_2 evolution rate, N_A is the Avogadro constant ($6.02 \times 10^{23}\ \mathrm{mol^{-1}}$), I is the light intensity, E_λ is the energy of one photon at 365 nm, and A is the incident area of the light (Lin *et al* 2020).

Here E_λ is the energy of one photon at 365 nm and can be calculated by

$$E_\lambda = hv = \frac{hc}{\lambda}, \tag{5.11}$$

where h is the Planck's constant ($6.626 \times 10^{-34}\ \mathrm{J\ Hz^{-1}}$), c is the speed of light ($3 \times 10^{8}\ \mathrm{m\ s^{-1}}$), and λ is the wavelength of photon.

Liu *et al* (2023) found that the presence of a co-catalyst (i.e. Pt) in the $CoS_2/Zn_3In_2S_6$ led to an increase in the AQE from 14.9% at a wavelength of 370 nm to 66.2%, which corresponds to an increase in the H_2 yield from 5.69 to 24.17 mmol $(\mathrm{h \cdot g})^{-1}$. Wang *et al* (2023) reported that 99.2% of AQE was achieved when using a sandwich type $PtCu/TiO_2$ photocatalyst, which produces an H_2 production rate of 2383.9 μmol $\mathrm{h^{-1}}$. In addition to the factors described above, several reaction parameters play a crucial role in affecting the photocatalytic water splitting and H_2 evolution. Section 5.3 will focus on the influences of a sacrificial agent type, pH of the reaction medium, photocatalyst concentration, and light source and intensity on photocatalytic water splitting for producing H_2.

5.2.2 Reaction variables

5.2.2.1 *Sacrificial agent*

As discussed in previous sections, several key factors of photocatalysts that affect the water splitting efficiency can be improved by modifying the semiconductor using different techniques. Aside from the enhancement of photocatalyst properties, sacrificial agents have been applied to promote the water splitting reaction (Yasuda *et al* 2018). Specifically, the sacrificial agent is able to act as electron donors and scavenge holes and thus prevent the recombination of electrons and holes. Since O_2 is not produced in the presence of a sacrificial agent, the reverse reaction to form H_2O is suppressed, thereby increasing H_2 yield (Schneider and Bahnemann 2013).

As illustrated in figure 5.3(A), the energy needed to complete the water splitting reaction when no sacrificial agent is high, and the reverse reaction is rapid. On the other hand, in the presence of a sacrificial agent (i.e. CH_3OH), even though the water splitting reaction is still an up-hill process, the energy change is significantly smaller. In early work regarding photocatalytic water splitting, a net *al*cohol reaction mixture such as methanol (Prahov *et al* 1984), ethanol (Ohtani *et al* 1993), or 2-propanol (Ohtani *et al* 1993) was used. The mixture of alcohol and water has been utilized in recent years. Like CH_3OH, the overall reaction (equation (5.12)) for the water splitting reaction when using TiO_2 and ethanol as the sacrificial agent is to co-generate H_2 and CO_2 instead of H_2 and O_2 and limit the reverse reaction between H_2 and O_2 to form H_2O:

$$C_2H_5OH + 3H_2O \xrightarrow{TiO_2,\ h\nu} 6H_2 + 2CO_2. \tag{5.12}$$

In general, sacrificial agents can be categorized as organic and inorganic electron donors. For the organic electron donors, various organics such as alcohols, organic acids, and hydrocarbons have been utilized in the water splitting reaction as the hole scavengers or electron donors. Among them, CH_3OH is the most used sacrificial agent. For a practical application, CH_3OH is required to be extracted from biomass or other organic waste. CH_3OH can conduct an irreversible reaction with the photogenerated holes in the VB to enhance the separation efficiency of photogenerated charge carriers and quantum efficiency. Nevertheless, it should be noted that CH_3OH is continuously consumed in the photocatalytic reaction and thus a sustained addition of CH_3OH to the reaction system is needed to maintain H_2 production.

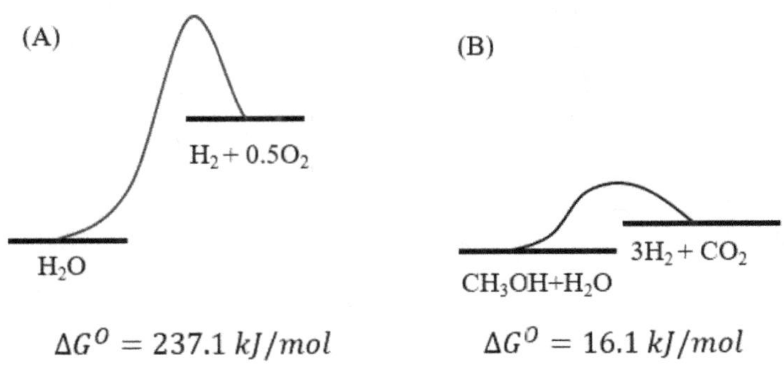

Figure 5.3. Energy diagrams for (A) large up-hill and (B) small up-hill process when adding CH_3OH as the sacrificial agent. (Adapted with permission from (Yasuda *et al* 2018). Copyright 2018 Elsevier.)

In a previous study, the decomposition mechanism (equations (5.13)–(5.15)) of CH_3OH in the TiO_2-driven photocatalytic reaction has been explored:

$$CH_3OH(l)\xrightarrow{TiO_2,h\nu} HCHO(g) + H_2(g) \quad \Delta G^o = 64.1 \text{ kJ mol}^{-1} \tag{5.13}$$

$$HCHO(g)\xrightarrow{TiO_2,h\nu} HCOOH(l) + H_2(g) \quad \Delta G^o = 47.8 \text{ kJ mol}^{-1} \tag{5.14}$$

$$HCOOH(l)\xrightarrow{TiO_2,h\nu} CO_2(g) + H_2(g) \quad \Delta G^o = -95.8 \text{ kJ mol}^{-1}. \tag{5.15}$$

Given equations (5.13) and (5.14), the Gibbs free energy is a positive value and therefore both reactions are not thermodynamically favorable at room temperatures. The photon energy absorbed by the TiO_2 can then drive the reaction toward the products' side. Equation (5.15) has a negative value of Gibbs free energy and hence it can be found that intrinsically the undesirable reverse consumption of H_2 is not favorable.

In the photocatalytic H_2 production from an aqueous alcoholic solution, at least more than half of the generated H_2 is formed through the action of holes rather than the action of electrons. Consequently, it is fair enough to claim that the yield of H_2 evolved from such systems cannot be called the water splitting efficiency. The possible mechanism is depicted in figure 5.4. Initially, the photon energy excites the TiO_2 and makes the electrons move from VB to CB thereby leaving holes in the VB. In the CB, the electron will be trapped by Ti^{4+}. Then, the CH_3OH is oxidized to form CH_2OH and H^+. Through a process called 'current doubling', HCHO is formed by injecting electrons into the CB. Then Ti^{4+} is reduced to form Ti^{3+}, and the final step is the recombination channel. A similar mechanism is also applicable to other alcohols including ethanol, 2-propanol, butanol, and polyvinyl alcohol (Schneider and Bahnemann 2013).

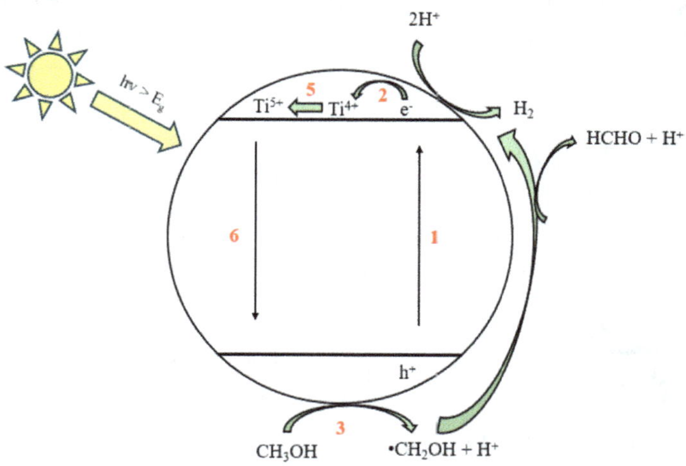

Figure 5.4. Possible mechanisms of photocatalytic H_2 production from a methanol-water mixture. (Adapted with permission from (Schneider and Bahnemann 2013). Copyright 2013 American Chemical Society.)

Regarding the inorganic electron donor, both sulfide (S^{2-}) and sulfite (SO_3^{2-}) are frequently employed since they are efficient hole acceptors and therefore enhance the separation efficiency of charge carriers. The relating possible reaction mechanisms are shown in the following equations:

$$2H_2O + 2e^- \rightarrow H_2 + 2OH^- \tag{5.16}$$

$$SO_3^{2-} + 2OH^- + 2h^+ \rightarrow SO_4^{2-} + H_2O \tag{5.17}$$

$$2SO_3^{2-} + 2h^+ \rightarrow S_2O_6^{2-} \tag{5.18}$$

$$2S^{2-} + 2h^+ \rightarrow S_2^{2-} \tag{5.19}$$

$$SO_3^{2-} + S^{2-} + 2h^+ \rightarrow S_2O_3^{2-} \tag{5.20}$$

$$S_2^{2-} + SO_3^{2-} \rightarrow S_2O_3^{2-} + S^{2-}. \tag{5.21}$$

There are two major advantages of sulfide and sulfite sacrificial agents over alcohols, including (i) the sulfide and sulfite are more oxidizable than alcohols and (ii) when using CdS (a common photocatalyst especially when using either sulfide or sulfite as the sacrificial agent), the dissolved Cd^{2+} can react with S^{2-} to reform CdS. However, in the alcoholic reaction medium, the reaction does not occur. In particular, the energy level of VB in the CdS is more positive than NHE (normal hydrogen electrode at pH 0) to accelerate the oxidation of S^{2-} and SO_3^{2-}. The oxidation of S^{2-} and SO_3^{2-} can be achieved either by a two-electron transfer process (equations (5.17)–(5.19)) or one electron oxidation (less thermodynamically favorable).

In short, the majority of past literature concerns organic and inorganic sacrificial electron donors. A limited number of studies have employed electron acceptors as the sacrificial agent such as Ag^+ (Schneider and Bahnemann 2013).

5.2.2.2 Photocatalyst concentration

Past literature has suggested that the maximum H_2 evolution rate (mol $(h{\cdot}g)^{-1}$) is independent of the dosage of photocatalyst, as shown in figure 5.5. The saturation of light absorption occurs at the optimized photocatalyst concentration. Specifically, an excessive amount of photocatalyst affects the penetration depth of the incident light and the degree of light scattering, which could result in no change or even a decrease in the photocatalytic activity at a higher dosage of photocatalyst (Rahman *et al* 2020).

Enzweiler *et al* (2020) observed that an increase in the H_2 production rate with increasing photocatalyst concentration up to an optimized photocatalyst concentration, followed by a reduction in the H_2 production rate when further increasing photocatalyst concentration. A similar trend between photocatalyst concentration and H_2 production was reported by Edelmannová *et al* (2021), in which the highest H_2 yield was obtained at 0.25 g l^{-1} but an increase in photocatalyst concentration to 1 g l^{-1} lowered the H_2 yield.

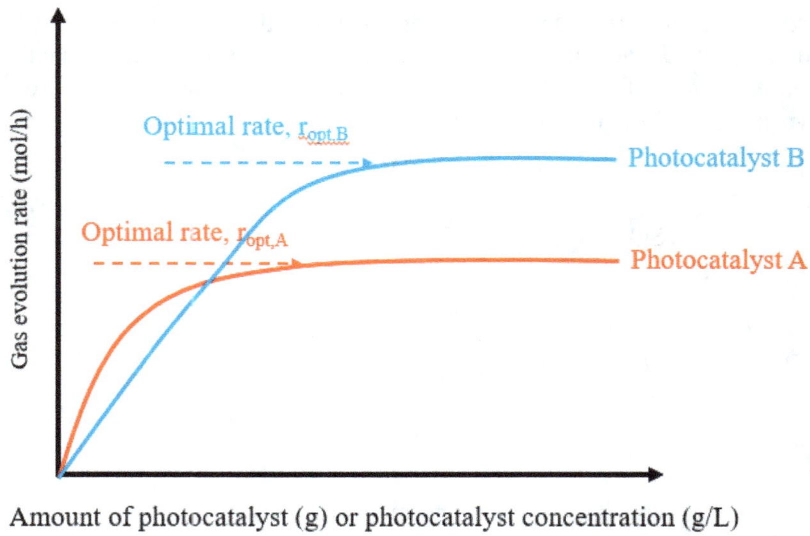

Figure 5.5. Influence of photocatalyst concentration on gas evolution rate.

Zhao *et al* (2022) also studied the effect of MoS_2/CdS concentration on photocatalytic water splitting for producing H_2. The relationship between photocatalyst concentration and H_2 yield was found to be similar.

On the contrary, Wei *et al* (2018) found that double or triple the concentration of photocatalyst led to an improvement in H_2 productivity by 91.63% and 201.97%, respectively. The differences in the photocatalytic activity results could be related to variations in the light source and intensity and pH of the reaction medium. These are other factors influencing the H_2 evolution rate and will be discussed in the following sections.

5.2.2.3 Light intensity

Light intensity is an important factor for a photocatalytic reaction by affecting the reaction rate and the utilization of energy. Owing to the small amounts of H_2 per unit area and the challenges in harvesting and light and containing gaseous products in the system, a concentrated light system might be promising. Bell *et al* (2013) highlighted the importance of light intensity for a photocatalytic water splitting reaction, and the results showed that TiO_2 was not saturated until 52 suns. Kerzig and Wenger (2019) controlled the reactivity of a photocatalytic system by altering the light intensity. This approach allowed the regulation of redox potential and switch of the electron transfer reactions. In particular, by changing the light intensity per area using beam collimation, the authors achieved a switching from one to two photon substrate activation chemistry. Different major products were selectively formed under similar reaction conditions.

In the field of photocatalytic water splitting, few studies have explored the effects of light intensity on the water splitting and H_2 evolution rate. However, some previous studies on photocatalytic degradation have observed that a loss of light energy has been found in the heterogenous photocatalytic reaction. The loss in the light energy could be due to (i) photon energy absorbed lower than the bandgap; (ii) photon energy absorbed that is higher than that required to create charge carriers; (iii) energy of reflected light; (iv) energy of transmitted light; and (v) energy of electron–hole recombination. This loss in the light energy can be reduced by the selection of the appropriate photocatalyst, the particle size of the photocatalyst, and light source and intensity (Yang and Liu 2007). The understanding of the influences of light intensity on water splitting efficiency and H_2 evolution are topics of future research.

5.3 Bio-oil production

5.3.1 Hydrothermal liquefaction (HTL)

In comparison, HTL is preferred over pyrolysis when high temperature is a constraint due to its ability to be performed at lower temperatures and utilization of wet biomass as a feedstock (Aliyu *et al* 2021, Alhassan *et al* 2016). The HTL process is known for its production of a large number of value added by-products along with its key product (bio-oil) such as aromatics and phenolic derivatives, carboxylic acid, esters, nitrogenous ring structures and branched aliphatic compounds (Xiu *et al* 2010). The yield of bio-oil from HTL usually varied in the range to 24–64 wt.% (Atelge *et al* 2020). Bio-oil produced via HTL has a high heteroatom content, significantly large acid value, highly viscous in nature and lower higher heating value (Saber *et al* 2016). The HTL of biomass for producing bio-oil consists of a series of reactions such as decarboxylation, degradation, hydrolysis, aromatization, and repolymerization (Leng *et al* 2020). HTL is known to be a more energy efficient process as it does not require drying of biomass, hence saving up to 75% of the total energy consumed by other processes such as pyrolysis (Xu and Savage 2017).

Water is also known to be an essential component of HTL, as it is capable of providing multiple roles in bio-oil production. Initially, water appears in the form of a reactant and hence supports the production of bio-oil. Water acts like a catalyst, which can promote acidic/basic reactions. Furthermore, water is a benign solvent and could offer a suitable environment for the reactants to undergo reactions. Both homogeneous and heterogeneous catalysts can be used for HTL to increase the quality and yield of bio-oil. Certain catalysts which have been used in the past literature are zeolites (Duan and Savage 2011), KOH (Ross *et al* 2010), and CH_3COOH (Ross *et al* 2010). Although HTL can be performed over a wide range of biomass, in recent years algae or the marine biomass remains a primary focus of researchers. Certain micro and macro algae such as *Dunaliella*, *Scenedesmus*, and *Spirulina* have been well mentioned in the literature. Recent studies have also indicated the use of two stage HTL using microalgae, which is pre-treated with an NaOH/urea solution, resulted in higher quality and quantity of bio-crude produced as compared to conventional single stage HTL in terms of reduced oxygen and nitrogen content. Bio-oil was also found to have a higher energy content (Hu *et al* 2017). A comprehensive list of recent studies regarding HTL of biomass for bio-oil production at different reaction conditions is given in table 5.3.

Table 5.3. Recent studies on bio-oil production via HTL of biomass.

Feedstock	Operating conditions	Major properties of bio-oil	References
Laminaria	350 °C; 16.5 bar	The yield was 79 wt.%. HHV was 35.97 MJ kg^{-1}	Bach *et al* (2014)
Nannochloropsis oceanica	240 °C–300 °C; 32–43 bar	Addition of alcohol co-solvents increased yield	Caporgno *et al* (2016)
Pal mesocarp fiber, palm kernel shell and raw fruit bunch	330 °C–390 °C; 25–35 MPa	Optimal conditions for maximizing yield were 390 °C and 25 MPa	Chan *et al* (2015)
One water INC Indianapolis feedstock	260 °C–320 °C; 1 MPa	HHV was 33.3 MJ kg^{-1} Chemicals identified in bio-oil were hydrocarbons, fatty acids and cyclic amines	Chen *et al* (2014)
Spirulina and *Tetraselmis*	300 °C–350 °C; 30 bar	HHV was 40 MJ kg^{-1}	Eboibi *et al* (2014)
Oil mill waste water	200 °C–300 °C; 190 bar	Phenols were the major compound The yield was 58 wt.%	Hadhoum *et al* (2016)
Chlamydomonas reinhardtii	220 °C–310 °C; 10 MPa	Pyrolysis produced a higher yield, but a lower energy recovery ratio compared to HTL	Hognon *et al* (2015)
Landscape waste leaves	300 °C; 2 MPa	Ketones, esters, acids, and alcohol were found in bio-oil	Cao *et al* (2016)
Chlorella, Scenedesmus obliques, Tetrasemis	350 °C; 20.7 MPa	Acetic acid, propionic acid, pyridine, acetone, ethanol were detected in bio-oil	Maddi *et al* (2016)
Kenaf and wheat straw biomass	200–350 °C; 5 MPa	Bio-oil yield increased with temperature till 300 °C	Meryemoğlu *et al* (2014)
Nannochloropsis gaditana and *Chlorella spp*	300 °C; 14 bar	HHV was 39 MJ kg^{-1}	Reddy *et al* (2016)
Water hyacinth	Temperature: 250 °C–300 °C; Pressure: 1300 Psi	Highest yield was obtained using KOH and K_2CO_3 as catalysts, i.e. 23%, with lesser oxygen utilization	Singh *et al* (2015a)
Rice straw	280 °C–320 °C; 6–9 MPa	Biomass conversion was 78% and bio-oil yield was 17 wt.%	Singh *et al* (2015b)
Algal blooms of Dianchi lake	260 °C –340 °C; 3 MPa	High ash content reduced the bio-oil yield	Tian *et al* (2015)

A current limitation of HTL of biomass is limited understanding on how the presence of inorganics in the biomass affect the biomass conversion routes. Inorganic materials can act either as a catalyst to promote the reaction or an inhibitor deterring the biomass degradation. Unlike woody biomass, a high content of inorganics can be normally identified in seaweed and municipal solid waste. However, few studies have considered the development of a more comprehensive kinetic model to include the effect of ash content and composition on the HTL of biomass. A common reaction pathway of inorganics in HTL is the partition into an aqueous phase. Previous studies have suggested that this inorganic fraction can be transferred to the bio-oil and solid phases. This is a promising area of future research on HTL of biomass.

5.3.2 Pyrolysis

Pyrolysis usually involves the utilization of lignocellulosic biomass such as agricultural waste (rice husk, sugarcane bagasse and hazelnut glume, etc), marine waste (algae, etc), municipal sludge waste (sanitary napkins, paints, syringes and batteries), and forestry waste (sawdust, wood chips and wood barks, etc). Biomass pyrolysis is further divided into various types, differentiating mainly on two parameters, temperature and heating rate. Such differentiation leads to various mechanisms of slow, conventional, fast, and flash pyrolysis. Slow pyrolysis is performed in the temperature range of 300 °C–700 °C and a very slow heating rate of 0.1–1 °C s^{-1}. Fast pyrolysis usually occurs in the temperature range of 400 °C–800 °C with a minimal vapor residence time of 2 s. A high heating rate is a key feature of fast pyrolysis, typically between 10–200 °C s^{-1}. Fast pyrolysis and its short residence time is responsible for about 70 wt.% bio-oil yield (Fahmy *et al* 2020). Flash pyrolysis occurs at the highest temperature range, which is 1000 °C with a high heating rate and small vapor retention time, i.e. 2500 °C s^{-1} and 0.5 s, respectively. Lastly, conventional or intermediate pyrolysis typically occurs between 400 °C and 650 °C with a vapor residence time of 0.5–20 s (Kazawadi *et al* 2021). The heating rate in the literature for intermediate pyrolysis is 1–10 °C s^{-1}.

Pyrolysis is usually performed in three major steps. The first step is drying to remove surface moisture without making in change in the molecular structure of the biomass. The second step is the initiation of the devolatilization. During this stage, hemicellulose decomposition begins with unstable polymers to generate volatiles and some gases such as CO_2 and CO (Qiao *et al* 2018). Third, a charring process occurs which consists of decarboxylation, decarbonylating and radical reactions. Thermogravimetric analysis–Fourier transform infrared spectrometry–mass spectrometry (TG–FTIR–MS) and pyrolysis–gas chromatography/mass spectrometry (Py–GC/MS) are two techniques which are frequently used to study the thermal behavior of volatiles and kinetic mechanism of pyrolysis (Gai *et al* 2013).

In terms of the heating mode during pyrolysis, an alternative heating method is microwave heating. Microwaves are electromagnetic waves that consist of electric and magnetic fields with wavelengths ranging from one millimeter to one meter and frequencies between 300 MHz and 300 GHz (Naterer 2022). The use of microwave heating in biomass conversion and processing has several benefits, including a higher heating rate, and selective heating. Thus it can limit the unwanted side reactions. It is easier to control as a non-contact and volumetric heating mechanism. Another major advantage of microwave heating over traditional electrical heating is that the heating profile varies from the core of the biomass particle which is initially heated by the microwave and the temperature is decreasing from the core to the surface. In microwave pyrolysis, due to the low microwave absorbing ability of biomass, it is typical to add a microwave adsorber together with biomass, such as silica carbide and activated carbon.

5.4. Conclusions

Current worldwide energy dependency on fossil fuels can be dramatically reduced using liquid and gaseous fuels produced from clean and renewable resources. This chapter has focused on the production of H_2 fuel by photocatalytic water splitting and liquid fuel (i.e. bio-oil) production by HTL and pyrolysis. The main conclusions in this book chapter are: (i) the significance of synthesizing a semiconductor-based photocatalyst with a suitable bandgap, crystallinity, crystallite size, Fermi level, dimensionality, and apparent quantum efficiency; (ii) HTL could be more favorable for processing high water containing feedstock due to the non-requirement for pre-drying when compared with pyrolysis; and (iii) bio-oil can be produced from either HTL or pyrolysis and subsequent upgrading is needed to improve the quality of bio-oil. Overall, this chapter has provided new perspectives and understanding of the key factors in selecting and synthesizing semiconductor photocatalysts and their applications in water splitting reactions for H_2 production as well as the utilization of HTL and pyrolysis of biomass.

Acknowledgments

The authors are grateful for the financial support of the Natural Sciences and Engineering Research Council (NSERC) of Canada.

References

Abdulkadir B A, Teh L P, Abidin S Z, Setiabudi H D and Jusoh R 2023 Advancements in silica-based nanostructured photocatalysts for efficient hydrogen generation from water splitting *Chem. Eng. Res. Des.* **199** 541–68

Abdullah F H, Bakar N H H A and Bakar M A 2022 Current advancements on the fabrication, modification, and industrial application of zinc oxide as photocatalyst in the removal of organic and inorganic contaminants in aquatic systems *J. Hazard. Mater.* **424** 127416

Abdulyekeen K A, Umar A A, Patah M F A and Daud W M A W 2021 Torrefaction of biomass: production of enhanced solid biofuel from municipal solid waste and other types of biomass *Renew. Sustain. Energy Rev.* **150** 111436

Al Abri R, Al Marzouqi F, Kuvarega A T, Meetani M A, Al Kindy S M Z, Karthikeyan S, Kim Y and Selvaraj R 2019 Nanostructured cerium-doped ZnO for photocatalytic degradation of pharmaceuticals in aqueous solution *J. Photochem. Photobiol.* A **384** 112065

Alhassan Y, Kumar N and Bugaje I M 2016 Hydrothermal liquefaction of de-oiled *Jatropha curcas* cake using deep eutectic solvents (DESs) as catalysts and co-solvents *Bioresour. Technol.* **199** 375–81

Aliyu A, Lee J G M and Harvey A P 2021 Microalgae for biofuels: a review of thermochemical conversion processes and associated opportunities and challenges *Bioresour. Technol. Rep.* **15** 100694

Anandan S, Ohashi N and Miyauchi M 2010 ZnO-based visible-light photocatalyst: band-gap engineering and multi-electron reduction by co-catalyst *Appl. Catal.* B **100** 502–9

Atelge M R *et al* 2020 A critical review of pretreatment technologies to enhance anaerobic digestion and energy recovery *Fuel* **270** 117494

Ayub Y, Ren J, He C and Azzaro-Pantel C 2024 Co-gasification of biomass and plastic waste for green and blue hydrogen production: novel process development, economic, exergy, advanced exergy, and exergoeconomics analysis *Chem. Eng. J.* **480** 148080

Aziz Aljar M A, Zulqarnain M, Shah A, Akhter M S and Iftikhar F J 2020 A review of renewable energy generation using modified titania for photocatalytic water splitting *AIP Adv.* **10** 70701

Bach Q-V, Sillero M V, Tran K-Q and Skjermo J 2014 Fast hydrothermal liquefaction of a Norwegian macro-alga: screening tests *Algal Res.* **6** 271–6

Baig N 2023 Two-dimensional nanomaterials: a critical review of recent progress, properties, applications, and future directions *Composites* A **165** 107362

Bell S, Will G and Bell J 2013 Light intensity effects on photocatalytic water splitting with a titania catalyst *Int. J. Hydrogen Energy* **38** 6938–47

Benkö G, Skårman B, Wallenberg R, Hagfeldt A, Sundström V and Yartsev A P 2003 Particle size and crystallinity dependent electron injection in fluorescein 27-sensitized TiO_2 films *J. Phys. Chem.* B **107** 1370–5

Bo L, He K, Tan N, Gao B, Feng Q, Liu J and Wang L 2017 Photocatalytic oxidation of trace carbamazepine in aqueous solution by visible-light-driven $Znln_2S_4$: performance and mechanism *J. Environ. Manage.* **190** 259–65

Byakodi M, Shrikrishna N S, Sharma R, Bhansali S, Mishra Y, Kaushik A and Gandhi S 2022 Emerging 0D, 1D, 2D, and 3D nanostructures for efficient point-of-care biosensing *Biosens. Bioelectron.* X **12** 100284

Cao L, Luo G, Zhang S and Chen J 2016 Bio-oil production from eight selected green landscaping wastes through hydrothermal liquefaction *RSC Adv.* **6** 15260–70

Caporgno M P, Pruvost J, Legrand J, Lepine O, Tazerout M and Bengoa C 2016 Hydrothermal liquefaction of *Nannochloropsis oceanica* in different solvents *Bioresour. Technol.* **214** 404–10

Chan Y H, Yusup S, Quitain A T, Tan R R, Sasaki M, Lam H L and Uemura Y 2015 Effect of process parameters on hydrothermal liquefaction of oil palm biomass for bio-oil production and its life cycle assessment *Energy Convers. Manage.* **104** 180–8

Chen W-H, Lee J E, Jang S-H, Lam S-S, Rhee G H, Jeon K-J, Hussain M and Park Y-K 2022 A review on the visible light active modified photocatalysts for water splitting for hydrogen production *Int. J. Energy Res.* **46** 5467–77

Chen W-T, Zhang Y, Zhang J, Yu G, Schideman L C, Zhang P and Minarick M 2014 Hydrothermal liquefaction of mixed-culture algal biomass from wastewater treatment system into bio-crude oil *Bioresour. Technol.* **152** 130–9

Deng Y, Shu S, Fang N, Wang R, Chu Y, Liu Z and Cen W 2023 One-pot synthesis of $SrTiO_3$–$SrCO_3$ heterojunction with strong interfacial electronic interaction as a novel photocatalyst for water splitting to generate H_2 *Chin. Chem. Lett.* **34** 107323

Duan P and Savage P E 2011 Hydrothermal liquefaction of a microalga with heterogeneous catalysts *Ind. Eng. Chem. Res.* **50** 52–61

Eboibi B E-O, Lewis D M, Ashman P J and Chinnasamy S 2014 Hydrothermal liquefaction of microalgae for biocrude production: improving the biocrude properties with vacuum distillation *Bioresour. Technol.* **174** 212–21

Edelmannová M, de los Milagros Ballari M, Přibyl M and Kočí K 2021 Experimental and modelling studies on the photocatalytic generation of hydrogen during water-splitting over a commercial TiO_2 photocatalyst P25 *Energy Convers. Manage.* **245** 114582

Eidsvåg H, Bentouba S, Vajeeston P, Yohi S and Velauthapillai D 2021 TiO_2 as a photocatalyst for water splitting—an experimental and theoretical review *Molecules* **26** 1687

Enzweiler H, Yassue-Cordeiro P H, Schwaab M, Barbosa-Coutinho E, Olsen Scaliante M H N and Fernandes N R C 2020 Catalyst concentration, ethanol content and initial pH effects on hydrogen production by photocatalytic water splitting *J. Photochem. Photobiol.* A **388** 112051

Fahmy T Y A, Fahmy Y, Mobarak F, El-Sakhawy M and Abou-Zeid R E 2020 Biomass pyrolysis: past, present, and future *Environ., Dev. Sustain.* **22** 17–32

Gai C, Dong Y and Zhang T 2013 The kinetic analysis of the pyrolysis of agricultural residue under non-isothermal conditions *Bioresour. Technol.* **127** 298–305

Gollakota A R K, Kishore N and Gu S 2018 A review on hydrothermal liquefaction of biomass *Renew. Sustain. Energy Rev.* **81** 1378–92

Grafmüller J *et al* 2022 Wood ash as an additive in biomass pyrolysis: effects on biochar yield, properties, and agricultural performance *ACS Sustain. Chem. Eng.* **10** 2720–9

Grushevskaya S, Belyanskaya I and Kozaderov O 2022 Approaches for modifying oxide-semiconductor materials to increase the efficiency of photocatalytic water splitting *Materials* **15** 4915

Hadhoum L, Balistrou M, Burnens G, Loubar K and Tazerout M 2016 Hydrothermal liquefaction of oil mill wastewater for bio-oil production in subcritical conditions *Bioresour. Technol.* **218** 9–17

Han H *et al* 2017 α-Fe_2O_3/TiO_2 3D hierarchical nanostructures for enhanced photoelectrochemical water splitting *Nanoscale* **9** 134–42

Han X, Liu P, Ran R, Wang W, Zhou W and Shao Z 2022 Non-metal fluorine doping in Ruddlesden–Popper perovskite oxide enables high-efficiency photocatalytic water splitting for hydrogen production *Mater. Today Energy* **23** 100896

Hognon C, Delrue F, Texier J, Grateau M, Thiery S, Miller H and Roubaud A 2015 Comparison of pyrolysis and hydrothermal liquefaction of *Chlamydomonas reinhardtii*. Growth studies on the recovered hydrothermal aqueous phase *Biomass Bioenergy* **73** 23–31

Hu M, Shu J, Xu L, Liu C and Feng Q 2022 A novel nonmetal intercalated high crystalline g-C_3N_4 photocatalyst for efficiency enhanced H_2 evolution *Int. J. Hydrogen Energy* **47** 11841–52

Hu Y, Feng S, Xu C (C) and Bassi A 2017 Production of low-nitrogen bio-crude oils from microalgae pre-treated with pre-cooled NaOH/urea solution *Fuel* **206** 300–6

Jawhari A H 2023 One-pot facile synthesis of ZrO_2–$CdWO_4$: a novel nanocomposite for hydrogen production via photocatalytic water splitting *Appl. Sci.* **13** 300–6

Juliya A P, Mujeeb V M A, Sreenivasan K P and Muraleedharan K 2021 Enhanced H_2 evolution via photocatalytic water splitting using mesoporous $TiO_2/RuO_2/CuO$ ternary nanomaterial *J. Photochem. Photobiol.* **8** 100076

Kazawadi D, Ntalikwa J and Kombe G 2021 A review of intermediate pyrolysis as a technology of biomass conversion for coproduction of biooil and adsorption biochar *J. Renew. Energy* **2021** 5533780

Kerzig C and Wenger O S 2019 Reactivity control of a photocatalytic system by changing the light intensity *Chem. Sci.* **10** 11023–9

Khalid N R, Israr Z, Tahir M B and Iqbal T 2020 Highly efficient Bi_2O_3/MoS_2 p–n heterojunction photocatalyst for H_2 evolution from water splitting *Int J. Hydrogen Energy* **45** 8479–89

Khatun M, Mitra P and Mukherjee S 2023 Effect of band gap and particle size on photocatalytic degradation of $NiSnO_3$ nanopowder for some conventional organic dyes *Hybrid Adv.* **4** 100079

Kunthakudee N, Puangpetch T, Ramakul P, Serivalsatit K and Hunsom M 2022 Light-assisted synthesis of Au/TiO_2 nanoparticles for H_2 production by photocatalytic water splitting *Int. J. Hydrogen Energy* **47** 23570–82

Lachos-Perez D, César Torres-Mayanga P, Abaide E R, Zabot G L and De Castilhos F 2022 Hydrothermal carbonization and liquefaction: differences, progress, challenges, and opportunities *Bioresour. Technol.* **343** 126084

Leng L, Zhang W, Peng H, Li H, Jiang S and Huang H 2020 Nitrogen in bio-oil produced from hydrothermal liquefaction of biomass: a review *Chem. Eng. J.* **401** 126030

Li X, Xiong J, Gao X, Huang J, Feng Z, Chen Z and Zhu Y 2019 Recent advances in 3D g-C_3N_4 composite photocatalysts for photocatalytic water splitting, degradation of pollutants and CO_2 reduction *J. Alloys Compd.* **802** 196–209

Lin H, Zhang K, Yang G, Li Y, Liu X, Chang K, Xuan Y and Ye J 2020 Ultrafine nano 1T-MoS_2 monolayers with NiO_x as dual co-catalysts over TiO_2 photoharvester for efficient photocatalytic hydrogen evolution *Appl. Catal.* B **279** 119387

Liu S, Jaffrezic N and Guillard C 2008 Size effects in liquid-phase photo-oxidation of phenol using nanometer-sized TiO_2 catalysts *Appl. Surf. Sci.* **255** 2704–9

Liu T, Xiong Y, Wang X, Xue Y, Liu W and Tian J 2023 Dual cocatalysts and vacancy strategies for enhancing photocatalytic hydrogen production activity of $Zn_3In_2S_6$ nanosheets with an apparent quantum efficiency of 66.20% *J. Colloid Interface Sci.* **640** 31–40

Liu Z, Yu Y, Zhu X, Fang J, Xu W, Hu X, Li R, Yao L, Qin J and Fang Z 2022 Semiconductor heterojunctions for photocatalytic hydrogen production and Cr(VI) reduction a review *Mater. Res. Bull.* **147** 111636

Lu Q, Abdelgawad A, Li J and Eid K 2022 Non-metal-doped porous carbon nitride nanostructures for photocatalytic green hydrogen production *Int. J. Mol. Sci.* **23** 15129

Maddi B, Panisko E, Wietsma T, Lemmon T, Swita M, Albrecht K and Howe D 2016 Quantitative characterization of the aqueous fraction from hydrothermal liquefaction of algae *Biomass Bioenergy* **93** 122–30

Maeda K 2011 Photocatalytic water splitting using semiconductor particles: history and recent developments *J. Photochem. Photobiol.* C **12** 237–68

Maeda K and Domen K 2007 New non-oxide photocatalysts designed for overall water splitting under visible light *J. Phys. Chem.* C **111** 7851–61

Makuła P, Pacia M and Macyk W 2018 How to correctly determine the band gap energy of modified semiconductor photocatalysts based on UV–Vis spectra *J. Phys. Chem. Lett.* **9** 6814–7

Meryemoğlu B, Hasanoğlu A, Irmak S and Erbatur O 2014 Biofuel production by liquefaction of kenaf (*Hibiscus cannabinus* L.) biomass *Bioresour. Technol.* **151** 278–83

Molaei M J 2024 Recent advances in hydrogen production through photocatalytic water splitting: a review *Fuel* **365** 131159

Moniruddin M, Afroz K, Shabdan Y, Bizri B and Nuraje N 2017 Hierarchically 3D assembled strontium titanate nanomaterials for water splitting application *Appl. Surf. Sci.* **419** 886–92

Mursyalaat V, Variani V I, Arsyad W O S and Firihu M Z 2023 The development of program for calculating the band gap energy of semiconductor material based on UV–vis spectrum using Delphi 7.0 *J. Phys. Conf. Ser.* **2498** 012042

Naterer G F 2022 *Advanced Heat Transfer* 3rd edn (Boca Raton, FL: CRC Press)

Ohtani B, Kakimoto M, Nishimoto S and Kagiya T 1993 Photocatalytic reaction of neat alcohols by metal-loaded titanium(IV) oxide particles *J. Photochem. Photobiol.* A **70** 265–72

Prahov L T, Disdier J, Herrmann J-M and Pichat P 1984 Room temperature hydrogen production from aliphatic alcohols over UV-illuminated powder Ni/TiO_2 catalysts *Int. J. Hydrogen Energy* **9** 397–403

Qiao Y, Xu F, Xu S, Yang D, Wang B, Ming X, Hao J and Tian Y 2018 Pyrolysis characteristics and kinetics of typical municipal solid waste components and their mixture: analytical TG-FTIR study *Energy Fuels* **32** 10801–12

Rahman M, Tian H and Edvinsson T 2020 Revisiting the limiting factors for overall water-splitting on organic photocatalysts *Angew. Chem. Int. Ed.* **59** 16278–93

Reddy H K *et al* 2016 Temperature effect on hydrothermal liquefaction of *Nannochloropsis gaditana* and *Chlorella* sp *Appl. Energy* **165** 943–51

Reli M, Ambrožová N, Valášková M, Edelmannová M, Čapek L, Schimpf C, Motylenko M, Rafaja D and Kočí K 2021 Photocatalytic water splitting over CeO_2/Fe_2O_3/Ver photocatalysts *Energy Convers. Manage.* **238** 114156

Ross A B, Biller P, Kubacki M L, Li H, Lea-Langton A and Jones J M 2010 Hydrothermal processing of microalgae using alkali and organic acids *Fuel* **89** 2234–43

Saber M, Nakhshiniev B and Yoshikawa K 2016 A review of production and upgrading of algal bio-oil *Renew. Sustain. Energy Rev.* **58** 918–30

Samsudin E M and Abd Hamid S B 2017 Effect of band gap engineering in anionic-doped TiO_2 photocatalyst *Appl. Surf. Sci.* **391** 326–36

Sanchez-Martinez A, Ortiz-Beas J P, Huerta-Flores A M, López-Mena E R, Pérez-Álvarez J and Ceballos-Sanchez O 2021 ZnSe nanoparticles prepared by coprecipitation method for photocatalytic applications *Mater. Lett.* **282** 128702

Schneider J and Bahnemann D W 2013 Undesired role of sacrificial reagents in photocatalysis *J. Phys. Chem. Lett.* **4** 3479–83

Singh R, Balagurumurthy B, Prakash A and Bhaskar T 2015a Catalytic hydrothermal liquefaction of water hyacinth *Bioresour. Technol.* **178** 157–65

Singh R, Chaudhary K, Biswas B, Balagurumurthy B and Bhaskar T 2015b Hydrothermal liquefaction of rice straw: effect of reaction environment *J. Supercrit. Fluids* **104** 70–5

Sivagami M and Asharani I V 2022 Phyto-mediated Ni/NiO NPs and their catalytic applications —a short review *Inorg. Chem. Commun.* **145** 110054

Tian C, Liu Z, Zhang Y, Li B, Cao W, Lu H, Duan N, Zhang L and Zhang T 2015 Hydrothermal liquefaction of harvested high-ash low-lipid algal biomass from Dianchi Lake: effects of operational parameters and relations of products *Bioresour. Technol.* **184** 336–43

Ullah S, Shabir M, Rasheed M A, Ahmad I, Ahmed E, Ahmad M, Khalid N R and Khan W Q 2023 Silver and yttrium co-doped ZnO nanoparticles as a potential water splitting photocatalyst for the H_2 evolution reaction *J. Sol-Gel Sci. Technol.* **108** 756–67

University of Utah 2024 Percent crystallinity by the XRD integration method *Materials Characterization Lab, University of Utah* https://mcl.mse.utah.edu/xrd-crystallinity-by-integration/

Wang G, van den Berg R, de Mello Donega C, de Jong K P and de Jongh P E 2016 Silica-supported Cu_2O nanoparticles with tunable size for sustainable hydrogen generation *Appl. Catal.* B **192** 199–207

Wang H *et al* 2023 High quantum efficiency of hydrogen production from methanol aqueous solution with PtCu–TiO_2 photocatalysts *Nat. Mater.* **22** 619–26

Wang J and Wang S 2022 A critical review on graphitic carbon nitride (g-C_3N_4)-based materials: preparation, modification and environmental application *Coord. Chem. Rev.* **453** 214338

Wang X, Sø L, Su R, Wendt S, Hald P, Mamakhel A, Yang C, Huang Y, Iversen B B and Besenbacher F 2014 The influence of crystallite size and crystallinity of anatase nanoparticles on the photo-degradation of phenol *J. Catal.* **310** 100–8

Wei Q, Yang Y, Liu H, Hou J, Liu M, Cao F and Zhao L 2018 Experimental study on direct solar photocatalytic water splitting for hydrogen production using surface uniform concentrators *Int. J. Hydrogen Energy* **43** 13745–53

Xiu S, Shahbazi A, Shirley V and Cheng D 2010 Hydrothermal pyrolysis of swine manure to bio-oil: effects of operating parameters on products yield and characterization of bio-oil *J. Anal. Appl. Pyrolysis* **88** 73–9

Xu D and Savage P E 2017 Effect of temperature, water loading, and Ru/C catalyst on water-insoluble and water-soluble biocrude fractions from hydrothermal liquefaction of algae *Bioresour. Technol.* **239** 1–6

Xu J, Wang W, Sun S and Wang L 2012 Enhancing visible-light-induced photocatalytic activity by coupling with wide-band-gap semiconductor: a case study on Bi_2WO_6/TiO_2 *Appl. Catal.* B **111–2** 126–32

Yang L and Liu Z 2007 Study on light intensity in the process of photocatalytic degradation of indoor gaseous formaldehyde for saving energy *Energy Convers. Manage.* **48** 882–9

Yasuda M, Matsumoto T and Yamashita T 2018 Sacrificial hydrogen production over TiO_2-based photocatalysts: polyols, carboxylic acids, and saccharides *Renew. Sustain. Energy Rev.* **81** 1627–35

Zhang Y, Zhang N, Tang Z-R and Xu Y-J 2012 Graphene transforms wide band gap ZnS to a visible light photocatalyst. The new role of graphene as a macromolecular photosensitizer *ACS Nano* **6** 9777–89

Zhao H, Fu H, Yang X, Xiong S, Han D and An X 2022 MoS_2/CdS rod-like nanocomposites as high-performance visible light photocatalyst for water splitting photocatalytic hydrogen production *Int. J. Hydrogen Energy* **47** 8247–60

IOP Publishing

Renewable Energy Systems
The way forward
David S-K Ting and Jacqueline A Stagner

Chapter 6

A proposal for the typological categorization of architectural integration approaches for wind energy in tall towers

Hüseyin Emre Ilgın and Mehmet Halis Gunel

A notable contribution to the energy demands of urban infrastructure is being made by wind turbines, which harness wind for electricity generation. Currently, wind energy is harnessed through wind turbines, predominantly situated on tall structures within wind farms, and occasionally incorporated into lofty edifices due to the heightened wind speeds at greater elevations. In the foreseeable future, wind energy is poised to be harnessed through wind turbines seamlessly integrated into the architectural fabric of both newly constructed and existing tall buildings. This evolving trend underscores the growing significance of wind-powered architectural design as a prominent embodiment of renewable technology. This chapter endeavors to delve into the future prospects of wind energy integration within urban settings, particularly from an architectural perspective. Within this framework, an all-encompassing and distinctive typological framework is introduced for the integration of wind energy into tall buildings. This proposed classification system is anticipated to be more encompassing and harmonious compared to alternative categorizations. Furthermore, it is expected that this proposed framework will offer architectural practitioners guidance and encouragement in the conceptualization and execution of building projects that seamlessly incorporate wind energy technology.

6.1 Introduction

In the current era, the global population is increasing steadily, necessitating greater energy for routine activities such as household appliances, agriculture, industries, and transportation. Energy plays a pivotal role in the socio-economic advancement of a nation, enhancing efficiency in human endeavors. However, the conventional

doi:10.1088/978-0-7503-6179-8ch6

energy generation from fossil fuels such as coal and oil not only poses environmental concerns but also presents other adverse impacts.

Thus, the significance of renewable energy has increased considerably. Architects and engineers are compelled to engage in thorough investigations regarding the potential utilization of renewable energy within urban settings. In this sphere, the integration of wind energy into architectural design has emerged as a burgeoning trend. However, it is imperative that the local wind climate (wind resource) of the region and the building's environmental context align suitably with the efficacy of the wind energy system. Under such circumstances, the incorporation of wind turbines into building design becomes a focal point, ensuring the viable and effective deployment of wind turbines.

In anticipation of future energy requirements, wind turbines stand as a promising and economically viable proposition for integrating power sources into buildings. This potential offers the advantage of diminishing dependence on grid-provided electricity while preserving the integrity of the natural surroundings [1, 2]. The incorporation of wind energy into the constructed environment holds appeal, as wind turbines, leveraging a highly predictable and environmentally friendly energy source—wind—could operate for approximately 85% of the year, maintaining operational effectiveness for a minimum service life of 20 years [3]. Typically, the placement strategy for wind turbines within the constructed environment tends to prioritize aesthetic, architectural, or regulatory factors over a performance-centered methodology grounded in the accessible wind energy resource.

The seamless integration of wind turbines into tall buildings presents a multifaceted challenge, requiring a sophisticated interplay of engineering, environmental science, and architectural design to achieve a harmonious synergy between urban infrastructure and sustainable energy solutions [4–7]. Noise reduction strategies transcend traditional methods, encompassing meticulous choices of sound-absorbing materials and the optimization of aerodynamic profiles [8]. Going beyond these fundamental aspects, state-of-the-art technology comes into play with the integration of data-driven algorithms [9]. These algorithms enable turbines to adjust their operation dynamically in real-time, responding to the ever-changing noise environment in their vicinity. This adaptive approach not only maximizes the turbines' operational efficiency but also minimizes potential disturbances to nearby communities, exemplifying a forward-thinking balance between technological innovation and the preservation of the urban living experience. The successful execution of such integrated systems reflects a transformative step towards environmentally conscious urban development, where renewable energy infrastructure seamlessly coexists with architectural aesthetics and community well-being.

Addressing the challenge of vibration in the integration of wind turbines into tall buildings involves deploying advanced structural engineering solutions that go beyond conventional methods [10–12]. These solutions include the incorporation of sophisticated dampening systems and the utilization of smart materials specifically designed to dissipate and absorb vibrations. This comprehensive approach not only ensures the preservation of the structural integrity of the building but also enhances the durability and longevity of the turbines themselves. The incorporation of predictive maintenance systems further underscores the commitment to safety and sustainability. By leveraging sensors and artificial intelligence, these systems continuously monitor the health and performance of the turbines, allowing for the early detection of potential issues before they escalate into significant problems [13–15]. This predictive maintenance strategy not only minimizes the risk of operational disruptions and malfunctions but also maximizes the operational lifespan of the turbines, contributing substantially to the overall long-term sustainability and efficiency of the integrated wind energy system.

Designing turbines for tall buildings necessitates a meticulous equilibrium in blade sizing, accounting for critical factors such as aerodynamic efficiency, structural load distribution, and adaptability to fluctuating wind conditions [16–20]. Achieving this balance is paramount for optimizing energy capture while ensuring that the building structure is not unduly stressed. The integration of advanced materials, such as carbon composites, plays a pivotal role by enhancing the strength-to-weight ratio of turbine blades, thereby improving overall efficiency. This not only allows for increased energy production but also contributes to the structural integrity of the building. Beyond performance considerations, the design must incorporate foresight for maintenance requirements. Features such as automated inspection platforms and modular components are essential, streamlining and facilitating routine servicing. By prioritizing maintenance considerations in the design phase, the turbines become more resilient, ensuring prolonged operational life, minimized downtime, and ultimately enhancing the overall sustainability of the integrated wind energy system within tall buildings.

The assessment of wind speed and availability in the integration of wind turbines into tall buildings requires a profound comprehension of local wind patterns and microclimates [21–23]. Achieving this understanding is crucial for optimizing energy harnessing. Advanced meteorological modeling, coupled with sensor networks, emerges as a cornerstone in this process, providing precise data that enables accurate predictions of wind behavior. This ensures that the turbines can capitalize on the maximum available energy in their specific location. Concurrently, addressing wind shadow and turbulence becomes paramount in mitigating aerodynamic challenges introduced by the building's presence. The utilization of computational fluid dynamics simulations and real-world measurements becomes instrumental in refining turbine placement and design. By systematically analysing these aerodynamic factors, engineers can fine-tune the positioning and characteristics of the turbines, maximizing their efficiency and overall energy output while minimizing the impact of the building on the aerodynamic performance of the turbines. This holistic approach underscores the importance of merging cutting-edge technology and scientific understanding to optimize the integration of wind turbines into urban landscapes.

Ultimately, the successful integration of wind turbines into tall buildings transcends mere technological innovation; it demands a holistic and collaborative approach that converges expertise from diverse fields. By bringing together professionals in engineering, environmental science, architecture, and more, a synergistic effort can be orchestrated to craft sustainable, efficient, and environmentally conscious energy solutions that harmoniously enhance the urban landscape. This comprehensive integration not only transforms buildings into active contributors to energy generation but also sets a precedent and blueprint for future sustainable urban development. Beyond the immediate benefits of renewable energy production, such integrated solutions pave the way for urban environments that prioritize environmental stewardship, community well-being, and energy resilience. The collaborative fusion of expertise becomes a catalyst for a new era in urban design, where the integration of renewable technologies seamlessly becomes part of the fabric of sustainable city planning, offering a model for others to emulate in the pursuit of a greener and more resilient future.

In the intricate nexus of turbine designs, tall buildings, and urban context, the integration of wind energy stands as a catalytic force driving transformative advancements in the pursuit of sustainable urban development. Scientifically, this integration transcends mere reduction in carbon footprint and clean energy generation, evolving into a dynamic field that orchestrates a symphony of multidimensional opportunities. Analysing the intricacies of wind patterns, structural engineering, and urban climatology, researchers delve into optimizing the synergy between turbines and tall structures, not only to maximize energy output but also to influence the microclimates of urban environments. This scientific exploration extends into architectural innovation, considering the aerodynamic efficiency of turbine designs and their harmonious integration within the urban landscape. Moreover, from a socio-economic perspective, this integration is investigated for its potential to stimulate economic growth, foster community engagement, and redefine urban living standards. Thus, the scientific discourse surrounding the confluence of turbine designs, tall buildings, and urban context becomes a pioneering venture, revealing the multifaceted and interconnected dimensions through which wind energy integration holds the promise of reshaping the sustainable future of urban life.

The architectural integration of turbines into tall buildings represents a pivotal shift in design thinking, compelling architects and engineers to reconsider traditional paradigms. This paradigmatic shift goes beyond the mere addition of functional elements; it necessitates a holistic approach that harmonizes form and function. In the pursuit of optimal structural stability and the efficient capture of wind energy, designers are driven to explore innovative architectural solutions that seamlessly integrate renewable energy technologies. This exploration extends to the aerodynamic refinement of turbine placement, the integration of cutting-edge materials, and the development of responsive systems that adapt to dynamic wind conditions. As architects reimagine the visual language of tall buildings with integrated turbines, a transformative design evolution unfolds—one that challenges conventional notions of aesthetics. This evolution serves as a catalyst for a new era in urban design, where the marriage of sustainability and beauty becomes not only feasible but also essential. The tall building, once perceived as a static element in the urban landscape, is now envisaged as a dynamic, energy-generating structure that contributes to the visual and environmental identity of the cityscape. In this way, the integration of turbines into tall buildings becomes a powerful agent in redefining urban aesthetics and fostering a sustainable and visually striking built environment.

Expanding on the intricate impact of wind turbines on local microclimates, the integration of tall buildings into this dynamic relationship adds a layer of complexity that extends well beyond the immediate realm of energy considerations. These tall structures, acting as both wind concentrators and deflectors, wield a profound influence on the surrounding atmospheric conditions, thereby presenting a unique opportunity for holistic environmental enhancement. The potential to modulate temperature variations and air quality within the immediate vicinity of these structures holds significant implications for urban climate management. Notably, the strategic deployment of wind turbines alongside tall buildings can contribute to the mitigation of urban heat islands, addressing a prevalent challenge in densely populated urban areas. Furthermore, the combined effect of these architectural elements opens avenues for enhancing natural ventilation patterns, promoting air circulation, and potentially contributing to localized climate regulation. The scientific exploration of these possibilities necessitates a nuanced understanding of meteorological patterns, fluid dynamics, and the intricacies of urban heat dispersion. Through this multidisciplinary lens, researchers can decipher the intricate interplay between tall buildings, wind turbines, and the urban microclimate, paving the way for innovative solutions that transcend conventional boundaries and foster sustainable urban environments.

Within the expansive realm of urban development, the integration of wind energy serves as a critical linchpin within the overarching discourse on smart cities. Scientifically, this integration represents a convergence of wind turbines with sophisticated technologies such as energy storage systems, smart grids, and Internet of Things (IoT) devices. The synergy created by this amalgamation establishes a foundation for a highly interconnected and adaptive urban infrastructure, thereby constituting a paradigm shift in the management and utilization of energy resources. Scientific investigations in this realm delve into the optimization of energy storage mechanisms, employing advanced materials and innovative designs to enhance efficiency and durability. Simultaneously, the scientific exploration of smart grids involves the application of data analytics, machine learning algorithms, and real-time monitoring to intelligently balance energy supply and demand. Additionally, the integration of IoT devices into this scientific framework necessitates rigorous research into communication protocols, edge computing capabilities, and cybersecurity measures, aiming to create a resilient and secure network of interconnected devices. This comprehensive scientific approach not only improves energy efficiency but also contributes to the creation of urban environments that are not only resilient but also responsive, capable of dynamically adapting to the ever-evolving needs and challenges of contemporary society.

Navigating the intricate terrain of the intersectionality between turbine designs, tall buildings, and urban context requires a collaborative approach that serves as a cornerstone for unlocking the full potential of this integrated endeavor. Scientifically, the complexity of this intersectionality necessitates cross-disciplinary partnerships that transcend traditional boundaries, bringing together the diverse expertise of architects, engineers, environmental scientists, urban planners, economists, and local communities.

The collaborative synergy that emerges from such partnerships is essential for developing a comprehensive understanding of the multifaceted challenges and opportunities inherent in the integration of wind energy into urban landscapes. The scientific exploration within these collaborative frameworks extends beyond the technical realm, encompassing in-depth analyses of social dynamics, economic viability, and cultural considerations that intricately shape the fabric of urban life.

From a scientific perspective, these collaborative efforts facilitate a more holistic approach to sustainable urban development. The study of turbine designs, their interaction with tall buildings, and their impact on the urban context requires not only technical expertise but also a nuanced understanding of the socio-economic and cultural aspects. Engineers and architects work in tandem to optimize turbine efficiency and integration with building structures, while environmental scientists assess the ecological impact and potential benefits. Urban planners contribute insights into the spatial considerations and the implications for the overall cityscape, and economists evaluate the economic viability and long-term sustainability of such projects.

Scientific exploration within these collaborative frameworks extends beyond the technical realm, encompassing in-depth analyses of social dynamics, economic viability, and cultural considerations that intricately shape the fabric of urban life. Through these collaborative scientific endeavors, a holistic approach to sustainable urban development emerges—one that not only addresses the technical intricacies of turbine integration but also accounts for the broader social, economic, and cultural dimensions, fostering a more inclusive and resilient urban future.

In conclusion, the intersectionality of turbine designs, tall buildings, and urban context heralds a new era in sustainable urban development. Beyond its immediate impact on energy generation, it sparks a renaissance in design thinking, fosters climate-responsive urban environments, and propels the evolution of smart cities. As we stand at the precipice of this transformative journey, the integration of wind energy into the urban landscape becomes not merely a technological endeavor but a collective aspiration towards a more harmonious and sustainable future for our cities. The collaborative and interdisciplinary nature of these scientific efforts ensures that the integration of wind energy is not only technically efficient but also socially, economically, and culturally resonant with the diverse needs of urban communities.

Within this chapter, an extensive typological categorization for the incorporation of wind energy within architectural structures is put forward. This proposed classification is deemed to exhibit greater intricacy and comprehensiveness in contrast to alternative classifications. It is notable that this synthesis should primarily emphasize the establishment of an aesthetic and conceptually coherent integration of buildings with wind turbines, surpassing the mere fulfillment of technical, financial, or environmental considerations.

6.2 Literature review

Within the existing literature, a prevailing inclination among researchers, particularly within the engineering domain, is to predominantly emphasize the technical functionality and economic viability of wind turbines. In contrast, only a restricted subset of studies has delved extensively into the interrelated considerations pertaining to the architectural assimilation of wind turbines into the constructed environment, encompassing their aesthetic attributes and spatial implications.

Campbell *et al* [24] concentrated on strategies for enhancing and incorporating wind energy to elevate the annual energy production. This encompassed considerations of visual aspects, aerodynamics, architectural compatibility, environmental factors, and structural aspects. The practical manifestation of these principles materialized through experimentation involving a two-story prototype building featuring an integrated wind turbine. The research yielded guidelines encompassing the conceptual phases of design, methodologies for energy yield projection, as well as the classification and evaluation of environmental repercussions and financial outlays. Additionally, three overarching integration techniques emerged within urban landscapes, which encompassed:

1. Full integration, whereby wind turbines profoundly influence the architectural configuration.
2. Retrofitting of wind turbines onto pre-existing structures.
3. Strategic placement of standalone wind turbines within urban surroundings.

This aforementioned typology was also referenced within the investigation by Degrassi *et al* [25]. Their study featured a retrospective examination of the incorporation of small-scale wind turbines into architectural systems, predominantly tailored for urban contexts, with the presentation of representative case studies.

In the conclusive document titled 'Assessing the feasibility of wind turbines integrated and mounted onto buildings: realizing their potential for mitigating carbon emissions,' authored by Dutton *et al* [26], an evaluation of diverse configurations of wind turbines integrated and mounted onto buildings was conducted to ascertain their technical viability and energy generation capabilities. The study yielded the deduction that wind energy could hold a substantial capacity for meeting energy demands within the constructed environment. Furthermore, by employing a dual strategy involving both innovative structures equipped with custom-designed wind energy apparatus and the retrofitting of turbines onto pre-existing edifices, a prospective annual energy generation of up to 5.0 TWh from these integrated wind turbines in urban surroundings was envisioned by 2020.

In this comprehensive analysis, the term 'building-integrated/mounted wind turbine' was established as an encompassing nomenclature, encompassing any wind turbine seamlessly assimilated within the built environment. Within this classification, 'building-integrated wind turbines' referred to turbines strategically situated in close proximity to buildings, capitalizing on any synergies fostered by the edifice. Additionally, 'building-mounted wind turbines' denoted turbines intricately connected to the structural framework of the building. Lastly, 'building-augmented wind turbines' indicated instances where the building's design was intentionally harnessed to modify and enhance wind flow into the turbine system.

In the publication titled 'Harnessing wind energy within the constructed environment: concentration phenomena induced by buildings', authored by Mertens [27], the focus lay on energy preservation strategies within urban settings, with a specific emphasis on the symbiotic relationship between wind turbines and buildings. The core concept explored was the optimization of wind energy collection through three distinct approaches: 'proximal positioning of wind turbines to buildings', 'integration of wind turbines within a configuration of aerofoil-shaped buildings', and 'incorporation of wind turbines within ducts passing through buildings'.

A study by Abohela *et al* [28] conducted an extensive review of existing research regarding the incorporation of wind turbines into urban settings, focusing primarily on three distinct thematic areas: 'strategies for integrating wind turbines within built environments', 'assessment of wind patterns within constructed environments using various methodologies', and the third thematic strand encompassing 'viability of integrating wind turbines into built environments concerning ecological, economic, and societal dimensions.'

In the context of the first thematic strand, drawing from the insights of Aguiló *et al* [29], three techniques for the incorporation of wind turbines within built environments were outlined:

1. Building-integrated wind turbines, involving the placement of an independent wind turbine on a standalone tower separated from the building itself.
2. Building-mounted wind turbines, wherein the wind turbine is affixed to the structure of the building.
3. Building-augmented wind turbines, characterized by architectural modifications designed to channel wind flow towards the wind turbine.

Drawing insights from an architectural studio context, Poerschke *et al* [30] undertook an exploration of concepts concerning wind energy integration within mid-rise structures in northwest Pennsylvania. This inquiry encompassed a multidimensional analysis, incorporating technical, environmental, and visual considerations, to ascertain wind-optimized architectural configurations and facilitate the harmonious visual alignment between wind turbines and building design. The architectural studio's investigation yielded several distinct avenues for the incorporation of wind power within the constructed environment, encompassing the following approaches: 'wind turbines affixed to rooftop surfaces', 'wind turbines positioned atop roof parapets', 'adoption of a dual-roof configuration to channel wind', 'integration into building façades', and 'integration within the surrounding landscape'. Poerschke *et al* [30] ultimately deduced that the fusion of small-scale wind turbines into building structures presented an underexplored research domain in relation to both energy performance and the aesthetic aspects, thus accentuating the potential for interdisciplinary collaboration among architects and engineers.

Park *et al* [31] introduced an innovative concept encompassing a novel system for integrating wind turbines within building structures, aiming to harness wind pressure via utilization of the building façade or exterior wall as an energy-generating mechanism within urban environments. Their study delineated the categorization of building-integrated wind turbines of substantial dimensions, concerning plausible placements within high-rise structures, as outlined below: positioned upon the rooftop, situated amidst two contiguous buildings, and housed within an aperture within the building's interior.

The objective of the undertaking by Haase and Löfström [32] was to formulate comprehensive integrated solutions for small wind turbines within the Norwegian context. This was accomplished by an exhaustive analysis of contemporary technologies, energy performance metrics, costs, and salient practices related to these wind turbine advancements, encompassing their associated challenges and potential benefits. The findings underscored the utmost significance of wind turbine placement, alongside considerations of turbine type and dimensions. Furthermore, the study identified the subsequent wind turbine classifications and domains of application within the realm of building-associated wind turbines:

1. Rotor varieties (horizontal and vertical axis wind turbines).
2. Building-mounted wind turbines (directly attached to the building's structural framework).
3. Building-integrated wind turbines (incorporated into the building's design in some capacity).
4. Building-augmented wind turbines (strategically integrated to modify and enhance airflow into the turbine via the building's architecture).

The above-stated classification was also employed in the research conducted by Filipowicz *et al* [33]. Their study provided an overview and data pertaining to wind turbine operations integrated with the Center of Energy AGH building. The outcomes highlighted that the proximity of wind turbines to the building could potentially influence the building's structural integrity yet exhibited minimal interference with the surrounding environment. Moreover, appropriately installed small wind turbines were shown to play a pivotal role during unforeseen circumstances by furnishing requisite energy supplies.

In the study by Vita *et al* [34], the attributes of flow patterns were analysed through wind tunnel experiments for a representative high-rise structure, exploring diverse arrangements and wind orientations. The aim was to enhance comprehension regarding wind energy potential within the constructed milieu and furnish designers with pertinent insights into optimal wind turbine placement strategies, thereby enhancing overall performance. The investigation revealed that, when situated 10 m above the rooftop, the airflow exhibits characteristics akin to atmospheric turbulence, displaying heightened turbulence intensity exceeding 10%, accompanied by substantial length scales around 200 m. Findings further indicated that arrays of tall buildings could offer an advantageous layout for incorporating urban wind turbines. However, a notable divergence in wind turbine efficiency was observed between a placement at the roof's center and positions at the leeward or windward corners or edges, contingent upon the prevailing wind direction.

Kumar and Prakash [35] aimed to analyse the utilization of micro wind turbines on tall buildings, specifically high-rise structures, serving as tower installations for these turbines. The investigation involves an examination of India's wind map to determine suitable locations for wind turbine installation. Through the utilization of wind maps, the airflow patterns at various elevations have been investigated. The outcomes of this study provide insights into the magnitude of wind energy generation across different Indian states.

Additionally, research consensus predominantly affirms that the prominent attributes of the flow pattern involve the augmentation of velocity and the amplification of turbulence intensity [36, 37]. Conversely, a relatively restricted body of investigation has concentrated on refining the architectural form of buildings to facilitate the integration of wind turbines within the building's overall design [38].

The analysis of the literature presented earlier reveals a noticeable absence of research that broadly addresses the architectural integration of wind turbines into buildings. This study endeavors to fill this void within the existing body of literature by concentrating this topic in a comprehensive manner.

6.3 Architectural integration strategies of wind energy to tall buildings

In this chapter, while acknowledging the precedent scholarly inquiries highlighted in the preceding section, the authors formulated a comprehensive typological categorization encompassing the architectural integration tactics of wind energy into buildings (depicted in figure 6.1). This classification is envisaged to possess enhanced inclusivity and coherence, catering to both extant structures and edifices in the developmental phase:

- Building-independent wind turbine utilization (contextually dependent, in the form of wind farms).

- Building-dependent/augmented wind turbine utilization.
    - Building-mounted/retrofitted wind turbine utilization (architecturally independent):
        - architectural form concerned.
        - architectural form unconcerned.

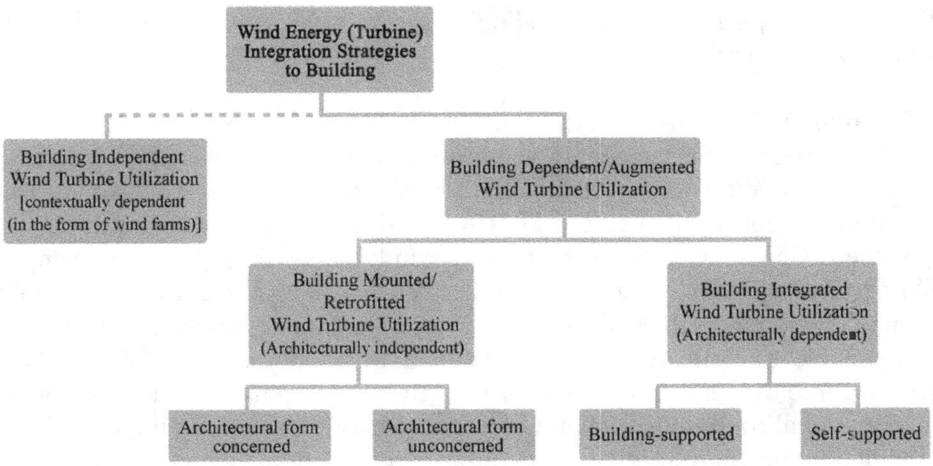

Figure 6.1. Architectural integration strategies of wind energy to buildings.

- Building-integrated wind turbine utilization (architecturally dependent):
  - building-supported.
  - self-supported (structurally independent).

This categorization hinges on the fundamental concept of architectural designs influenced by or propelling wind energy dynamics. Given that the linkage is predominantly contextual rather than constituting comprehensive or partial architectural integration, the cluster denoted as 'building-independent wind turbine utilization' is depicted as an auxiliary bubble connected to the principal diagram (depicted in figure 6.1) through dashed lines.

6.3.1 Building-independent wind turbine utilization

This category pertains to wind turbines that are situated in a manner that is both architecturally and structurally detached from the building, yet contextually linked to the built environment or landscape. This contextuality implies the placement of wind turbines in proximity to existing building(s) within the landscape, establishing a contextual association between the building and its surroundings. In this configuration, wind turbines do not leverage the building's structure to enhance wind flow into the turbine; instead, they remain self-supported and separate.

6.3.2 Building-dependent/augmented wind turbine utilization

Building-dependent/augmented wind turbine utilization encompasses the placement of wind turbines that are architecturally and/or structurally reliant on the building. This arrangement can be further divided into two categories:

- Utilization of building-mounted/retrofitted wind turbines.
- Utilization of building-integrated wind turbines.

In the former, augmentation is achieved through factors such as building height, as seen in building-mounted/retrofitted wind turbines, while in the latter, augmentation is accomplished by the building's design form, as observed in building-integrated wind turbines.

6.3.2.1 Building-mounted/retrofitted wind turbine utilization

This encompasses the placement of wind turbines that are structurally dependent or linked to the building, where the building's roof and side walls serve as supports for the turbines (depicted in figure 6.2). These turbines derive their operation from augmented wind flow, thereby harnessing wind energy through the influence of building height. Notably, building-mounted/retrofitted wind turbines find application in tall buildings or those situated at elevated locations. Moreover, these turbines are adaptable to both existing buildings and those in the design phase. When integrated into existing structures, measures are taken to ensure structural integrity and operational conditions, including aspects such as vibration and noise insulation.

Building-mounted/retrofitted wind turbines do not take advantage of potential enhancements in wind flow resulting from the aerodynamic configuration of the

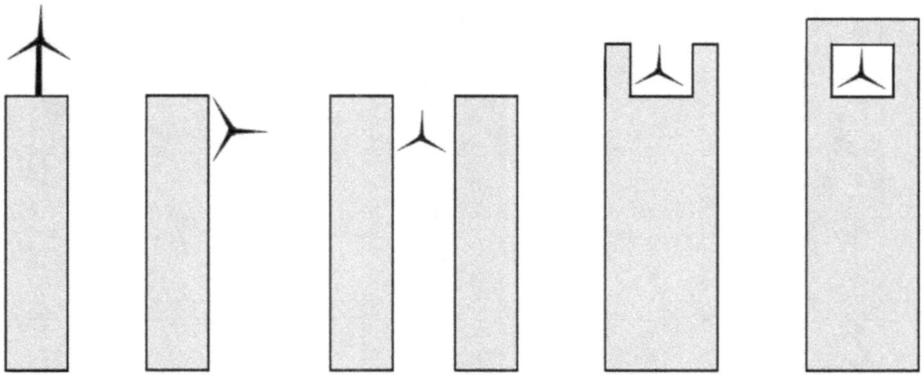

Figure 6.2. Multiple integration approaches for the building-mounted/retrofitted wind turbines.

building. This fundamental differentiation between building-mounted/retrofitted wind turbines and building-integrated wind turbines lies in their approach to harnessing wind energy through energy-oriented design principles.

Various integration tactics for these wind turbines are depicted in figure 6.2.

Building-mounted/retrofitted wind turbine utilization can be bifurcated into two categories:

- architectural form concerned, and
- architectural form unconcerned.

In the architectural form concerned subgroup, the wind turbine case is tailored to correspond with the building's architectural structure, considering aesthetic attributes, as seen in the instance of The Tower, One St George Wharf (London, 2014), a towering structure measuring 181 m in height (figure 6.3). The architectural layout of the tower is derived from the configuration of a Catherine wheel, characterized by a floor-plan design that encompasses five apartments on each floor, arranged around a central core. The structure features sky gardens, which offer inhabitants a partially external space extending from the central circular design. This arrangement introduces variations in the building's exterior, accentuating its height and adding intricacy to the otherwise minimalist cladding. The tower is categorized into three distinct segments: a lower portion housing communal amenities such as the lobby, business lounge, gym, spa, and swimming pool; a middle segment accommodating the majority of the apartments; and an upper section that narrows in diameter, allowing for 360° terraces and the incorporation of a wind turbine atop the edifice. This vertical axis wind turbine of approximately 10 m in height, designed to harmonize with the cylindrical building form, is utilized to power lighting in public areas [39].

Within the architectural form unconcerned subgroup, the wind turbine's integration is not factored into the architectural design. Instances of this can be observed in building façade integrated wind turbines (figure 6.4). These wind turbines are directly incorporated into the building's exterior surface, which remains under constant wind pressure. This innovative approach is exemplified by the work of [30, 31], who proposed this system that not only harnesses wind flow from predominant

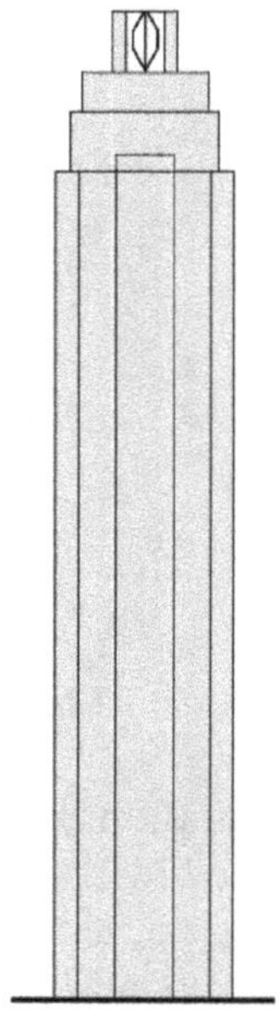

Figure 6.3. The Tower, One St George Wharf, London, 2014.

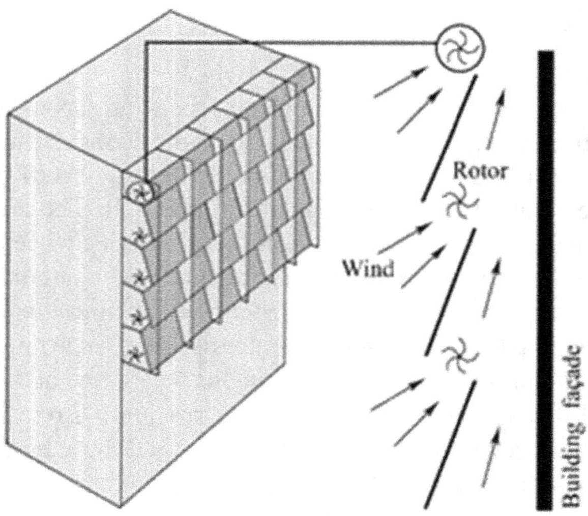

Figure 6.4. Building façade integrated wind turbines.

wind directions but also captures ascending airflow along the building's exterior. Due to the intricate nature of the wind patterns around the building, further investigation is warranted for this system.

6.3.2.2 Building-integrated wind turbine utilization

This category involves the placement of wind turbines that are architecturally linked to the building, although not mandatorily structurally dependent. Within this classification, the architectural design of the building(s) is fundamentally influenced by the utilization of wind energy. These wind turbines have the ability to leverage the aerodynamic building form, enhancing the energy generation efficiency of the turbine by directing and intensifying the wind flow into it.

To assess the potential of incorporating building-integrated wind turbines, the aerodynamic efficiency of building shapes is evaluated through methods such as wind tunnel testing, flow visualization techniques, and computational fluid dynamics simulations. In this context, the following key outcomes from these energy investigations have been highlighted [24]:

- Curved or smoothly contoured building forms can be advantageous for optimal performance.
- Buildings designed with a kidney or boomerang layout exhibit the highest wind augmentation potential.
- Aerodynamic wings or additions that create a three-dimensional aerodynamic duct structure around the turbine demonstrate significant efficiency, effectively preventing flow separation and wind augmentation loss.
- Notable wind enhancement is achievable within incidence angles of approximately 40°–50°, and substantial airflow through an aerodynamic ducted orifice can be achieved even at acute angles.

The category of building-integrated wind turbine utilization can be subdivided into two distinct groups:

- 'building-supported', both architecturally and structurally dependent to the building, and
- 'self-supported', architecturally dependent on the building but structurally independent from the building.

Building-supported wind turbines encompass wind turbines that are strategically positioned to leverage both architectural and structural aspects of the building. These turbines exploit the aerodynamic configuration of the building to channel and optimize wind flow, thereby enhancing their capacity for generating energy.

The rational arrangement of spaces with respect to their buffering function necessitates the inclusion of intermittently utilized or service areas (such as elevators, staircases, etc) in proximity to the turbine. This is crucial since the areas adjoining the turbine inherently possess lower value and lower utilization demands compared to other spaces.

A notable instance of building-supported wind turbines is exemplified by Bahrain World Trade Center (Manama, 2008) towering at 240 m (figure 6.5). This structure proudly stands as the world's 'first extensive electricity-generating wind turbine integrated tall building' [40, 41]. The core design concept was influenced by the aerodynamics of sails and traditional Arabian wind towers. In the initial stages of conception, the preliminary energy yield from three horizontal axis wind turbines, each with a diameter of 29 m, was projected to contribute around 15% of the building's total electrical energy consumption [42, 43]. The elliptical plan configuration effectively channels and intensifies the wind flow, enhancing wind speeds by

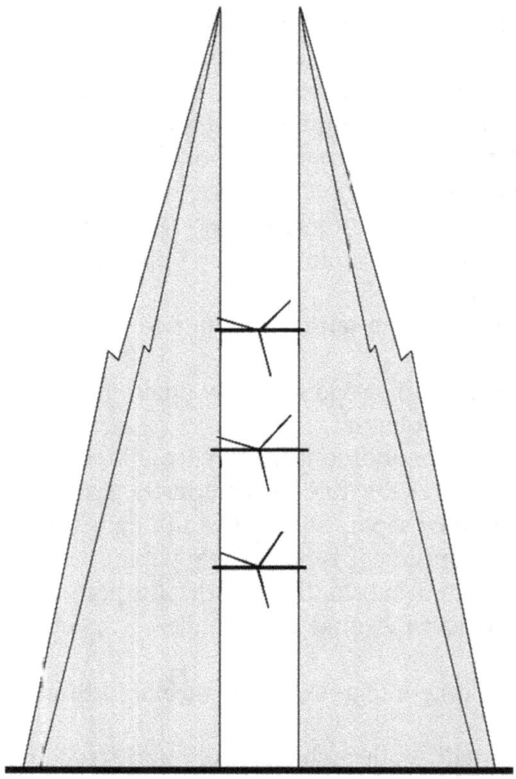

Figure 6.5. Bahrain World Trade Center, Manama, 2008.

up to 30% [44]. Furthermore, the tapered structure towards the upper section mitigates the impact of heightened wind flow. This combined effect, coupled with the lower wind speed at ground level, results in the nearly uniform rotation of all three wind turbines at comparable velocities, consequently yielding equivalent energy outputs.

Another remarkable illustration of building-supported wind turbines is exemplified by Strata, standing at a height of 148 m (figure 6.6), and widely recognized as the 'world's inaugural building featuring cladding-enclosed wind turbines' [39]. The tower encompasses a total of 408 apartments, complemented by four retail units situated at the ground level. Additionally, a pavilion structure is incorporated into the design, comprising three levels of shared ownership apartments and one level designated for housing plant facilities. These two buildings are linked by a singular basement, serving as a communal space for parking cars, motorcycles, and bicycles. The tower's triangular layout is characterized by two sides featuring outwardly curved, convex shapes adorned with aluminum and glass panels. In contrast, the third side displays a concave design, entirely glazed from top to bottom. As the structure ascends, its form is further emphasized by a distinctive architectural feature: the aluminum and glass panels on the outwardly curved surfaces are pulled back, unveiling a fully glazed façade. The architectural configuration and alignment

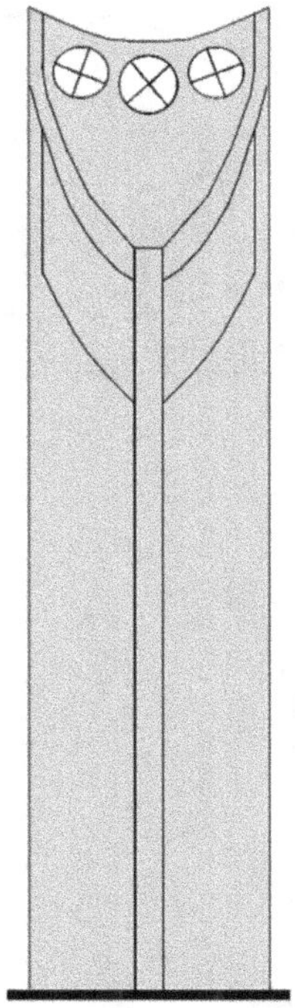

Figure 6.6. Strata, London, 2010.

of the structure strategically harness the predominant wind direction to optimal effect [45]. The trio of wind turbines situated atop the tower possess a maximum capacity of 19.5 kW each. Their anticipated output is no less than 50 MWh of electrical energy annually, constituting approximately 8% of the building's entire energy consumption [46].

An illustrative instance of building-supported wind turbines is embodied by Pearl River Tower, towering at a height of 309 m (figure 6.7), wherein its strikingly sculpted structure strategically channels wind through apertures situated at its mechanical floors, propelling turbines that generate energy for the edifice. The architectural design ingeniously facilitates the flow of air through these apertures, resulting in a twofold increase in wind velocity [47]. These wind turbines are projected to contribute approximately 1% of the building's overall energy requirements [48]. The dynamically curved configuration of the building, designed with a keen focus on performance, notably augments the efficacy of the turbines. Moreover, the external solar shading system and glass outer shell of the Pearl River Tower seamlessly incorporate a photovoltaic system. Within the cavity for solar shading and glare control, motorized sunshade devices are present. These devices are automated, responding to photocells that monitor the sun's position relative to the building's elevation. The Building Management Systems efficiently adjust the sunshade angles based on solar intensity, solar altitude angle, and solar azimuth angle, effectively minimizing solar heat. Similarly, a low-energy, highly efficient lighting system employs radiant panel geometry to facilitate light

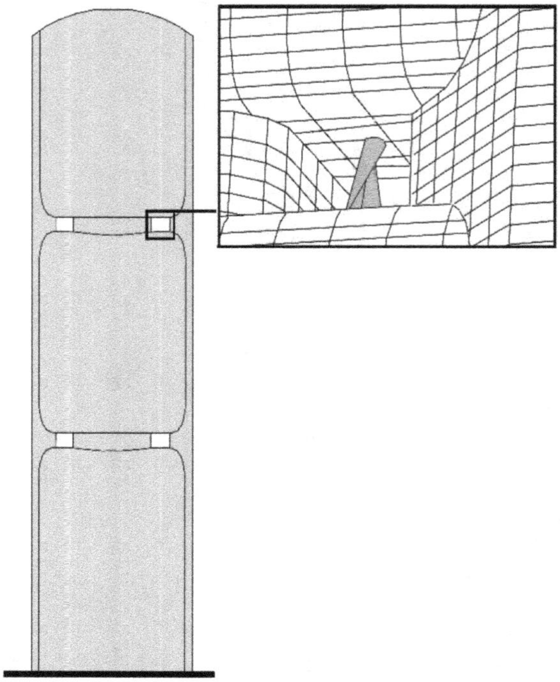

Figure 6.7. Pearl River Tower, Guangzhou, 2013.

distribution. Enhanced human thermal comfort, efficient heat exchange, and improved office acoustics are achieved through the radiant cooling system. Furthermore, the integration of daylight harvesting with automated blinds and daylight-responsive controls has led to substantial reductions in the building's mechanical, electrical, and lighting consumption, achieving a notable 45% reduction compared to the National Chinese Energy Code [49].

In pursuit of achieving its ambitious objective of attaining a 'zero energy target', Pertamina Energy Tower, soaring to a remarkable height of 523 m (figure 6.8), showcases a prominent illustration of building-supported wind turbines as a sustainable design strategy [50, 51]. This exceptional architectural endeavor transcends the conventional norms by harnessing wind power as a means to generate electricity, rather than merely mitigating or counteracting its effects, a characteristic that sets it apart. Functioning akin to an expansive wind farm, the tower incorporates vertical axis wind turbines that effectively harness the prevailing winds for power generation. The tower's design capitalizes on the prevailing wind patterns, transforming them into a renewable energy source. This innovative approach is manifested through a meticulously designed wind funnel situated at the gently tapering crown-shaped apex of the architecture. This design feature serves to amplify the flow of wind through the upper floors, optimizing wind capture over the turbines and consequently yielding a heightened energy output. Notably, this ingenious design enables the turbines to contribute up to 25% of the tower's total energy requirements, marking a substantial step toward its energy sustainability goals [52].

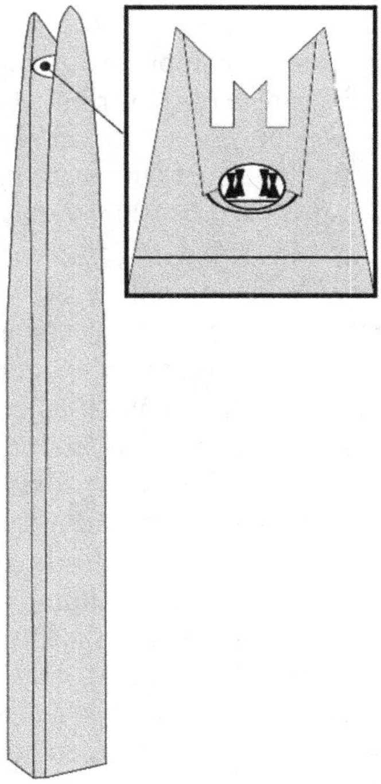

Figure 6.8. Pertamina Energy Tower, Jakarta, proposed, 2012.

It is worth noting that the incorporation of wind turbines into architectural structures such as the Bahrain World Trade Center, Strata, and Pearl River Tower serves a dual purpose, encompassing both aesthetic and symbolic elements. While these wind turbines contribute to the overall visual appeal of the buildings, their primary function is often symbolic rather than being a substantial source of power generation.

From a scientific perspective, the wind turbines integrated into these structures are designed with a focus on aesthetics, seamlessly blending into the architectural design to enhance the overall visual impact. The innovative incorporation of renewable energy elements into urban landscapes reflects a commitment to sustainability and environmental consciousness. The turbine's sleek and modern design, coupled with their strategic placement on the buildings, symbolizes a forward-thinking approach to energy solutions and a desire to showcase green technologies.

However, the practical contribution of these wind turbines to the actual power needs of the respective buildings is often limited. The architectural integration of wind turbines may face challenges such as inconsistent wind patterns, urban turbulence, and space limitations, which can hinder their efficiency in generating a substantial amount of electricity. In many cases, the energy generated by these turbines may be more symbolic than utilitarian, representing a gesture towards renewable energy rather than constituting a significant portion of the building's power supply.

Despite the limited power generation capacity of these integrated wind turbines, their symbolic value should not be underestimated. They serve as tangible representations of a commitment to sustainable practices and renewable energy sources, acting as educational tools for public awareness on environmental issues. Furthermore, these architectural endeavors contribute to a broader discourse on the integration of renewable energy solutions into urban planning, inspiring discussions on the feasibility and challenges associated with such implementations.

In conclusion, the integration of wind turbines into buildings like the Bahrain World Trade Center, Strata, and Pearl River Tower is a fascinating intersection of architecture and sustainability. While primarily serving aesthetic and symbolic purposes, these structures prompt important discussions about the potential and limitations of renewable energy integration in urban landscapes, fostering a broader awareness of the need for environmentally conscious design and energy solutions.

Self-supported wind turbines, a concept yet to be realized, encompass wind turbines strategically positioned in architectural proximity to buildings, exhibiting architectural reliance while maintaining structural independence from the building itself. These turbines operate in close proximity to the building, capitalizing on the accelerated wind speeds facilitated by the building's presence. Notably, these turbines derive support from their distinct structural framework rather than being reliant on the building's structural elements. Moreover, they exert a significant influence on the architectural design, particularly the aerodynamic configuration of the building. Figure 6.9 illustrates diverse representations of self-supported wind turbines, each accommodating various building mass compositions as an illustrative demonstration.

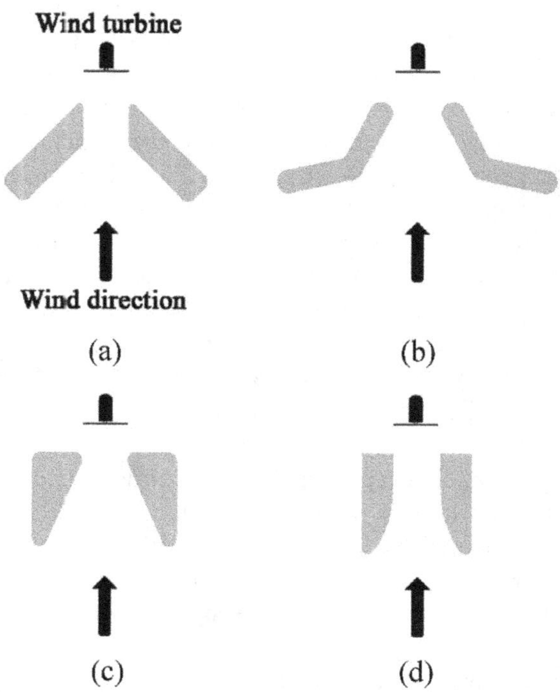

Figure 6.9. Self-supported wind turbines with different building mass compositions (plan view).

Expanding upon the scientific perspective, the diverse classifications of wind turbine utilization represent a nuanced exploration of the intricate relationship between architecture and renewable energy harvesting. The spectrum of options, each with its distinct characteristics, introduces a scientific dimension to the decision-making process for integrating wind energy systems into the built environment.

Building-independent turbines offer flexibility in placement, decoupling the turbine's location from the architectural structure. From a scientific standpoint, this flexibility allows for optimal positioning to capture prevailing wind patterns, potentially maximizing energy yield. The reduced constraints on architectural design simplify the integration process, making it an attractive option for projects emphasizing adaptability and minimal interference with existing structures.

Building-dependent approaches, including building-integrated and building-supported turbines, delve into the scientific intricacies of aerodynamics and structural interactions. Building-integrated turbines leverage the architectural form to enhance energy generation efficiency. Scientific evaluations through computational fluid dynamics simulations and wind tunnel testing become crucial to understand how the building's design influences wind flow and turbine performance. On the other hand, building-supported turbines strategically position themselves to exploit both architectural and structural aspects. Scientific considerations include the rational arrangement of spaces for buffering function and optimizing wind flow to enhance energy output.

The decision on classification adoption is guided by a scientific analysis of multiple factors. Local wind patterns, architectural nuances, and specific energy goals necessitate a comprehensive understanding of the project's context. Scientific methodologies, such as computational fluid dynamics simulations, provide insights into the complex aerodynamic interactions between buildings and turbines. This data-driven approach aids in making informed decisions to optimize energy efficiency while adhering to architectural and structural constraints.

Scientifically, a hybrid approach involves integrating multiple classifications based on the project's unique demands. This holistic strategy considers the synergies between building-independent and building-dependent systems, leveraging the strengths of each to create a harmonious and efficient energy ecosystem. Ongoing technological advancements, including innovations in materials and monitoring technologies, contribute to the continuous evolution of hybrid solutions, making them increasingly viable and effective.

Scientifically striking a balance between architectural aesthetics, structural feasibility, and energy efficiency is imperative for successful implementation. Advanced modeling techniques and simulations contribute to the optimization of wind energy systems, ensuring that they seamlessly integrate with the built environment without compromising visual appeal or structural integrity.

In conclusion, the scientific exploration of these classifications provides a foundation for informed decision-making in the pursuit of sustainable, energy-efficient, and aesthetically pleasing built environments. As technology advances, the scientific understanding of wind energy interactions within architectural landscapes will continue to evolve, opening new avenues for innovative and harmonious integration.

6.4 Conclusion

Throughout the past decade, there has been a consistent upward trajectory in the adoption of renewable energy sources, with a notable emphasis on wind power utilization. This globally recognized and continually expanding technological domain, centered around wind energy, holds significant potential for mitigating the impact of greenhouse gas emissions. It stands poised to contribute significantly to the realization of sustainable development goals while fostering economic advancement within low-carbon sectors, thereby garnering substantial interest and recognition.

The incorporation of wind energy within urban settings is an emerging concept. The integration of renewable energy resources into architectural structures not only addresses environmental considerations but also encompasses architectural aspects. Architects have innovatively shaped new architectural styles or building typologies that encourage the adoption of sustainable energy sources. The concept of harnessing wind energy through the placement of wind turbines atop tall buildings holds significant promise in fostering a more environmentally conscious built environment.

Within this framework, the fusion of wind energy with architectural design has emerged as a burgeoning architectural movement. Architectural concepts that seamlessly incorporate wind turbines have gained prominence, particularly in the designs of tall buildings worldwide. The anticipation is that urban buildings powered by wind energy will soon become a commonplace reality, enriching the lives of numerous households in the foreseeable future.

In this chapter, the focus pertains to harnessing wind energy from tall buildings, entailing an exploration of architectural integration methodologies. A thorough and distinctive typological categorization for the implementation of wind energy within building structures is introduced, deemed to be more comprehensive and cohesive than existing alternatives. Within the framework of this classification, the incorporation of wind turbines into the architectural design of tall buildings and the built environment can assume the following manifestations:

- Building-independent wind turbine utilization (contextually dependent, in the form of wind farms).
- Building-dependent/augmented wind turbine utilization:
    1. ○ Building-mounted/retrofitted wind turbine utilization (architecturally independent):
        - architectural form concerned.
        - architectural form unconcerned.
    2. ○ Building-integrated wind turbine utilization (architecturally dependent):
        - building-supported.
        - self-supported (structurally independent).

The choice among these classifications depends on various factors, including building design, location, and energy goals. Building-dependent approaches, especially building-integrated and building-supported turbines, offer opportunities for enhanced energy efficiency through aerodynamic design. Building-independent turbines, on the other hand, provide more flexibility in placement. The most effective solution may involve a combination of these approaches based on specific project requirements and objectives.

All the wind turbine implementations falling under the category of 'building-dependent/augmented wind turbines' exhibit significant potential, with particular emphasis on the subset of building-mounted/retrofitted turbines. This subset holds promise for utilization in pre-existing structures, although careful consideration is essential to mitigate potential challenges such as noise emissions, induced vibrations, and shadow flickering. It is noteworthy that these technical complexities appear amenable to resolution.

Among these categories, building-integrated wind turbines hold a distinctive significance. This is attributed to their ability to empower architects in manifesting their commitment to a sustainable energy trajectory, where the architectural design assumes a pivotal role in optimizing wind power utilization.

In the future, it is anticipated that architects will conceive structures capable of autonomously generating their complete energy requirements, even yielding surplus energy to supplement the city's power grid. Wind turbines are poised to become an integral facet of virtually every building, especially those of considerable height. At this juncture, the authors contend that a synergistic collaboration between architects and engineers, operating as an iterative feedback loop, should be purposefully oriented towards conceiving cohesive concepts encompassing renewable energy integration, building functionality, and the exploration of architectural forms, all coalescing to yield harmonious and favorable outcomes.

References

[1] Taylor D 1998 Using buildings to harvest wind energy *Build. Res. Inf.* **26** 199–202

[2] Smith P F 2003 *Sustainability at the Cutting Edge: Emerging Technologies for Low Energy Buildings* (Oxford: Architectural Press)

[3] Nwagu C N, Ujah C O, Kallon D V and Aigbodion V S 2024 Integrating solar and wind energy into the electricity grid for improved power accessibility *Unconven. Resour.* **5** 100129

[4] Rezaeiha A, Montazeri H and Blocken B 2020 A framework for preliminary large-scale urban wind energy potential assessment: roof-mounted wind turbines *Energy Convers. Manage.* **214** 112770

[5] Chagas C C M, Pereira M G, Rosa L P, da Silva N F, Freitas M A V and Hunt J D 2020 From megawatts to kilowatts: a review of small wind turbine applications, lessons from the US to Brazil *Sustainability* **12** 2760

[6] Veers P *et al* 2019 Grand challenges in the science of wind energy *Science* **366** eaau2027

[7] Tchakoua P, Wamkeue R, Ouhrouche M, Slaoui-Hasnaoui F, Tameghe T A and Ekemb G 2014 Wind turbine condition monitoring: state-of-the-art review, new trends, and future challenges *Energies* **7** 2595–630

[8] Wang T, Han Q, Chu F and Feng Z 2019 Vibration based condition monitoring and fault diagnosis of wind turbine planetary gearbox: a review *Mech. Syst. Sig. Process.* **126** 662–85

[9] Kisvari A, Lin Z and Liu X 2021 Wind power forecasting—a data-driven method along with gated recurrent neural network *Renew. Energy* **163** 1895–909

[10] Škvorc P and Kozmar H 2021 Wind energy harnessing on tall buildings in urban environments *Renew. Sustain. Energy Rev.* **152** 111662

[11] Malliotakis G, Alevras P and Baniotopoulos C 2021 Recent advances in vibration control methods for wind turbine towers *Energies* **14** 7536

[12] Xie F and Aly A M 2020 Structural control and vibration issues in wind turbines: a review *Eng. Struct.* **210** 110087

[13] Garcia Marquez F P and Peinado Gonzalo A 2022 A comprehensive review of artificial intelligence and wind energy *Arch. Comput. Meth. Eng.* **29** 2935–58

[14] Cao X, Xiong Y, Sun J, Zhu X, Sun Q and Wang Z L 2021 Piezoelectric nanogenerators derived self-powered sensors for multifunctional applications and artificial intelligence *Adv. Funct. Mater.* **31** 2102983

[15] Hsu J Y, Wang Y F, Lin K C, Chen M Y and Hsu J H Y 2020 Wind turbine fault diagnosis and predictive maintenance through statistical process control and machine learning *IEEE Access* **8** 23427–39

[16] Zhang D, Liu Z, Li W and Hu G 2023 LES simulation study of wind turbine aerodynamic characteristics with fluid–structure interaction analysis considering blade and tower flexibility *Energy* **282** 128840

[17] Anup K C, Whale J and Urmee T 2019 Urban wind conditions and small wind turbines in the built environment: a review *Renew. Energy* **131** 268–83

[18] Torres-Madroñero J L, Alvarez-Montoya J, Restrepo-Montoya D, Tamayo-Avendaño J M, Nieto-Londoño C and Sierra-Pérez J 2020 Technological and operational aspects that limit small wind turbines performance *Energies* **13** 6123

[19] Ali Q S and Kim M H 2020 Unsteady aerodynamic performance analysis of an airborne wind turbine under load varying conditions at high altitude *Energy Convers. Manage.* **210** 112696

[20] Guo S, Li Y and Chen W 2021 Analysis on dynamic interaction between flexible bodies of large-sized wind turbine and its response to random wind loads *Renew. Energy* **163** 123–37

[21] Arteaga-López E, Ángeles-Camacho C and Bañuelos-Ruedas F 2019 Advanced methodology for feasibility studies on building-mounted wind turbines installation in urban environment: applying CFD analysis *Energy* **167** 181–8

[22] Tasneem Z, Al Noman A, Das S K, Saha D K, Islam M R, Ali M F, Badal M F R, Ahamed M H, Moyeen S I and Alam F 2020 An analytical review on the evaluation of wind resource and wind turbine for urban application: prospect and challenges *Dev. Built Environ.* **4** 100033

[23] Manganhar A L, Rajpar A H, Luhur M R, Samo S R and Manganhar M 2019 Performance analysis of a savonius vertical axis wind turbine integrated with wind accelerating and guiding rotor house *Renew. Energy* **136** 512–20

[24] Campbell N, Stankovic S, Graham M, Parkin P, Duijvendijk M, Gruiter T, Behling S, Hieber J and Blanch M 2001 Wind energy for the built environment *Proc. European Wind Energy Conf. and Exhibition (Copenhagen)*

[25] Degrassi S, Castelli M R and Benini E 2013 Retrospective of wind turbine architectural integration in the built environment *World Acad. Sci., Eng. Technol., Int. J. Arch. Environ. Eng.* **7** 417–21

[26] Dutton A G, Halliday J A and Blanch M J 2005 The feasibility of building-mounted/integrated wind turbines (BUWTs) *Final Report* Energy Research Unit, CCLRC

[27] Mertens S 2006 *Wind Energy in the Built Environment* (London: Multi-Science)

[28] Abohela I, Hamza N and Dudek S 2011 Integration of wind turbines in the built form and environment. sustainable architecture and urban development *Forum E-J* **10** 23–39

[29] Aguiló A, Taylor D, Quinn A and Wiltshire D R 2004 Computational fluid dynamic modelling of wind speed enhancement through a building augmented wind concentration system *European Wind Energy Conf.*

[30] Poerschke U, Stewart S, Srebric J and Murtha T 2011 Design investigations on building-integrated wind energy: lessons from an architecture studio *Proc. 2011 ASES Conf (Raleigh, NC, 17–20 May)* (Boulder, CO: American Solar Energy Society) pp 955–62

[31] Park J, Jung H J, Lee S W and Park J 2015 New building-integrated wind turbine system utilizing the building *Energies* **8** 11846–70

[32] Haase M and Löfström E 2015 Building augmented wind turbines—BAWT *Integrated Solutions and Technologies of Small Wind Turbines* (Oslo: SINTEF Academic)

[33] Filipowicz M, Żołądek M, Goryl W and Sornek K 2018 Urban ecological energy generation on the example of elevation wind turbines located at Center of Energy AGH *E3S Web Conf.* **49** 00023

[34] Vita G, Šarkić-Glumac A, Hemida H, Salvadori S and Baniotopoulos C 2020 On the wind energy resource above high-rise buildings *Energies* **13** 3641

[35] Kumar N and Prakash O 2023 Analysis of wind energy resources from high rise building for micro wind turbine: a review *Wind Eng.* **47** 190–219

[36] Lu L and Ip K Y 2009 Investigation on the feasibility and enhancement methods of wind power utilization in high-rise buildings of Hong Kong *Renew. Sustain. Energy Rev.* **13** 450–61

[37] Ayhan D and Saglam S 2012 A technical review of building-mounted wind power systems and a sample simulation model *Renew. Sustain. Energy Rev.* **16** 1040–9

[38] Cho K P, Jeong S H and Sari D P 2011 Harvesting wind energy from aerodynamic design for building integrated wind turbines *Int. J. Technol.* **2** 189–98

[39] CTBUH Council on Tall Buildings and Urban Habitat Illinois Institute of Technology www.ctbuh.org (Accessed: 10 August 2023)

[40] WS Atkins and Partners 2008 Bahrain World Trade Center: the world's first large scale integration of wind turbines on a building *CTBUH Best Tall Building Awards* (London: WS Atkins and Partners) pp 1–21

[41] Gunel M H and Ilgın H E 2014 *Tall Buildings: Structural Systems and Aerodynamic Form* (London and New York: Routledge)

[42] Killa S and Smith R F 2008 Harnessing energy in tall buildings: Bahrain World Trade Center and beyond *Proc. of the 8th World Congress on Tall and Green: Typology for a Sustainable Urban Future (CTBUH'08) (2008, March)* pp 144–50

[43] Bellini O E and Daglio L 2009 *Bahrain World Trade Center. Building Arabia* (Vercelli: White Star) pp 36–40

[44] Smith R F and Killa S 2007 Bahrain World Trade Center (BWTC): the first large-scale integration of wind turbines in a building *Struct. Des. Tall Spec. Build.* **16** 429–39

[45] Bogle I 2011 Integrating wind turbines in tall buildings *CTBUH J.* 4 30–4

[46] Cowan D 2010 Strata SE1, London-propelling sustainable regeneration *Proc. Inst.-Civ. Eng.* **163** 56–63

[47] Hu Y 2022 Wind energy innovative application on Shanghai Tower and analytical contrast with other buildings *IOP Conf. Ser.* **1011** 012020

[48] Sharpe T 2010 The role of aesthetics, visual and physical integration in building mounted wind turbines—an alternative approach *Paths Sustain. Energy* (London: InTech)

[49] Selcuk S A and Ilgın H E 2017 Performative approaches in tall buildings: Pearl River Tower *Eurasian J. Civ. Eng. Archit.* **1** 11–20

[50] Besjak C, Biswas P, Petrov G I, Meinschein G and Jordan A 2015 New heights in sustainability—Pertamina Energy Tower *Structures Congress* pp 945–60

[51] Wolny P 2019 *21st Century Skyscrapers (Feats of 21st Century Engineering)* (Berkeley Heights, NJ: Enslow)

[52] Charles J 2016 *Sustainable Construction, Green Building Design and Delivery* 4th edn (Kibert: Wiley) pp 25–34

Chapter 7

Meeting the energy needs of the urban and rural poor in the Congo Basin through renewables: challenges and prospects

Nyong Princely Awazi

The Congo Basin, rich in renewable energy resources, faces challenges in meeting the energy needs of its urban and rural poor. Despite abundant biomass, solar, and hydroelectric potential, wind, tidal, and geothermal energy remain untapped. Countries such as the Democratic Republic of Congo, Central African Republic, and Congo-Brazzaville struggle with poor governance, weak policies, high transition costs, and inadequate infrastructure. However, favorable prospects exist, with national strategies promoting renewable energy, green growth, and Sustainable Development Goal 7. Harnessing renewable energy could significantly address energy deficits and provide a sustainable solution for the region's energy needs.

7.1 Introduction

The Congo Basin harbors rich biodiversity and serves as a vital ecological asset for the planet (Anthony *et al* 2015, Heimpel *et al* 2024). However, within its dense forests and sprawling urban centers, many communities struggle with a fundamental necessity: access to reliable and sustainable energy (Kusakana 2016, Kenfack *et al* 2017). For both urban and rural populations in this region, the provision of electricity remains a significant challenge, impeding socio-economic development and perpetuating cycles of poverty (Aquilas and Atemnkeng 2022). Urban centers across the Congo Basin, such as Kinshasa, Yaoundé, Douala, Bamenda, Libreville, Bafoussam, Malabo, Bangui and Brazzaville, grapple with inadequate infrastructure and limited access to electricity (Bishoge *et al* 2020). High costs, unreliable grids, and a burgeoning population exacerbate the energy deficit, hindering businesses, education, healthcare, and overall quality of life (Aquilas and Atemnkeng 2022). Moreover, reliance on fossil fuels contributes to environmental degradation and

doi:10.1088/978-0-7503-6179-8ch7

exacerbates climate change, posing further risks to the region's ecological integrity and human well-being.

In rural areas, the energy dilemma takes on a different guise, with off-grid communities bearing the brunt of energy poverty (Shure *et al* 2014). Remote villages lack basic electricity services, forcing residents to rely on inefficient and often hazardous alternatives such as kerosene lamps and biomass fuels (Tongele 2021). This not only restricts productive activities but also jeopardizes health and safety, particularly for women and children exposed to indoor air pollution. Amidst these challenges, however, lie opportunities for transformative change. The Congo Basin boasts immense renewable energy potential, including abundant sunlight, hydroelectric resources, and biomass feedstock (Dalder *et al* 2024). Harnessing these indigenous energy sources presents a pathway toward energy access that is both sustainable and inclusive (Mukandala *et al* 2020). By investing in decentralized renewable energy solutions, such as solar panels, micro-hydro systems, and biomass technologies, communities can leapfrog the constraints of traditional grid infrastructure and unlock newfound opportunities for development (Enongene and Fobissie 2016, Bishoge *et al* 2020). Moreover, initiatives that empower local stakeholders, foster innovation, and prioritize environmental stewardship can pave the way for a more resilient and equitable energy future in the Congo Basin. In light of the forgoing, this chapter examines the specific challenges and prospects of meeting the energy needs of the urban and rural poor in the Congo Basin through renewables.

7.2 Renewable energy potential in the Congo Basin

The Congo Basin stands as a reservoir of untapped renewable energy potential, offering a beacon of hope for addressing the energy needs of its urban and rural populations. With its vast expanse of lush forests, abundant sunlight, and mighty rivers, this region possesses the natural resources necessary to power sustainable development.

7.2.1 Biomass

Biomass energy, derived from organic materials such as agricultural residues, forest biomass, and animal waste, plays a significant role in meeting the energy needs of communities across the Congo Basin (Gond *et al* 2016, Migolet *et al* 2022, Yufenyuy *et al* 2023). With its dense tropical forests and rich biodiversity, the region boasts abundant biomass resources, offering a locally available and renewable source of energy. The Congo Basin is home to a wealth of biomass resources, including agricultural residues such as crop residues and palm oil waste, forest biomass from logging residues and wood processing, and organic waste from livestock farming and food production (Maniatis 2011, Behrendt *et al* 2013). This diversity of biomass sources provides ample opportunities for energy production across urban and rural areas. Biomass energy has been utilized in the Congo Basin for centuries, with traditional practices such as woodfuel for cooking and heating prevalent in many communities (Bilonda 2020). However, modern technologies are increasingly being employed to harness biomass energy more efficiently, including biogas digesters, biomass boilers, and gasification systems for electricity generation (Nfah and Ngundam 2009, Mayala *et al* 2022). Biomass energy plays a crucial role in addressing energy poverty in both urban and rural areas of the Congo Basin. In off-grid communities, biomass fuels such as firewood and charcoal are often the primary sources of energy for cooking and heating, providing essential services where access to electricity is limited.

While biomass energy offers a renewable alternative to fossil fuels, its unsustainable harvesting and use can have adverse environmental impacts, including deforestation, habitat loss, and air pollution (Behrendt *et al* 2013, Schure 2014, Tyukavina *et al* 2018). Sustainable biomass management practices, such as reforestation, agroforestry, and efficient cookstove programs, are essential for mitigating these environmental concerns and ensuring the long-term viability of biomass energy in the Congo Basin (Tegegne *et al* 2016, Tongele 2021). Policy frameworks and investment incentives play a crucial role in promoting the sustainable development of biomass energy in the Congo Basin. Governments, international organizations, and private sector stakeholders can support the adoption of clean biomass technologies, improve biomass supply chains, and facilitate community engagement to maximize the socio-economic benefits of biomass energy while minimizing its environmental footprint. Biomass energy holds significant promise for meeting the energy needs of the Congo Basin while contributing to sustainable development goals.

7.2.2 Solar

Solar energy represents a transformative opportunity for meeting the energy needs of communities across the Congo Basin, leveraging the region's abundant sunlight to drive sustainable development (Kenfack *et al* 2017). The Congo Basin benefits from a wealth of solar irradiation, with ample sunlight available throughout the year. This abundant solar resource presents a renewable and reliable source of energy for both urban and rural communities across the region. In remote and off-grid areas where traditional electricity infrastructure is lacking, solar energy offers a viable solution for powering essential services such as lighting, telecommunications, and water pumping. Solar photovoltaic (PV) systems, ranging from small-scale solar lanterns to larger solar home systems and mini-grids, provide off-grid communities with access to clean and reliable electricity (Abanda 2012). In urban centers and areas with existing electricity infrastructure, solar energy can complement traditional power sources and help alleviate pressure on the grid. Utility-scale solar PV installations can contribute to the diversification of the energy mix, reducing reliance on fossil fuels and enhancing energy security. The adoption of solar energy technologies in the Congo Basin can generate significant socio-economic benefits, including job creation, income generation, and improved access to education and healthcare services (Tongele 2021). Solar entrepreneurship and local capacity building initiatives can empower communities to harness solar energy for economic development and poverty alleviation.

Solar energy is inherently clean and renewable, emitting no greenhouse gases or harmful pollutants during operation. By promoting the widespread adoption of solar technologies, the Congo Basin can reduce its carbon footprint, mitigate climate change impacts, and safeguard the region's fragile ecosystems and biodiversity (Mukandala *et al* 2020). Despite its potential, the widespread adoption of solar energy in the Congo Basin faces several challenges, including upfront costs, financing constraints, technical capacity limitations, and regulatory barriers. Addressing these challenges requires concerted efforts from governments, international organizations, and private sector stakeholders to create enabling environments for solar investment and deployment. Solar energy holds immense promise for powering sustainable development in the Congo Basin, offering a clean, reliable, and locally available energy solution (Kusakana 2016).

7.2.3 Hydroelectric power

Hydroelectric power (HEP) holds immense potential in the Congo Basin, fueled by the region's extensive network of rivers and waterways (Deng *et al* 2020, Kalonda *et al* 2023). The Congo Basin is endowed with a vast hydrological network, including the mighty Congo River and its tributaries. These abundant water resources offer significant potential for hydroelectric power generation, providing a renewable and reliable source of electricity for the region. Several large-scale hydroelectric projects have been developed or proposed in the Congo Basin, harnessing the power of its rivers to generate electricity for domestic consumption and export. Projects such as the Inga dams in the Democratic Republic of Congo (DRC), and the Lom Panga, Mekim, Memve'ele, Edea and Lagdo dams in Cameroon represent ambitious efforts to tap into the region's hydroelectric potential and drive economic development. Hydroelectricity is a clean and renewable energy source, producing minimal greenhouse gas emissions during operation. By leveraging hydroelectric power, the Congo Basin can reduce its reliance on fossil fuels, mitigate climate change impacts, and contribute to global efforts to transition to a low-carbon economy (Warner *et al* 2019). Hydroelectricity has the potential to significantly improve energy access in the Congo Basin, particularly in rural and remote areas where grid electrification is challenging. Large hydroelectric projects can provide reliable electricity to urban centers, while smaller-scale hydro installations and microgrids can serve off-grid communities, powering homes, schools, and healthcare facilities.

The development of hydroelectricity infrastructure in the Congo Basin can stimulate economic growth, create jobs, and drive industrialization. Additionally, revenue generated from hydroelectric power generation can be reinvested in social infrastructure, education, and healthcare, improving livelihoods and reducing poverty in surrounding communities (Showers 2009). While hydroelectricity is considered a clean energy source, large-scale hydroelectric projects can have significant environmental and social impacts. These include habitat alteration, displacement of communities, and changes to river ecosystems. It is essential to conduct thorough environmental and social impact assessments, engage with local stakeholders, and implement mitigation measures to minimize adverse effects.

Many rivers in the Congo Basin are shared across multiple countries, highlighting the importance of transboundary cooperation in the development and management of hydroelectric projects. Collaborative efforts among riparian states can promote sustainable water resource management, address shared challenges, and foster regional integration and stability. Hydroelectricity power holds great promise for driving sustainable development in the Congo Basin, offering a clean, reliable, and locally available energy solution.

7.2.4 Wind

Wind energy presents a promising but relatively untapped resource in the Congo Basin, offering a clean and renewable alternative to traditional fossil fuels (Abanda 2012). While the Congo Basin is not traditionally known for its strong winds, certain areas within the region, such as highlands and coastal regions, experience consistent wind patterns that are conducive to wind energy generation. Preliminary assessments suggest that there is untapped wind energy potential in specific locations across the Congo Basin. Wind energy in the Congo Basin is primarily utilized in small-scale applications, such as off-grid installations for rural electrification and powering remote infrastructure. Small wind turbines, ranging from a few kilowatts to a few hundred kilowatts in capacity, can be deployed in isolated communities to supplement existing energy sources and improve energy access. The integration of utility-scale wind farms into the existing electricity grid presents challenges in the Congo Basin. Limited grid infrastructure, transmission constraints, and variability in wind patterns may pose obstacles to the widespread adoption of large-scale wind energy projects. However, advancements in grid technology and strategic planning can help mitigate these challenges. Wind energy is considered environmentally friendly, emitting no greenhouse gases or air pollutants during operation. However, the construction and operation of wind turbines can have localized environmental impacts, such as habitat disturbance and bird collisions. Careful siting and environmental assessments are essential to minimize these potential negative effects.

Accurate assessment of wind resources and strategic planning are critical for the successful development of wind energy projects in the Congo Basin (Manjong *et al* 2021, Aquilas and Atemnkeng 2022). Wind mapping studies, meteorological data collection, and feasibility assessments can help identify suitable locations for wind farm development and optimize project performance. Policy support and investment incentives play a crucial role in promoting the deployment of wind energy in the Congo Basin. Governments, international organizations, and private sector stakeholders can collaborate to establish favorable regulatory frameworks, provide financial incentives, and facilitate technology transfer to accelerate the adoption of wind energy solutions. Building local capacity and fostering knowledge sharing are essential for the sustainable development of wind energy in the Congo Basin. Training programs, technology transfer initiatives, and partnerships with academic institutions can empower local communities and institutions to actively participate in the planning, implementation, and operation of wind energy projects. Wind energy holds significant potential as a clean and renewable energy source in the

Congo Basin, offering opportunities for energy access, economic development, and environmental sustainability.

7.2.5 Tidal

Tidal energy, derived from the natural rise and fall of ocean tides, represents a promising but relatively unexplored renewable energy resource in the Congo Basin (Vallaeys *et al* 2021). The Congo Basin encompasses a stretch of coastline along the Atlantic Ocean, characterized by estuaries, bays, and tidal channels. These coastal features create favorable conditions for tidal energy generation, as the ebb and flow of tides result in significant water movement and energy potential. While tidal ranges in the Congo Basin may not be as pronounced as those in other regions, such as the Bay of Fundy or the Severn Estuary, tidal currents can still be strong and predictable in certain areas. These tidal currents can be harnessed using tidal stream turbines or other marine energy technologies to generate electricity. Tidal energy in the Congo Basin has the potential to provide clean and reliable electricity for coastal communities, industrial facilities, and offshore installations. Additionally, tidal energy can complement other renewable energy sources, such as solar and wind, to create a diversified and resilient energy mix. The development of tidal energy projects in the Congo Basin faces several technological challenges, including the design and deployment of robust marine energy devices capable of withstanding harsh marine environments and variable tidal conditions. Research and development efforts are needed to advance tidal energy technologies and reduce costs.

Tidal energy projects can have environmental impacts on marine ecosystems, including alterations to sediment transport, changes in water quality, and disturbances to marine habitats and species (Abanda 2012). Environmental impact assessments and monitoring programs are essential to minimize these impacts and ensure the sustainable deployment of tidal energy technologies. The development of tidal energy projects in the Congo Basin requires clear regulatory and permitting frameworks to address issues related to site selection, environmental protection, stakeholder engagement, and project financing. Governments and regulatory authorities can play a key role in facilitating the permitting process and providing incentives for tidal energy development. Collaboration among stakeholders, including government agencies, research institutions, industry partners, and local communities, is essential for the successful deployment of tidal energy projects in the Congo Basin. Knowledge sharing, capacity building, and technology transfer initiatives can accelerate the development and adoption of tidal energy solutions. Tidal energy holds promise as a clean and renewable energy source in the Congo Basin, offering opportunities for energy diversification, economic development, and environmental sustainability along the region's coastline.

7.2.6 Geothermal energy

Geothermal energy, derived from the heat stored beneath the Earth's surface, represents a promising yet largely untapped renewable energy resource in the Congo Basin (Mukandala *et al* 2020). The Congo Basin sits atop the East African Rift

System, a geologically active region characterized by tectonic activity, volcanic eruptions, and geothermal heat flux. This geological setting creates favorable conditions for the occurrence of geothermal resources, including hot springs, geysers, and volcanic activity (Deogratias *et al* 2022, Feumoe *et al* 2022). The heat beneath the Earth's surface in the Congo Basin can be harnessed through geothermal technologies, such as drilling wells to access hot water and steam reservoirs. This heat can then be used for various applications, including electricity generation, district heating, greenhouse heating, and industrial processes. Geothermal energy has the potential to provide baseload electricity generation in the Congo Basin, complementing intermittent renewable sources such as solar and wind. Geothermal power plants can be constructed to utilize high-temperature reservoirs, where steam or hot water is extracted to drive turbines and generate electricity (Demissie 2010, Kusakana 2016). In addition to electricity generation, geothermal energy can be directly utilized for heating and cooling applications in the Congo Basin. District heating systems can supply hot water for residential and commercial buildings, while geothermal heat pumps can provide space heating and cooling in homes and facilities. Despite its potential, the exploration and assessment of geothermal resources in the Congo Basin are still in the early stages (Odhipio *et al* 2020). Geological surveys, geophysical studies, and exploration drilling are needed to identify and characterize geothermal reservoirs, assess resource potential, and determine feasibility for development.

The development of geothermal energy projects in the Congo Basin faces technical challenges, including drilling deep wells in challenging geological formations, managing reservoir fluids, and maintaining reservoir integrity (Muanza 2015). Innovative drilling techniques, reservoir modeling, and engineering solutions are required to overcome these challenges. Geothermal energy projects require significant upfront investment in exploration, drilling, and infrastructure development. Access to financing, investment incentives, and risk mitigation mechanisms are essential to attract private sector investment and support the development of geothermal projects in the Congo Basin. While geothermal energy is considered a clean and renewable energy source, the development of geothermal projects can have environmental impacts, including land use changes, water consumption, and induced seismicity. Environmental impact assessments and mitigation measures are necessary to minimize these impacts and ensure sustainable development. Geothermal energy holds great promise as a clean, reliable, and locally available energy source in the Congo Basin.

7.3 The potentials of renewable energy meeting the energy needs of the rural and urban poor in the Congo Basin

The Congo Basin harbors vast renewable energy potential that holds promise for meeting the energy needs of both rural and urban communities, lifting them out of poverty and driving sustainable development (Dountio *et al* 2016, Mboumboue and Njomo 2016, Tamba 2017, Muh *et al* 2018, Flora *et al* 2019, Enow-Arrey 2020, Nsafon *et al* 2020, Tangka 2021, Yimen *et al* 2022). With its abundant sunlight, vast biomass resources, mighty rivers, and geothermal hotspots, the Congo Basin is rich in renewable energy sources waiting to be harnessed. These indigenous resources offer a locally available, clean, and sustainable alternative to traditional fossil fuels, presenting a pathway towards energy access and socio-economic empowerment for the region's marginalized populations. In both rural and urban areas, energy poverty persists as a barrier to development, hindering access to essential services, economic opportunities, and a higher quality of life. However, the adoption of renewable energy solutions, ranging from solar panels and micro-hydro systems to biomass and geothermal technologies, holds the promise of transforming energy access in the Congo Basin. Through investment in decentralized renewable energy infrastructure, empowering local communities, and fostering innovation, stakeholders can unlock the potentials of renewable energy to bridge the energy gap, alleviate poverty, and promote resilience in the face of climate change.

7.3.1 Access to clean cooking facilities in the Congo Basin

Access to clean cooking facilities remains a critical challenge for many communities across the Congo Basin, with significant implications for public health, gender equality, and environmental sustainability (Masamba *et al* 2023). In both urban and rural areas of the Congo Basin, traditional cooking methods such as open fires and rudimentary stoves fueled by biomass (wood, charcoal, agricultural residues) are prevalent. These inefficient and polluting cooking practices contribute to indoor air pollution, respiratory diseases, deforestation, and climate change. The use of traditional cooking methods exposes households to high levels of indoor air pollution, resulting in respiratory illnesses, cardiovascular diseases, and adverse pregnancy outcomes. Women and children, who are often responsible for cooking and spending extended periods of time indoors, are disproportionately affected by indoor air pollution. The widespread reliance on biomass fuels for cooking contributes to deforestation, habitat loss, and soil degradation in the Congo Basin. Unsustainable harvesting practices exacerbate environmental degradation, threatening biodiversity, water resources, and ecosystem services essential for human well-being. Limited access to clean cooking facilities perpetuates energy poverty in the Congo Basin, particularly in rural and marginalized communities. Dependence on biomass fuels consumes significant time and resources, constraining opportunities for education, income generation, and socio-economic development. Several barriers hinder the transition to clean cooking facilities in the Congo Basin, including affordability, availability, awareness, and cultural preferences. High upfront costs, limited distribution networks, and a lack of government support further impede the adoption of clean cooking technologies.

Addressing the challenge of access to clean cooking facilities requires a multifaceted approach that integrates technological, policy, and behavioral interventions. Promoting improved cookstoves, biogas digesters, and solar cookers can enhance efficiency, reduce emissions, and improve health outcomes (Piergallini 2019). Governments in the Congo Basin can play a crucial role in facilitating the transition to clean cooking facilities through supportive policies, regulations, and incentives. This includes subsidies for clean cooking technologies, tax exemptions, and awareness campaigns to promote behavior change. Engaging local communities, particularly women and marginalized groups, is essential for the successful adoption of clean cooking solutions in the Congo Basin. Empowering communities through education, training, and participatory decision-making processes can enhance ownership and sustainability. Improving access to clean cooking facilities is essential for promoting public health, environmental sustainability, and inclusive development in the Congo Basin.

7.3.2 Energy needs of rural and urban populations in the Congo Basin

The energy needs of both rural and urban populations in the Congo Basin represent a pressing challenge and a pivotal opportunity for sustainable development (Reed and Miranda 2007, Gamaidandi 2021). With its sprawling urban centers and remote rural communities, the Congo Basin is a microcosm of energy disparity, where access to reliable and clean energy sources remains a privilege rather than a right. In urban areas such as Kinshasa and Brazzaville, inadequate infrastructure, high costs, and unreliable grids exacerbate energy poverty, hindering economic growth and perpetuating social inequalities. Meanwhile, in rural villages scattered throughout the region's dense forests and river basins, off-grid communities struggle with limited access to modern energy services, relying instead on inefficient and polluting biomass fuels for cooking, heating, and lighting. This reliance not only compromises public health but also contributes to environmental degradation and perpetuates cycles of poverty. In both rural and urban contexts, the energy landscape of the Congo Basin is shaped by a complex interplay of socio-economic, environmental, and political factors. From the challenges of infrastructure development and energy access to the opportunities presented by indigenous renewable resources and innovative technologies, the quest for sustainable energy solutions in the Congo Basin is multi-faceted and urgent.

7.3.2.1 Energy for production

Energy for production plays a crucial role in driving economic growth, industrialization, and job creation in the Congo Basin (De Wit *et al* 2015). Energy is a fundamental input for productive activities in the Congo Basin, powering agricultural machinery, industrial processes, and mining operations. Reliable and affordable energy sources are essential for enhancing productivity, value addition, and competitiveness across various sectors of the economy. Agriculture is a key driver of the economy in the Congo Basin, supporting livelihoods and food security for millions of people. Access to energy for irrigation, mechanization, and post-harvest processing is critical for increasing agricultural productivity, reducing post-harvest losses, and promoting agro-industrial development. The manufacturing sector in the Congo Basin is characterized by a diverse range of industries, including food processing, textiles, and construction materials. Access to reliable energy sources, such as electricity and thermal energy, is essential for powering manufacturing processes, enhancing efficiency, and expanding industrial capacity. The Congo Basin is endowed with rich mineral resources, including copper, cobalt, and gold, which contribute significantly to export earnings and foreign investment. Energy-intensive mining operations require reliable and affordable sources of electricity and fuel for extraction, processing, and transportation of minerals (Reed and Miranda 2007). Despite the importance of energy for production, the Congo Basin faces several challenges in meeting the energy needs of productive sectors. These include inadequate energy infrastructure, unreliable electricity supply, high energy costs, and limited access to modern energy services, particularly in rural and remote areas.

The Congo Basin possesses abundant renewable energy resources, including hydroelectric, solar, biomass, and geothermal potential, which can be harnessed to meet the energy needs of productive sectors sustainably (Abanda 2012). Investments in renewable energy infrastructure, grid expansion, and energy efficiency measures present opportunities for enhancing energy access and promoting inclusive growth. Governments in the Congo Basin can play a key role in promoting energy for production through supportive policies, regulations, and institutional frameworks. This includes establishing conducive investment environments, providing incentives for renewable energy development, and promoting public–private partnerships. Private sector engagement is essential for driving investments in energy infrastructure and promoting innovation in energy technologies (Tangka 2021). Public–private partnerships, market-based approaches, and value chain interventions can mobilize private sector resources and expertise to address energy challenges and unlock opportunities for sustainable development. Energy for production is a critical enabler of economic development and poverty reduction in the Congo Basin.

7.3.2.2 Energy for households

Access to reliable and clean energy for households is essential for improving living standards, promoting health, and enhancing overall quality of life in the Congo Basin. In both rural and urban areas of the Congo Basin, households rely on a mix of traditional and modern energy sources for cooking, lighting, heating, and other domestic activities. Traditional biomass fuels such as firewood, charcoal, and agricultural residues are commonly used for cooking, while kerosene lamps and candles are prevalent for lighting (Kusakana 2016). Despite abundant energy resources in the Congo Basin, energy poverty persists, with millions of households lacking access to modern energy services. Limited access to electricity, inefficient cooking technologies, and reliance on polluting fuels contribute to poor health outcomes, environmental degradation, and economic constraints for households (Schure *et al* 2014, Ngbolua *et al* 2018). The use of traditional cooking methods and fuels in households results in indoor air pollution, leading to respiratory diseases, eye problems, and other health ailments, particularly among women and children, who are disproportionately affected (Schure 2012). Access to clean cooking facilities is essential for reducing indoor air pollution and improving public health. While access to electricity has improved in urban areas of the Congo Basin, rural electrification rates remain low, with many households lacking access to grid-connected electricity. Off-grid solutions such as solar home systems, microgrids, and mini-hydro systems can provide decentralized electricity access to rural communities, enhancing energy access and promoting socio-economic development. High upfront costs and limited affordability pose significant barriers to energy access for households in the Congo

Basin. Many low-income households are unable to afford clean energy technologies and rely on inefficient and expensive alternatives, perpetuating energy poverty and exacerbating socio-economic disparities.

The Congo Basin is rich in renewable energy resources, including solar, hydro, biomass, and geothermal potential, which can be harnessed to provide clean and affordable energy for households (Flora *et al* 2019, Nsafon *et al* 2020). Investments in renewable energy infrastructure, technology innovation, and financing mechanisms can expand access to clean energy solutions and promote sustainable development. Governments in the Congo Basin can play a crucial role in promoting energy access for households through supportive policies, regulations, and incentives. This includes prioritizing rural electrification, promoting clean cooking technologies, and implementing targeted subsidy programs to make clean energy solutions more affordable and accessible to households. Engaging local communities and empowering households to participate in decision-making processes is essential for ensuring the success and sustainability of energy access initiatives in the Congo Basin. Community-led approaches, capacity building programs, and awareness campaigns can mobilize support and foster ownership of energy projects at the grassroots level. Improving access to reliable, clean, and affordable energy for households is essential for promoting sustainable development, enhancing livelihoods, and achieving energy justice in the Congo Basin (Tongele 2021).

7.3.2.3 Energy for agriculture

Energy plays a pivotal role in agricultural productivity, food security, and rural livelihoods in the Congo Basin. Energy is a critical input for agricultural activities in the Congo Basin, powering irrigation systems, machinery, processing facilities, and post-harvest storage (Kusakana 2016). Access to reliable and affordable energy sources is essential for increasing productivity, enhancing value addition, and improving resilience to climate change in the agricultural sector. The majority of agricultural practices in the Congo Basin are still labor-intensive and rely on traditional methods, limiting productivity and efficiency. However, modernizing agricultural practices through mechanization, irrigation, and agro-processing can significantly enhance productivity and income generation for farmers (Baumüller 2021). In rural areas of the Congo Basin, where grid electricity is limited or non-existent, households and farmers rely on biomass fuels such as firewood, charcoal, and agricultural residues for cooking, heating, and lighting (Malabo Montpellier Panel 2018). Off-grid renewable energy solutions, including solar pumps, biomass gasifiers, and biogas digesters, can provide decentralized energy access for agricultural activities (Falchetta *et al* 2022). Irrigation is crucial for ensuring crop yields and mitigating the impacts of climate variability in the Congo Basin. Energy-intensive irrigation systems, such as diesel and electric pumps, are commonly used in areas with access to grid electricity. However, solar-powered irrigation technologies offer a sustainable and cost-effective alternative for off-grid and remote agricultural areas.

Post-harvest losses due to inadequate processing and storage facilities are a significant challenge for farmers in the Congo Basin. Access to reliable energy sources for drying, milling, and preservation can reduce post-harvest losses, improve food quality, and increase market access for agricultural produce (Woomer *et al* 2023). Despite the importance of energy for agriculture, the Congo Basin faces several challenges in meeting the energy needs of farmers. These include inadequate energy infrastructure, high energy costs, limited access to modern energy services, and lack of awareness about energy-efficient technologies among farmers. The Congo Basin is endowed with abundant renewable energy resources, including solar, hydro, and biomass potential, which can be harnessed to meet the energy needs of agriculture sustainably. Investments in renewable energy infrastructure, technology adoption, and capacity building can enhance energy access and promote agricultural development in the region. Governments in the Congo Basin can play a vital role in promoting energy for agriculture through supportive policies, regulations, and incentives (Kenfack *et al* 2017). This includes prioritizing rural electrification, promoting energy-efficient technologies, and providing financial incentives for renewable energy adoption in the agricultural sector (Varela Pérez *et al* 2022). Energy is a vital enabler of agricultural development and food security in the Congo Basin.

7.3.2.4 Energy for transport

Energy for transport is a crucial component of economic development and connectivity in the Congo Basin, facilitating trade, mobility, and access to essential services (Tongele 2021). The Congo Basin encompasses a diverse network of transport infrastructure, including roads, railways, waterways, and air transport. These transportation arteries serve as lifelines for the movement of goods, people, and resources within and across the region's vast landscapes. Diesel and gasoline are the primary fuels used for road transport in the Congo Basin, powering trucks, buses, and motorcycles. In remote and rural areas, where access to petroleum fuels may be limited or expensive, alternative fuels such as biofuels and liquefied petroleum gas (LPG) can provide cleaner and more affordable energy options. The transport sector in the Congo Basin faces several challenges, including inadequate infrastructure, poor road conditions, high transportation costs, and limited access to modern and efficient vehicles (Ulimwengu *et al* 2009, Sogena *et al* 2018). These challenges hinder trade flows, increase transport times, and contribute to economic inefficiencies. The reliance on fossil fuels for transport in the Congo Basin results in significant environmental impacts, including air pollution, greenhouse gas emissions, and deforestation. Improving energy efficiency, promoting cleaner fuels, and adopting sustainable transport practices are essential for mitigating these environmental consequences. Water transport, particularly along the Congo River and its tributaries, plays a vital role in facilitating trade and commerce in the region (Munyangeyo 2021). Promoting inland waterway transport and investing in port infrastructure can enhance connectivity, reduce transportation costs, and alleviate pressure on road networks.

Electric mobility presents an emerging opportunity for reducing dependence on fossil fuels and promoting sustainable transport in the Congo Basin. Electric vehicles (EVs), including buses, motorcycles, and bicycles, powered by renewable energy sources such as solar and hydroelectricity, offer a cleaner and more efficient mode of transport for urban and peri-urban areas. Governments in the Congo Basin can play a critical role in promoting energy-efficient and sustainable transport through supportive policies, regulations, and investments. This includes investing in transport infrastructure, promoting renewable energy adoption, and implementing fuel efficiency standards for vehicles (Dountio *et al* 2016). Enhancing transport connectivity and facilitating cross-border trade are essential for regional integration and economic development in the Congo Basin. Regional cooperation initiatives, such as the Trans-African Highway network and the Central African Economic and Monetary Community (CEMAC), can promote harmonization of transport policies and infrastructure development. Energy for transport is a linchpin of socio-economic development and regional integration in the Congo Basin. Addressing infrastructure deficits, promoting sustainable transport practices, and leveraging renewable energy solutions are key to building a more efficient, resilient, and environmentally sustainable transport system for the benefit of its people and economies (Dominguez and Foster 2011, Kevin *et al* 2017).

7.3.2.5 Energy for education

Energy plays a vital role in facilitating access to education, improving learning outcomes, and enhancing educational opportunities in the Congo Basin (Mboumboue and Njomo 2016). Energy is a fundamental enabler of education, powering schools, classrooms, and educational facilities in the Congo Basin. Access to reliable and affordable energy sources is essential for providing adequate lighting, heating, cooling, and technology-enabled learning environments conducive to effective teaching and learning. While progress has been made in expanding access to education in the Congo Basin, many schools, particularly in rural and remote areas, still lack access to grid electricity. Off-grid renewable energy solutions, such as solar panels and mini-grids, can provide decentralized electricity access for schools, powering lights, computers, and other educational equipment. With the increasing integration of technology in education, access to electricity and digital infrastructure is becoming increasingly important for facilitating digital learning initiatives in the Congo Basin. Internet connectivity, computers, tablets, and e-learning platforms require reliable energy sources to function effectively, bridging the digital divide and expanding access to quality education. Access to electricity extends study hours for students, enabling evening classes, after-school activities, and adult education programs in the Congo Basin. Reliable lighting provided by solar-powered lanterns or electrified classrooms creates a conducive learning environment and enhances educational opportunities for learners of all ages.

Energy access also contributes to the health and safety of students and teachers in schools. Reliable lighting reduces the risk of accidents and improves security on school premises, particularly during evening hours. Access to clean cooking facilities powered by modern energy sources can also improve indoor air quality and reduce health risks associated with traditional cooking methods. Despite the importance of energy for education, the Congo Basin faces several challenges in meeting the energy needs of schools. These include inadequate energy infrastructure, high energy costs, limited financial resources for electrification projects, and a lack of awareness about the benefits of energy access for education. The Congo Basin is endowed with abundant renewable energy resources, including solar, hydro, biomass, and wind potential, which can be harnessed to meet the energy needs of schools sustainably. Investments in renewable energy infrastructure, technology adoption, and capacity building can enhance energy access and support educational development in the region. Governments in the Congo Basin can play a crucial role in promoting energy for education through supportive policies, regulations, and investments. This includes prioritizing electrification of schools, providing incentives for renewable energy adoption, and integrating energy access into national education strategies and plans. Energy access is essential for ensuring inclusive, equitable, and quality education for all in the Congo Basin.

7.3.2.6 Energy for health

Energy plays a critical role in supporting healthcare delivery, promoting public health, and ensuring access to essential medical services in the Congo Basin. Access to reliable and affordable energy sources is essential for powering healthcare facilities, including hospitals, clinics, and health centers, in the Congo Basin. Electricity is needed to operate medical equipment, refrigerate vaccines and medicines, provide lighting for surgeries and consultations, and support telemedicine and digital health initiatives (Mboumboue and Njomo 2016). Many medical devices and diagnostic tools used in healthcare settings, such as x-ray machines, ultrasound scanners, and laboratory equipment, require a constant and uninterrupted power supply to function effectively. Access to reliable electricity is essential for ensuring accurate diagnosis, treatment, and patient care in the Congo Basin. Maintaining the cold chain for vaccines is critical for preserving vaccine potency and preventing disease outbreaks in the Congo Basin. Refrigeration equipment powered by electricity or solar energy is essential for storing vaccines at the required temperatures from production to delivery at healthcare facilities, particularly in remote and off-grid areas. Energy access is vital for supporting emergency medical services, including ambulance operations, emergency rooms, and intensive care units, in the Congo Basin. Reliable electricity ensures the availability of life-saving medical interventions, equipment, and lighting during emergencies and natural disasters.

The integration of technology in healthcare delivery, such as telemedicine, electronic medical records, and mobile health applications, relies on access to electricity and digital infrastructure. Reliable energy sources power computers, Internet connectivity, and telecommunications equipment, enabling remote consultations, medical diagnosis, and health information exchange. Access to clean cooking facilities powered by modern energy sources, such as liquefied petroleum gas (LPG) or electric stoves, contributes to improved indoor air quality and reduces the risk of respiratory diseases among households in the Congo Basin. Promoting clean cooking solutions can enhance public health outcomes and reduce the burden of disease. Despite the importance of energy for health, the Congo Basin faces several challenges in meeting the energy needs of healthcare facilities. These include inadequate energy infrastructure, unreliable electricity supply, high energy costs, and limited access to modern energy services, particularly in rural and remote areas. The Congo Basin is endowed with abundant renewable energy resources, including solar, hydro, and biomass potential, which can be harnessed to meet the energy needs of healthcare facilities sustainably. Investments in renewable energy infrastructure, technology adoption, and capacity building can enhance energy access and support healthcare delivery in the region. Governments in the Congo Basin can play a crucial role in promoting energy for health through supportive policies, regulations, and investments. This includes prioritizing electrification of healthcare facilities, providing incentives for renewable energy adoption, and integrating energy access into national health strategies and plans. Energy access is essential for ensuring effective healthcare delivery, promoting public health, and achieving universal health coverage in the Congo Basin.

7.3.2.7 Energy for water

Energy plays a crucial role in water management, sanitation, and access to clean water in the Congo Basin (Omole and Ndambuki 2014, De Angelis *et al* 2021). Access to reliable and affordable energy sources is essential for pumping, treating, and distributing water to communities, households, and industries in the Congo Basin. Electricity and diesel-powered pumps are commonly used to extract groundwater from boreholes, wells, and surface water sources for domestic, agricultural, and industrial use. Energy is required to power water treatment plants and facilities that purify and disinfect water to meet drinking water standards in the Congo Basin. Treatment processes such as filtration, chlorination, and desalination rely on electricity, diesel generators, or solar energy to remove contaminants and pathogens from water sources (Mboumboue and Njomo 2016). Energy access is essential for supporting sanitation services, including wastewater treatment, sewage disposal, and solid waste management, in the Congo Basin. Treatment plants, pumping stations, and sewer networks require electricity or alternative energy sources to operate effectively and prevent environmental pollution and public health risks. Energy-intensive irrigation systems, such as electric and diesel pumps, play a vital role in supporting agricultural productivity and food security in the Congo Basin. Access to reliable energy sources for irrigation can enhance crop yields, reduce water stress, and mitigate the impacts of climate variability on agriculture. The Congo Basin is endowed with abundant hydropower potential, with numerous rivers, waterfalls, and hydroelectric sites suitable for power generation. Hydropower projects can provide clean and renewable energy for water pumping, treatment, and distribution, as well as electricity for households, industries, and irrigation schemes.

In remote and off-grid areas of the Congo Basin, where access to grid electricity is limited or non-existent, off-grid renewable energy solutions, such as solar-powered pumps and water treatment systems, can provide decentralized water supply and sanitation services for communities and households. Despite the importance of energy for water management, the Congo Basin faces several challenges in meeting the energy needs of the water sector. These include inadequate energy infrastructure, unreliable electricity supply, high energy costs, and limited access to modern energy services, particularly in rural and underserved areas. The Congo Basin is endowed with abundant renewable energy resources, including solar, hydro, and biomass potential, which can be harnessed to meet the energy needs of the water sector sustainably. Investments in renewable energy infrastructure, technology adoption, and capacity building can enhance energy access and support water management in the region. Governments in the Congo Basin can play a crucial role in promoting energy for water through supportive policies, regulations, and investments. This includes prioritizing the electrification of water infrastructure, providing incentives for renewable energy adoption, and integrating energy access into national water and sanitation strategies and plans. Energy access is essential for ensuring sustainable water management, sanitation, and access to clean water in the Congo Basin. Addressing energy challenges, promoting renewable energy solutions, and fostering collaboration among stakeholders are vital to harnessing the transformative potential of energy for water and improve the well-being of its population (Nagabhatla *et al* 2021, Nsah 2022).

7.3.2.8 Energy for women

Energy plays a significant role in shaping the lives of women in the Congo Basin, influencing their health, economic opportunities, and social empowerment. Women in the Congo Basin often bear the primary responsibility for household chores, including cooking, cleaning, and childcare. Access to reliable and clean energy sources for cooking and lighting can alleviate the burden on women, freeing up time for education, income-generating activities, and community participation (Mboumboue and Njomo 2016). Traditional cooking methods, such as open fires and rudimentary stoves fueled by biomass, pose significant health risks to women and their families due to indoor air pollution. Access to clean cooking solutions, such as improved cookstoves, biogas digesters, and solar cookers, can improve indoor air quality, reduce respiratory illnesses, and empower women to lead healthier lives. Energy access can create economic opportunities for women in the Congo Basin, particularly in rural and underserved areas. Women entrepreneurs can benefit from off-grid renewable energy solutions, such as solar-powered irrigation systems, microgrids, and energy-efficient appliances, to start businesses, generate income, and improve livelihoods. Reliable electricity access at home and in schools can support women and girls' education in the Congo Basin. Lighting powered by solar energy enables extended study hours, access to digital learning resources, and participation in educational programs, enhancing literacy rates and educational outcomes among women and girls.

Access to reliable energy sources is essential for supporting women's health and well-being in healthcare facilities. Electricity powers medical equipment, lighting, and refrigeration units for storing vaccines and medicines, ensuring access to quality maternal and reproductive health services, prenatal care, and childbirth facilities (World Health Organization 2016). Inadequate lighting in public spaces and households increases the risk of gender-based violence against women and girls in the Congo Basin. Access to reliable electricity and street lighting can enhance safety and security, reduce the fear of violence, and promote women's mobility and participation in public life. Energy access can empower women to participate in decision-making processes and community development initiatives in the Congo Basin (Casati *et al* 2023). Women's involvement in energy planning, project implementation, and leadership roles can ensure that their voices are heard, needs are addressed, and rights are respected. Governments and stakeholders in the Congo Basin can promote gender-responsive energy policies and programs that prioritize women's needs, rights, and participation. This includes integrating gender considerations into energy planning, providing financial incentives for women entrepreneurs, and implementing gender-sensitive energy access initiatives. Energy access is central to promoting gender equality and women's empowerment in the Congo Basin.

7.3.2.9 Energy for the environment

Energy production and consumption have profound implications for the environment in the Congo Basin, influencing biodiversity, ecosystem health, and climate change (Yufenyuy *et al* 2023). The Congo Basin is vulnerable to the impacts of climate change, including changes in temperature, rainfall patterns, and extreme weather events. Energy-related activities, such as deforestation, fossil fuel combustion, and industrial emissions, contribute to greenhouse gas emissions, exacerbating climate change impacts in the region (Ngbolua *et al* 2018). Energy production from biomass fuels, such as firewood and charcoal, contributes to deforestation and habitat loss in the Congo Basin. Unsustainable harvesting practices for fuelwood and land clearing for energy crops threaten biodiversity, ecosystem services, and the resilience of forest ecosystems in the region. Traditional cooking methods, such as open fires and rudimentary stoves fueled by biomass, generate indoor and outdoor air pollution, leading to respiratory illnesses and environmental degradation. Access to clean cooking solutions, renewable energy technologies, and energy-efficient appliances can reduce emissions and improve air quality in the Congo Basin (Mboumboue and Njomo 2016). Energy production and hydropower development in the Congo Basin can impact water resources, river ecosystems, and aquatic biodiversity. Large-scale hydropower projects may alter river flows, sedimentation patterns, and aquatic habitats, affecting fish populations, water quality, and downstream ecosystems. Energy infrastructure development, such as roads, transmission lines, and power plants, can lead to land use change, fragmentation of habitats, and loss of biodiversity in the Congo Basin. Sustainable siting, planning, and management of energy projects are essential for minimizing environmental impacts and protecting sensitive ecosystems.

The Congo Basin is endowed with abundant renewable energy resources, including solar, hydro, biomass, and geothermal potential, which can be harnessed to meet energy needs sustainably. Investments in renewable energy infrastructure, technology innovation, and policy support can reduce dependence on fossil fuels and mitigate environmental impacts. Improving energy efficiency in the Congo Basin can reduce energy consumption, lower greenhouse gas emissions, and promote sustainable development. Energy-efficient technologies, building codes, and appliance standards can enhance resource efficiency, minimize waste, and support a transition to a low-carbon economy. Engaging local communities, indigenous peoples, and civil society organizations is essential for promoting sustainable energy practices and environmental stewardship in the Congo Basin. Participatory approaches, community-based natural resource management, and awareness-raising campaigns can empower communities to conserve biodiversity and protect the environment. Energy production and use have significant environmental implications for the Congo Basin, impacting biodiversity, ecosystem health, and climate resilience.

7.3.3 How renewable energy can meet the needs of the rural and urban poor in the Congo Basin

Renewable energy holds significant potential to meet the energy needs of both rural and urban poor communities in the Congo Basin through decentralized energy access, off-grid solutions, clean cooking technologies, income generation, agricultural productivity, health benefits, climate resilience, as well as policy support and financing (Dountio *et al* 2016 Nsafon *et al* 2020, Yimen *et al* 2022). Many rural and peri-urban areas in the Congo Basin lack access to centralized electricity grids. Renewable energy technologies, such as solar panels, small-scale hydro, and biomass systems, offer decentralized solutions that can provide reliable electricity access to off-grid communities. Off-grid renewable energy solutions, such as solar home systems, mini-grids, and micro-hydro systems, can provide clean and affordable electricity to rural households and communities in the Congo Basin. These solutions empower individuals to meet their energy needs for lighting, cooking, and powering appliances without relying on expensive and polluting fossil fuels. Traditional biomass fuels, such as firewood and charcoal, are commonly used for cooking in rural and urban poor households, contributing to indoor air pollution and deforestation. Clean cooking technologies, such as improved cookstoves, biogas digesters, and solar cookers, offer sustainable alternatives that reduce fuelwood consumption, improve indoor air quality, and mitigate environmental impacts (Kenfack *et al* 2017). Renewable energy technologies can create income-generating opportunities for rural and urban poor communities in the Congo Basin. Small-scale solar installations, for example, can provide households with access to electricity for charging mobile phones, running small businesses, and selling surplus energy back to the grid through net metering arrangements.

Renewable energy solutions, such as solar-powered irrigation systems and biomass gasifiers, can enhance agricultural productivity and income generation for rural farmers in the Congo Basin (Abanda 2012). Access to clean and reliable energy for irrigation, processing, and post-harvest handling can increase crop yields, improve food security, and stimulate rural development. Transitioning to renewable energy sources for cooking and lighting can improve public health outcomes in the Congo Basin. Clean cooking technologies reduce indoor air pollution, respiratory illnesses, and premature deaths associated with traditional biomass use, particularly among women and children in rural households (Mboumboue and Njomo 2016). Renewable energy technologies offer climate-resilient solutions that mitigate the impacts of climate change and enhance community resilience in the Congo Basin.

Solar, hydro, and wind power are abundant and sustainable energy sources that can reduce reliance on fossil fuels, lower greenhouse gas emissions, and promote climate adaptation and mitigation efforts. Governments, development agencies, and international organizations can support the deployment of renewable energy solutions for rural and urban poor communities in the Congo Basin through supportive policies, regulatory frameworks, and financing mechanisms. Incentives such as subsidies, tax breaks, and microfinance initiatives can make renewable energy technologies more accessible and affordable for underserved populations (Flora *et al* 2019).

7.4 Challenges to harnessing the renewable energy potentials in the Congo Basin to meet the needs of the rural and urban poor

Harnessing the renewable energy potential in the Congo Basin to meet the needs of the rural and urban poor presents a promising pathway towards sustainable development and energy access in the region. However, numerous challenges hinder the realization of this potential.

7.4.1 Limited infrastructure

The lack of adequate energy infrastructure, including transmission lines, distribution networks, and storage facilities, poses a significant barrier to harnessing renewable energy resources in the Congo Basin (Kenfack *et al* 2017). Insufficient grid connectivity and infrastructure gaps hinder the deployment of renewable energy solutions, particularly in remote and underserved areas. Limited infrastructure poses a significant challenge to harnessing the renewable energy potentials in the Congo Basin mainly in the aspects of grid connectivity, transmission and distribution networks, storage and backup systems, transportation and logistics, access to water resources, rural electrification, and technology deployment and maintenance. Many areas in the Congo Basin, particularly rural and remote communities, lack access to centralized electricity grids. The limited grid connectivity hinders the deployment of large-scale renewable energy projects such as solar and wind farms, as well as the integration of distributed generation systems such as mini-grids and microgrids. Weak transmission and distribution networks in the Congo Basin impede the efficient and reliable delivery of electricity generated from renewable energy sources to end-users (Maupin 2017). Inadequate infrastructure for transmitting and distributing power from renewable energy projects to urban centers, industrial zones, and rural communities limits the accessibility and affordability of clean energy solutions. Limited infrastructure for energy storage and backup systems in the Congo Basin poses challenges for integrating intermittent renewable energy sources such as solar and wind power into the grid. Inadequate storage capacity, battery technology, and backup systems hinder the ability to store excess renewable energy during periods of low demand and ensure grid stability and reliability.

Weak transportation infrastructure and logistical challenges in the Congo Basin impede the delivery, installation, and maintenance of renewable energy equipment and components (Dominguez and Foster 2011). Poor road conditions, limited access to ports, and customs procedures delay the importation and distribution of renewable energy technologies, increasing project costs and timelines. Hydropower, a significant renewable energy source in the Congo Basin, relies on access to water resources and the availability of suitable sites for hydroelectric development. Limited infrastructure for water management, dam construction, and reservoir operations may constrain the development of hydropower projects and the utilization of this renewable energy resource. Limited infrastructure for rural electrification poses challenges for extending electricity access to remote and underserved areas in the Congo Basin. The high costs of extending the grid to rural communities, coupled with geographical barriers and terrain challenges, hinder efforts to provide clean and reliable energy services to off-grid populations. Inadequate infrastructure for technology deployment, installation, and maintenance hampers the scalability and sustainability of renewable energy projects in the Congo Basin. Limited access to skilled labor, spare parts, and repair services undermines the performance and reliability of renewable energy systems, increasing downtime and operational costs. Addressing limited infrastructure requires strategic investments in energy infrastructure development, grid expansion, transportation networks, and technology deployment initiatives.

7.4.2 Financial constraints and high transition costs

Financing renewable energy projects and investments in the Congo Basin remains a challenge, especially for rural and urban poor communities with limited access to capital and credit. This is mainly through high initial investment costs, limited access to financing, uncertain return on investment, transition costs and economic impacts, subsidies and incentives, limited revenue streams, as well as capacity building and technical assistance (Onyeji-Nwogu 2017). The upfront costs of renewable energy infrastructure, such as solar panels, wind turbines, and hydropower plants, can be substantial, particularly for large-scale projects. High capital investment requirements deter investors, developers, and governments from initiating renewable energy projects in the Congo Basin, especially in rural and underserved areas where energy demand is high but financial resources are limited. Financial constraints, including limited access to capital, credit, and investment, pose barriers to funding renewable energy projects in the Congo Basin (Kidmo *et al* 2021). Domestic financial markets may lack the liquidity, expertise, and risk appetite to support renewable energy investments, while international financing sources may have stringent eligibility criteria or impose high interest rates and repayment terms. Investors and financiers may perceive renewable energy projects in the Congo Basin as high-risk ventures with uncertain returns on investment. Factors such as political instability, regulatory uncertainties, and currency fluctuations contribute to investor skepticism and risk aversion, making it challenging to attract private sector investment and mobilize capital for renewable energy initiatives. Transitioning from traditional energy sources, such as fossil fuels and biomass, to renewable energy technologies entails significant transition costs and economic impacts for the Congo Basin (Ijoma *et al* 2022). Industries, communities, and households reliant on conventional energy sources may face job losses, income disruptions, and adjustment challenges during the transition to renewable energy, exacerbating socio-economic inequalities and livelihood vulnerabilities.

Inadequate government subsidies, incentives, and support mechanisms for renewable energy development in the Congo Basin hinder market competitiveness and hinder the transition to clean energy solutions. Lack of fiscal incentives, feed-in tariffs, tax breaks, and investment guarantees limit the attractiveness of renewable energy investments and hinder the growth of the renewable energy sector. Generating revenue streams from renewable energy projects in the Congo Basin may be challenging due to factors such as low electricity tariffs, limited off-take agreements, and unreliable payment mechanisms. Insufficient revenue generation potential undermines the financial viability and sustainability of renewable energy projects, discouraging investment and hindering market growth (Amir and Khan 2022). Limited capacity and technical expertise in project development, financing, and risk management constrain efforts to overcome financial constraints and mobilize resources for renewable energy initiatives in the Congo Basin. Capacity building initiatives, technical assistance programs, and knowledge sharing platforms are needed to enhance financial literacy, investment readiness, and project bankability in the renewable energy sector (Niyibizi 2015). Addressing financial constraints and high transition costs requires a multi-faceted approach that involves governments, development partners, financiers, and stakeholders collaborating to mobilize resources, mitigate risks, and create enabling environments for renewable energy investment and deployment (Kenfack *et al* 2017).

7.4.3 Policy and regulatory barriers

Inconsistent or inadequate policy frameworks, regulations, and administrative procedures create uncertainties and barriers for renewable energy development in the Congo Basin. This is mainly through inconsistent policies and regulations, lack of supportive legal frameworks, bureaucratic procedures and red tape, limited capacity for policy implementation, lack of transparency and accountability, policy inertia and resistance to change, as well as lack of long-term planning and vision (Ouedraogo 2019). The lack of coherent and consistent policies and regulations for renewable energy development in the Congo Basin creates uncertainties for investors, developers, and stakeholders. Inconsistent policy frameworks, conflicting regulations, and overlapping mandates among government agencies and ministries hinder the development, financing, and implementation of renewable energy projects (Bishoge *et al* 2020). Inadequate legal frameworks and regulatory frameworks for renewable energy development limit the attractiveness of the sector to investors and financiers. Absence of clear laws, regulations, and guidelines governing renewable energy projects, such as feed-in tariffs, power purchase agreements, and grid access rules, impedes market competitiveness and inhibits private sector participation in the renewable energy sector. Lengthy bureaucratic procedures, complex permitting processes, and administrative delays pose barriers to project development and implementation in the Congo Basin. Cumbersome regulatory requirements, multiple approvals, and overlapping jurisdictional responsibilities increase transaction costs, project timelines, and risks for renewable energy investors and developers.

Weak institutional capacity for policy implementation and enforcement undermines the effectiveness of renewable energy policies and regulations in the Congo Basin. Inadequate resources, technical expertise, and monitoring mechanisms hinder the enforcement of renewable energy standards, compliance with regulatory requirements, and oversight of renewable energy projects (Njoh *et al* 2019). Transparency and accountability deficits in policy-making processes and regulatory decision-making undermine public trust and confidence in renewable energy governance in the Congo Basin. Lack of transparency in licensing, permitting, and procurement processes may lead to corruption, rent-seeking behavior, and regulatory capture, impeding fair competition and market development in the renewable energy sector (Labordena *et al* 2017). Resistance to policy reforms, inertia in government institutions, and vested interests in the status quo may hinder the adoption of renewable energy policies and regulations in the Congo Basin. Political opposition, industry lobbying, and institutional barriers to change limit the willingness of policymakers to enact transformative policies that promote renewable energy deployment and transition away from fossil fuels. Short-term planning horizons, lack of strategic vision, and political uncertainties contribute to policy and regulatory instability in the Congo Basin. Absence of long-term energy planning, sectoral coordination, and multi-stakeholder engagement hinder the development of integrated, holistic approaches to renewable energy governance and sustainable development in the region. Addressing policy and regulatory barriers requires concerted efforts to strengthen legal frameworks, enhance policy coherence, and promote good governance principles in the renewable energy sector. By fostering political will, building institutional capacity, and fostering stakeholder engagement, the Congo Basin can overcome policy and regulatory barriers and unlock its renewable energy potential to achieve sustainable development goals, energy access targets, and poverty reduction objectives (Onyeji-Nwogu 2017).

7.4.4 Technical capacity and skills

The shortage of technical expertise, trained personnel, and skilled workforce in renewable energy technologies and project management limits the implementation and scalability of renewable energy initiatives in the Congo Basin mainly through lack of qualified personnel, limited access to training and education, brain drain and talent migration, limited research and development (R&D) capacities, inadequate technical support services, skills mismatch and training needs, and gender disparities in technical fields (Ijoma *et al* 2022). The Congo Basin faces a shortage of qualified professionals with expertise in renewable energy technologies, project management, and engineering. Limited availability of skilled personnel, including engineers, technicians, and project managers, hampers the planning, implementation, and maintenance of renewable energy projects in the region. Inadequate access to training programs, vocational schools, and higher education institutions offering renewable energy courses restricts the development of technical skills and expertise in the Congo Basin. Lack of formal training opportunities and specialized curricula for renewable energy disciplines limits the pool of qualified professionals entering the renewable energy workforce. Brain drain, talent migration, and emigration of skilled workers to other regions or countries exacerbate skills shortages and capacity gaps in the renewable energy sector in the Congo Basin. Limited career opportunities, low salaries, and inadequate professional development prospects may drive qualified professionals to seek employment opportunities abroad, further depleting the local talent pool.

Weak R&D capacities, innovation ecosystems, and technology transfer mechanisms impede the development and adaptation of renewable energy technologies in the Congo Basin. Limited funding, infrastructure, and collaboration networks for R&D activities constrain innovation, technology diffusion, and knowledge creation in the renewable energy sector. Limited availability of technical support services, consulting firms, and engineering firms specializing in renewable energy projects restricts access to expertise and resources for project development and implementation in the Congo Basin. Insufficient capacity for feasibility studies, project design, and technical assessments undermines the quality and effectiveness of renewable energy initiatives. The mismatch between skills demand and supply in the renewable energy sector exacerbates capacity constraints and skills shortages in the Congo Basin. Lack of alignment between educational curricula, industry requirements, and job market demands results in skills mismatches, underemployment, and labor market inefficiencies in the renewable energy workforce. Gender disparities in technical fields, such as engineering and renewable energy, further exacerbate capacity gaps and skills shortages in the Congo Basin. Limited access to education, training, and employment opportunities for women in STEM (science, technology, engineering, and mathematics) disciplines hinders gender equality and women's participation in the renewable energy workforce. Addressing technical capacity and skills challenges requires multi-faceted approaches that involve investing in education, training, and professional development initiatives, fostering innovation and entrepreneurship, and promoting collaboration and knowledge exchange networks in the renewable energy sector.

7.4.5 Intermittency and reliability

The intermittent nature of renewable energy sources, such as solar and wind power, poses challenges for ensuring reliable and consistent energy supply in the Congo Basin mainly via the intermittent nature of renewable energy sources, grid integration challenges, energy storage limitations, backup generation requirements, hydropower variability, demand-side management, and grid infrastructure investment (Abanda 2012). Renewable energy sources such as solar and wind power are inherently intermittent, meaning their availability fluctuates based on weather conditions and time of day. In the Congo Basin, where solar and wind resources vary seasonally and geographically, the intermittent nature of these energy sources poses challenges for ensuring reliable and consistent electricity supply. Integrating intermittent renewable energy sources into the electricity grid presents technical challenges related to grid stability, balancing supply and demand, and managing fluctuations in power output. The lack of grid infrastructure and grid management capabilities in the Congo Basin complicates the integration of variable renewable energy generation, leading to grid instability and reliability concerns. Limited energy storage capacity in the Congo Basin hampers efforts to mitigate the intermittency of renewable energy sources and ensure reliable electricity supply. Energy storage technologies such as batteries, pumped hydro storage, and thermal storage systems are essential for storing excess renewable energy during periods of high generation and releasing it when demand exceeds supply. However, the high costs and technical challenges associated with energy storage hinder widespread deployment in the region. Due to the intermittent nature of renewable energy sources, backup generation capacity is needed to ensure reliable electricity supply during periods of low renewable energy generation. In the Congo Basin, where reliance on backup generation from fossil fuel sources may be necessary to meet peak demand and maintain grid stability, the transition to renewable energy faces challenges related to the availability, affordability, and environmental impacts of backup generation technologies.

Hydropower, a significant renewable energy source in the Congo Basin, is subject to seasonal variations in river flows and rainfall patterns. Changes in hydrological conditions, such as droughts or floods, can affect hydropower generation capacity and reliability, posing challenges for energy planning, water management, and climate resilience in the region. Implementing demand-side management measures, such as energy efficiency programs, demand response initiatives, and flexible pricing schemes, can help mitigate the intermittency of renewable energy sources by adjusting electricity consumption patterns to match supply fluctuations. However, limited awareness, incentives, and regulatory frameworks for demand-side management in the Congo Basin hinder efforts to optimize energy use and enhance grid flexibility. Enhancing grid infrastructure and transmission networks is essential for accommodating higher levels of renewable energy penetration and ensuring grid reliability in the Congo Basin. Investments in grid expansion, grid modernization, and smart grid technologies are needed to strengthen the resilience and flexibility of the electricity grid and support the integration of renewable energy sources. Addressing intermittency and reliability challenges requires a comprehensive approach that combines investments in grid infrastructure, energy storage technologies, demand-side management measures, and policy and regulatory reforms to support renewable energy deployment in the Congo Basin.

7.4.6 Social and cultural factors

Socio-cultural norms, perceptions, and behaviors may influence the acceptance and adoption of renewable energy technologies among rural and urban poor communities in the Congo Basin through traditional energy practices, perceptions of modern energy technologies, gender dynamics, community engagement and participation, cultural heritage and environmental conservation, religious beliefs and practices, as well as community resilience and adaptation (Kenfack *et al* 2017). Many communities in the Congo Basin have longstanding traditions and cultural practices related to energy use, such as reliance on biomass fuels for cooking and heating. The transition to renewable energy technologies may face resistance or skepticism due to ingrained habits, cultural norms, and attachment to traditional energy sources. Social attitudes and perceptions towards modern energy technologies, such as solar panels and wind turbines, may vary among different cultural groups in the Congo Basin. Lack of awareness, misconceptions, and distrust of unfamiliar technologies may hinder acceptance and adoption of renewable energy solutions, particularly in rural and remote areas. Gender roles and dynamics influence energy access, decision-making, and participation in the Congo Basin. Women often bear the primary responsibility for household energy management, including cooking, lighting, and water heating, yet they may have limited access to and control over modern energy technologies. Addressing gender disparities and promoting women's empowerment are essential for ensuring inclusive and equitable access to renewable energy solutions.

Meaningful engagement and participation of local communities in renewable energy projects are crucial for their success and sustainability. However, social hierarchies, power dynamics, and lack of community involvement in decision-making processes may hinder community acceptance and ownership of renewable energy initiatives in the Congo Basin. The Congo Basin is rich in cultural heritage, biodiversity, and natural resources that are deeply intertwined with local identities and traditions. Balancing the promotion of renewable energy development with the protection of cultural landscapes, sacred sites, and environmental conservation areas requires careful consideration of cultural values, indigenous knowledge, and community preferences. Religious beliefs and practices may influence attitudes towards environmental conservation, sustainable development, and renewable energy adoption in the Congo Basin. Religious leaders and institutions can play a significant role in promoting environmental stewardship, advocating for renewable energy solutions, and raising awareness about the interconnectedness of social, environmental, and spiritual well-being. Building community resilience and adaptation capacities to climate change impacts, such as extreme weather events and natural disasters, is essential for promoting sustainable energy solutions in the Congo Basin. Integrating renewable energy technologies with community-based adaptation strategies, disaster risk reduction measures, and livelihood diversification initiatives can enhance resilience and promote sustainable development outcomes. Addressing social and cultural factors requires culturally sensitive approaches, community engagement strategies, and participatory decision-making processes that respect local knowledge, values, and aspirations. By fostering dialogue, promoting social inclusion, and incorporating cultural perspectives into renewable energy planning and implementation, the Congo Basin can overcome social and cultural barriers and harness its renewable energy potentials to achieve sustainable development goals and energy access targets.

7.4.7 Access to land and resources

Land tenure issues, competing land uses, and resource conflicts may hinder the development of renewable energy projects, such as hydropower dams and bioenergy plantations, in the Congo Basin largely due to land tenure issues, competition for land use, environmental and social impact assessments, protected areas and biodiversity conservation, community land rights and indigenous peoples' rights, infrastructure development and access, and policy and regulatory frameworks (Ouedraogo 2019). Land tenure insecurity, overlapping land claims, and competing land uses pose challenges for siting renewable energy projects in the Congo Basin. Unclear land tenure arrangements, informal land tenure systems, and unresolved land disputes may hinder access to suitable sites for renewable energy development, leading to project delays and conflicts with local communities. The Congo Basin is characterized by diverse land uses, including agriculture, forestry, conservation, and urban development. Competition for land use between renewable energy projects and other economic activities, such as agriculture and mining, may arise, particularly in areas with high renewable energy potential. Balancing competing land uses and interests requires effective land use planning and stakeholder engagement processes. Renewable energy projects in the Congo Basin must undergo rigorous environmental and social impact assessments (ESIAs) to assess potential environmental and social risks and impacts. Limited capacity for conducting ESIAs, a lack of data and expertise, and challenges in stakeholder consultation may delay project approvals and increase regulatory compliance costs for renewable energy developers.

The Congo Basin is home to unique biodiversity hotspots, protected areas, and critical ecosystems that require special conservation considerations. Siting renewable energy projects in or near protected areas may conflict with biodiversity conservation objectives and trigger environmental concerns related to habitat loss, fragmentation, and wildlife disturbance. Indigenous peoples and local communities in the Congo Basin have customary land rights and traditional knowledge systems that are often not recognized or respected by formal legal frameworks. Ensuring free, prior, and informed consent (FPIC) and meaningful participation of indigenous peoples and local communities in renewable energy decision-making processes is essential for respecting land rights, safeguarding cultural heritage, and promoting social equity. Access to land and resources for renewable energy projects may be constrained by limited infrastructure and logistical challenges in the Congo Basin. Inadequate transportation networks, poor road conditions, and logistical constraints may hinder the delivery of renewable energy equipment, construction materials, and workforce to project sites, increasing project costs and timelines. Clear and transparent policy and regulatory frameworks governing land use, land tenure, and resource management are essential for facilitating renewable energy development in the Congo Basin. Strengthening land governance, clarifying land tenure rights, and streamlining permitting processes can reduce investment risks and promote sustainable land use practices compatible with renewable energy objectives. Addressing access to land and resources challenges requires participatory approaches, stakeholder engagement mechanisms, and multi-sectoral collaboration to ensure that renewable energy projects respect land rights, environmental safeguards, and social considerations in the Congo Basin.

7.4.8 Poor governance

Poor governance poses significant challenges to harnessing renewable energy resources in the Congo Basin, hindering sustainable development, energy access, and poverty alleviation efforts in the region. Poor governance affects the renewable energy sector through corruption and mismanagement, weak regulatory frameworks, political instability, lack of planning and coordination, limited capacity and expertise, resource misallocation, and exclusion of marginalized communities (Burnley 2011, Njoh *et al* 2019, Olanrewaju *et al* 2019). Corruption, bribery, and mismanagement of resources undermine efforts to develop and deploy renewable energy projects in the Congo Basin. Lack of transparency and accountability in procurement processes, licensing, and project implementation can lead to inefficiencies, cost overruns, and delays in renewable energy initiatives. Inadequate or poorly enforced regulatory frameworks for renewable energy development create uncertainties and barriers for investors, developers, and stakeholders in the Congo Basin. Absence of clear policies, regulations, and standards for renewable energy projects can deter investment, innovation, and market growth in the sector. Political instability, governance challenges, and conflicts in the Congo Basin contribute to an unpredictable investment climate and hinder the development of renewable energy infrastructure and projects. Political interference, lack of institutional capacity, and governance deficits undermine the stability and sustainability of renewable energy initiatives. Fragmented governance structures, overlapping mandates, and lack of coordination among government agencies, ministries, and stakeholders impede effective planning and implementation of renewable energy policies and projects in the Congo Basin. Incoherent policies, conflicting priorities, and institutional silos hamper progress towards achieving energy access goals and sustainable development objectives.

Weak institutional capacity, inadequate technical expertise, and skills gaps in renewable energy planning, regulation, and management constrain the development and deployment of renewable energy solutions in the Congo Basin. Capacity building, training programs, and knowledge exchange initiatives are needed to strengthen governance capacities and skills in the sector. Inefficient allocation of resources, budgetary constraints, and competing priorities within government budgets limit investments in renewable energy infrastructure and technology deployment in the Congo Basin. Resource misallocation, lack of prioritization, and short-term planning undermine long-term sustainability and resilience of the energy sector. Poor governance practices often result in the exclusion of marginalized communities, indigenous peoples, and vulnerable populations from decision-making processes, benefit sharing, and participation in renewable energy projects in the Congo Basin. Lack of consultation, participation, and representation exacerbates social inequalities and undermines social acceptance of renewable energy initiatives. Addressing poor governance requires concerted efforts to strengthen institutions, enhance transparency, promote accountability, and foster inclusive governance processes in the Congo Basin.

7.4.9 Weak institutional frameworks

Weak institutional frameworks pose significant challenges to harnessing renewable energy potentials in the Congo Basin, hindering sustainable development, energy access, and poverty alleviation efforts in the region. Weak institutional frameworks mainly affect the renewable energy sector through policy and regulatory uncertainty, ineffective planning and coordination, limited capacity and expertise, lack of financing and investment, poor project management and implementation, limited stakeholder engagement and participation, as well as lack of accountability and transparency (Njoh *et al* 2019). Weak institutional frameworks result in uncertain policy environments and inadequate regulatory frameworks for renewable energy development in the Congo Basin. Absence of clear policies, regulations, and standards creates uncertainties for investors, developers, and stakeholders, leading to a lack of confidence and investment in renewable energy projects. Weak institutional coordination and planning mechanisms among government agencies, ministries, and stakeholders impede the effective development and implementation of renewable energy policies and projects in the Congo Basin. Fragmented governance structures, overlapping mandates, and institutional silos hinder progress towards achieving energy access goals and sustainable development objectives. Institutions responsible for renewable energy planning, regulation, and management often lack the capacity, technical expertise, and skills needed to effectively promote and support renewable energy initiatives in the Congo Basin. Insufficient human resources, training programs, and knowledge sharing initiatives constrain institutional capacities to develop and implement renewable energy projects.

Weak institutional frameworks contribute to limited access to financing and investment for renewable energy projects in the Congo Basin. Inadequate mechanisms for project financing, risk mitigation, and public–private partnerships deter private sector investment, innovation, and market development in the renewable energy sector. Weak institutional capacities for project management and implementation lead to inefficiencies, delays, and cost overruns in renewable energy projects in the Congo Basin. Lack of effective project management systems, monitoring mechanisms, and performance indicators undermines the successful execution and delivery of renewable energy initiatives. Weak institutional frameworks often result in limited stakeholder engagement, consultation, and participation in decision-making processes related to renewable energy development in the Congo Basin. Exclusion of local communities, indigenous peoples, and civil society organizations from decision-making processes exacerbates social tensions, conflicts, and opposition to renewable energy projects. Weak institutional frameworks contribute to a lack of accountability and transparency in the management of renewable energy resources and projects in the Congo Basin. Inadequate mechanisms for oversight, monitoring, and evaluation of renewable energy initiatives increase the risk of corruption, mismanagement, and resource exploitation. Addressing weak institutional frameworks requires concerted efforts to strengthen institutions, enhance governance structures, and promote accountability and transparency in the renewable energy sector.

7.4.10 Limited technology transfer

Limited technology transfer presents a significant challenge to harnessing the renewable energy potentials in the Congo Basin. Limited technology transfer affects the renewable energy potential of Africa through dependency on imported technologies, high costs and affordability, lack of local capacity and expertise, infrastructure and logistics challenges, intellectual property rights and technology transfer agreements, capacity building and technology adaptation, as well as policy support and regulatory frameworks (Ouedraogo 2019, Bishoge *et al* 2020, Monyei *et al* 2022). The Congo Basin often relies on imported renewable energy technologies due to limited domestic manufacturing and research capabilities. However, the transfer of these technologies may be constrained by intellectual property rights, licensing agreements, and proprietary knowledge, leading to dependency on foreign suppliers and hindering local innovation and technology diffusion. Imported renewable energy technologies may be expensive, making them less accessible and affordable for rural and urban poor communities in the Congo Basin. High upfront costs, import tariffs, and currency fluctuations further exacerbate the financial barriers to technology adoption and deployment, limiting access to clean energy solutions. Limited local capacity and technical expertise in renewable energy technology development, installation, and maintenance hinder the effective transfer and utilization of renewable energy technologies in the Congo Basin. Inadequate training programs, skills development initiatives, and knowledge exchange platforms constrain the ability of local actors to adopt and adapt renewable energy solutions to local contexts.

Weak infrastructure, inadequate transportation networks, and logistical challenges in the Congo Basin may impede the transfer of renewable energy technologies from suppliers to end-users. Poor road conditions, limited access to ports, and customs procedures delay the delivery and installation of renewable energy equipment, prolonging project timelines and increasing costs. Intellectual property rights, technology transfer agreements, and licensing arrangements may restrict the transfer of proprietary renewable energy technologies to the Congo Basin. Limited access to patented technologies, proprietary knowledge, and research findings may impede innovation, research collaboration, and technology diffusion in the region. Effective technology transfer requires capacity building initiatives, technology adaptation efforts, and knowledge sharing mechanisms to ensure the successful deployment and utilization of renewable energy technologies in the Congo Basin. Collaborative partnerships, technology transfer agreements, and South–South cooperation initiatives can facilitate knowledge exchange, skills transfer, and technology diffusion among countries in the region. Governments in the Congo Basin can promote technology transfer and innovation in the renewable energy sector through supportive policies, regulatory frameworks, and incentive mechanisms. Measures such as tax incentives, research grants, and technology transfer agreements can encourage investment in renewable energy technology development, manufacturing, and deployment in the region. Addressing limited technology transfer requires a multi-faceted approach that involves strengthening local capacities, fostering collaboration and partnerships, and promoting supportive policies and regulatory frameworks.

7.5 Prospects of harnessing the renewable energy potentials in the Congo Basin to meet the needs of the rural and urban poor

The Congo Basin, a vast expanse of tropical forests and diverse ecosystems spanning several Central African countries, holds immense renewable energy potentials waiting to be harnessed. In recent years, there has been a growing recognition of the role that renewable energy can play in addressing the energy needs of both rural and urban populations in the region. This introduction explores the prospects of leveraging renewable energy resources to alleviate energy poverty and promote sustainable development in the Congo Basin. Despite its rich renewable energy resources, including solar, hydroelectric, biomass, wind, and geothermal, the Congo Basin continues to grapple with energy poverty, where a significant portion of its population lacks access to reliable and affordable electricity. In rural areas, communities rely heavily on traditional biomass fuels for cooking and heating, leading to deforestation, indoor air pollution, and adverse health outcomes. In urban centers, unreliable electricity supply from fossil fuel-based power systems contributes to energy insecurity and impedes socio-economic progress. However, the Congo Basin stands at a pivotal juncture, poised to harness its renewable energy potentials to address these pressing energy challenges. The region's abundant solar radiation, extensive river systems suitable for hydropower generation, and vast biomass resources offer promising opportunities for scaling up renewable energy deployment.

Renewable energy solutions, such as off-grid solar systems, mini-grids, and small-scale hydropower projects, can provide decentralized electricity access to remote and underserved communities, empowering them with reliable energy services for lighting, cooking, education, healthcare, and productive activities. Moreover, renewable energy can contribute to poverty reduction, job creation, and economic empowerment by stimulating local entrepreneurship, fostering innovation, and expanding energy-related livelihood opportunities across the region. Furthermore, renewable energy offers environmental sustainability benefits, aligning with the Congo Basin's conservation objectives and climate change mitigation efforts. By reducing reliance on fossil fuels and mitigating greenhouse gas emissions, renewable energy can contribute to forest conservation, biodiversity preservation, and climate resilience in the region. Additionally, renewable energy technologies, such as improved cookstoves and biogas digesters, offer cleaner and more sustainable alternatives to traditional biomass fuels, promoting environmental health and reducing deforestation pressures. However, realizing the full potential of renewable energy in the Congo Basin requires concerted efforts and multi-stakeholder collaboration. It demands supportive policy frameworks, institutional capacities, financial investments, and community engagement mechanisms to overcome barriers and facilitate the transition to clean energy pathways.

7.5.1 Existence of national development strategies that factor in renewable energy

National development strategies in the Congo Basin can play a crucial role in promoting renewable energy and driving sustainable development outcomes. These strategies can contribute to advancing renewable energy deployment in the region through the integration of renewable energy targets, policy and regulatory frameworks, institutional capacity building, investment promotion and facilitation, rural electrification and energy access, climate change mitigation and adaptation, as well as community engagement and social inclusion (Ouedraogo 2019, Bishoge *et al* 2020, Adenle 2020). National development strategies can incorporate specific targets and objectives for renewable energy deployment, such as capacity additions, electrification rates, and greenhouse gas emission reductions. By setting ambitious renewable energy targets aligned with international commitments, such as the Paris Agreement, countries in the Congo Basin can signal their commitment to transitioning towards low-carbon energy systems and attracting investment in renewable energy projects. National development strategies can provide the policy and regulatory frameworks necessary to support renewable energy development. This includes enacting legislation, regulations, and incentive mechanisms to promote renewable energy investment, streamline permitting processes, and provide financial support, such as feed-in tariffs, tax incentives, and concessional financing, for renewable energy projects. National development strategies can prioritize building institutional capacities and strengthening governance structures for renewable energy planning, regulation, and implementation. This involves enhancing the technical expertise, human resources, and coordination mechanisms within government agencies responsible for energy planning, environmental management, and rural development to effectively oversee and support renewable energy initiatives. National development strategies can create an enabling environment for renewable energy investment by promoting public–private partnerships, mobilizing domestic and international financing, and facilitating project development and financing mechanisms. This includes establishing investment promotion agencies, providing risk guarantees, and supporting project preparation facilities to attract private sector investment and leverage public resources for renewable energy projects.

National development strategies can prioritize rural electrification and energy access as key development priorities, recognizing the importance of renewable energy solutions in reaching remote and underserved communities. By targeting investments in off-grid and decentralized renewable energy systems, such as minigrids, solar home systems, and small-scale hydropower, national development strategies can expand electricity access, improve livelihoods, and reduce poverty in rural areas. National development strategies can integrate renewable energy as a central component of climate change mitigation and adaptation efforts. By reducing reliance on fossil fuels and promoting renewable energy technologies with low carbon footprints, countries in the Congo Basin can contribute to global climate goals while enhancing energy security, resilience to climate impacts, and sustainable development outcomes. National development strategies can prioritize community engagement, social inclusion, and gender equality in renewable energy planning and implementation. This involves consulting with local communities, indigenous peoples, and marginalized groups to ensure that renewable energy projects are socially acceptable, culturally appropriate, and beneficial to all stakeholders, particularly the rural and urban poor. Overall, national development strategies in the Congo Basin can serve as powerful tools for promoting renewable energy as a driver of sustainable development, economic growth, and poverty alleviation.

7.5.2 Increasing adoption of green growth strategies

The increasing adoption of green growth principles in the Congo Basin is contributing to the promotion of renewable energy in several ways including through policy alignment, investment prioritization, economic diversification, environmental conservation, community engagement and social inclusion, as well as climate resilience. Green growth strategies prioritize sustainability and environmental protection while promoting economic development (Warnock 2013, Milburn 2014, Schoneveld and Zoomers 2015, Enongene and Fobissie 2016, Marijnen and Schouten 2019, Corbino 2021). By aligning renewable energy promotion with green growth objectives, governments in the Congo Basin are integrating renewable energy targets and incentives into national development plans and policies. This alignment creates a conducive environment for renewable energy investment and deployment, signaling government commitment to transitioning towards low-carbon energy systems. Green growth strategies emphasize investment in sustainable infrastructure and technologies that reduce environmental impacts and enhance resource efficiency. In the Congo Basin, increasing adoption of green growth principles directs investment towards renewable energy projects, such as solar, hydroelectric, and biomass, that offer cleaner alternatives to fossil fuels and contribute to mitigating climate change. As a result, renewable energy projects receive greater attention and financial support from public and private investors aligned with green growth objectives. Green growth strategies promote economic diversification and job creation in renewable energy sectors. By investing in renewable energy technologies and value chains, the Congo Basin can capitalize on its abundant renewable energy resources to stimulate economic growth, create employment opportunities, and enhance energy security. Renewable energy projects offer potential for local job creation in construction, manufacturing, operation, and maintenance, contributing to poverty reduction and socio-economic development in the region.

Green growth principles emphasize the conservation and sustainable management of natural resources and ecosystems. Increasing adoption of green growth in the Congo Basin encourages the development of renewable energy projects that minimize environmental impacts and promote biodiversity conservation. For example, hydropower projects can be designed to minimize ecological disruption and incorporate environmental safeguards, while solar and wind projects have minimal land footprint and carbon emissions compared to fossil fuel-based energy generation. Green growth strategies prioritize community engagement, social inclusion, and participatory decision-making processes. In the Congo Basin, increasing adoption of green growth principles in renewable energy planning involves consulting with local communities, indigenous peoples, and other stakeholders to ensure that renewable energy projects are socially acceptable, culturally appropriate, and beneficial to all stakeholders. Community participation in renewable energy projects enhances project acceptance, reduces social conflicts, and promotes equitable distribution of benefits, particularly for rural and marginalized populations. Green growth strategies promote climate resilience and adaptation measures to address the impacts of climate change. Renewable energy technologies, such as solar and wind, offer decentralized and off-grid solutions that are less susceptible to climate impacts and can improve energy access in remote and vulnerable communities. Globally, the increasing adoption of green growth principles in the Congo Basin is creating synergies between sustainable development objectives and renewable energy promotion. Integrating renewable energy into green growth strategies is key to accelerating the transition towards a low-carbon, resilient, and inclusive energy future that promotes environmental sustainability, economic prosperity, and social equity.

7.5.3 Adherence to Sustainable Development Goal number 7 on affordable and clean energy

Adherence to Sustainable Development Goal (SDG) number 7 on affordable and clean energy is promoting renewable energy in the Congo Basin in different ways such as policy prioritization, energy access, cleaner energy sources, climate action, energy efficiency, partnerships and collaboration (Jagger *et al* 2019, Ntirumenyerwa Mihigo and Cliquet 2020, Monyei *et al* 2022, Egbende *et al* 2023). SDG 7 calls for ensuring access to affordable, reliable, sustainable, and modern energy for all. In line with this goal, governments in the Congo Basin are prioritizing renewable energy development as a means to achieve universal energy access. Renewable energy policies and strategies are being formulated and implemented to align with SDG 7 targets, driving investment and action towards clean energy solutions. SDG 7 aims to ensure universal access to electricity by 2030. In the Congo Basin, where a significant portion of the population lacks access to modern energy services, renewable energy plays a critical role in expanding energy access. Investments in renewable energy technologies such as solar, small-scale hydropower, and biomass are facilitating electrification efforts in remote and off-grid areas, providing affordable and reliable electricity to underserved communities. SDG 7 emphasizes the transition to cleaner energy sources, including renewables, to mitigate climate change and reduce air pollution. In the Congo Basin, where reliance on traditional biomass fuels contributes to deforestation and indoor air pollution, renewable energy offers cleaner alternatives for cooking, heating, and lighting. Increasing adoption of renewable energy technologies such as improved cookstoves, biogas digesters, and solar lanterns promotes cleaner energy use and improves environmental and public health outcomes.

SDG 7 aligns with SDG 13 on climate action, emphasizing the role of renewable energy in reducing greenhouse gas emissions and combating climate change. In the Congo Basin, where climate change poses significant risks to ecosystems, biodiversity, and livelihoods, promoting renewable energy contributes to climate mitigation and adaptation efforts. Hydropower, solar, and wind energy projects offer low-carbon alternatives to fossil fuels, reducing the region's carbon footprint and enhancing climate resilience. SDG 7 advocates for improving energy efficiency and promoting sustainable energy practices. Renewable energy technologies, such as solar PV systems and energy-efficient appliances, offer opportunities for enhancing energy efficiency and reducing energy consumption in the Congo Basin. By investing in energy-efficient solutions and promoting sustainable energy practices, the region can optimize energy use, reduce energy costs, and enhance energy security while advancing SDG 7 objectives. SDG 17 emphasizes the importance of partnerships and collaboration in achieving sustainable development goals. In the Congo Basin, partnerships between governments, international organizations, civil society, and the private sector are driving renewable energy initiatives aligned with SDG 7. Collaborative efforts, such as knowledge sharing, technology transfer, capacity building, and financing mechanisms, facilitate the scaling up of renewable energy projects and accelerate progress towards achieving SDG 7 targets. On the whole, adherence to SDG 7 on affordable and clean energy serves as a catalyst for promoting renewable energy in the Congo Basin, fostering sustainable development, environmental conservation, and socio-economic empowerment.

7.5.4 National Determined Contributions (NDCs) that factor in clean and renewable energy

National Determined Contributions (NDCs) that factor in clean and renewable energy are promoting renewable energy in the Congo Basin in various ways including through policy alignment, renewable energy targets, policy coherence, investment mobilization, technology transfer and capacity building, stakeholder engagement and participation, as well as monitoring, reporting, and verification (MRV) (Cabré and Sokona 2016, Fobissie *et al* 2019, Ozor *et al* 2020, Atyi 2021, Wiese *et al* 2021, Zahar 2023, Kohnert 2024). NDCs outline countries' commitments to climate action, including targets for reducing greenhouse gas emissions and increasing renewable energy deployment. By integrating clean and renewable energy targets into their NDCs, countries in the Congo Basin are aligning national climate policies and energy strategies with international climate objectives, signaling their commitment to transitioning towards low-carbon energy systems. NDCs often include specific targets and actions related to renewable energy deployment, such as increasing the share of renewables in the energy mix, expanding renewable energy capacity, and promoting energy efficiency measures. In the Congo Basin, NDCs provide a policy framework for setting ambitious renewable energy targets and implementing measures to accelerate the deployment of renewable energy technologies, such as solar, wind, hydro, and biomass. NDCs promote policy coherence by integrating climate and energy objectives into national development planning processes. By mainstreaming renewable energy considerations into NDC implementation strategies, countries in the Congo Basin can ensure that renewable energy promotion is prioritized across sectors, such as energy, environment, agriculture, and transport, and that synergies between climate and development goals are maximized.

NDCs serve as a basis for mobilizing domestic and international financing for renewable energy projects. By including renewable energy targets and priorities in their NDCs, countries in the Congo Basin can attract investment in renewable energy infrastructure, technologies, and capacity building initiatives. NDC-aligned renewable energy projects become eligible for climate finance, concessional loans, and support from international climate funds, facilitating project development and implementation. NDCs promote technology transfer and capacity building for renewable energy development. By including provisions for technology transfer, knowledge sharing, and capacity-building activities related to renewable energy in their NDCs, countries in the Congo Basin can access technical assistance, training programs, and expertise to enhance their renewable energy capabilities and accelerate the adoption of clean energy technologies. NDCs emphasize stakeholder engagement and participation in climate action planning and implementation processes. In the Congo Basin, involving relevant stakeholders, such as government agencies, civil society organizations, private sector actors, and local communities, in the development and implementation of NDC-aligned renewable energy initiatives promotes ownership, accountability, and transparency, leading to more effective and inclusive outcomes. NDCs require countries to establish robust MRV systems to track progress towards their climate and renewable energy targets. By strengthening MRV systems, countries in the Congo Basin can monitor renewable energy deployment, assess the effectiveness of policy measures, and enhance accountability and transparency in renewable energy governance. Generally, NDCs that factor in clean and renewable energy play a crucial role in promoting renewable energy in the Congo Basin by providing a policy framework, mobilizing investment, fostering technology transfer, enhancing stakeholder engagement, and ensuring accountability in renewable energy governance.

7.6 Conclusions

Meeting the energy needs of the urban and rural poor in the Congo Basin through renewables presents both challenges and prospects. While the region boasts abundant renewable energy resources, including solar, hydro, biomass, wind, and geothermal, significant barriers hinder the full realization of its renewable energy potential. Challenges such as poor governance, weak institutional frameworks, limited technology transfer, infrastructure constraints, financial constraints, policy and regulatory barriers, and technical capacity gaps pose obstacles to renewable energy deployment. Moreover, social and cultural factors, including traditional energy practices, gender dynamics, community engagement, and land tenure issues, further complicate efforts to harness renewable energy for poverty alleviation and sustainable development. However, amidst these challenges, there are promising prospects for advancing renewable energy in the Congo Basin. The increasing adoption of green growth principles, adherence to Sustainable Development Goal 7 on affordable and clean energy, and integration of clean and renewable energy into National Determined Contributions (NDCs) provide a policy framework and institutional support for promoting renewable energy. Additionally, advancements in technology, investment mobilization, stakeholder engagement, and capacity building offer opportunities to overcome barriers and accelerate renewable energy deployment. By addressing challenges and capitalizing on prospects, the Congo Basin can unlock its renewable energy potentials to meet the energy needs of the urban and rural poor, improve energy access, enhance energy security, and promote sustainable development. Renewable energy solutions can provide clean, reliable, and affordable electricity for lighting, cooking, heating, education, healthcare, and productive activities, empowering communities, fostering economic growth, and mitigating climate change impacts. Overall, achieving universal energy access and sustainable energy transition in the Congo Basin requires concerted efforts, multi-stakeholder collaboration, and innovative approaches that prioritize social equity, environmental sustainability, and inclusive development. By leveraging renewable energy as a catalyst for poverty reduction, climate resilience, and socio-economic empowerment, the Congo Basin can chart a path towards a more sustainable and equitable energy future for all its inhabitants.

References

Abanda F H 2012 Renewable energy sources in Cameroon: potentials, benefits and enabling environment *Renew. Sustain. Energy Rev.* **16** 4557–62

Adenle A A 2020 Assessment of solar energy technologies in Africa—opportunities and challenges in meeting the 2030 agenda and sustainable development goals *Energy Policy* **137** 111180

Amir M and Khan S Z 2022 Assessment of renewable energy: status, challenges, COVID-19 impacts, opportunities, and sustainable energy solutions in Africa *Energy Built. Environ.* **3** 348–62

Anthony N M, Atteke C, Bruford M W, Dallmeier F, Freedman A, Hardy O and Gonder M K 2015 Evolution and conservation of Central African biodiversity: priorities for future research and education in the Congo Basin and Gulf of Guinea *Biotropica* **47** 6–17

Aquilas N A and Atemnkeng J T 2022 Climate-related development finance and renewable energy consumption in greenhouse gas emissions reduction in the Congo basin *Energy Strat. Rev.* **44** 100971

Atyi R E A 2021 *State of the Congo Basin Forests in 2021: Overall Conclusions* **vol 2022** (Bogor: CIFOR) p 367

Baumüller H 2021 *From Potentials to Reality: Transforming Africa's Food Production: Investment and Policy Priorities for Sufficient, Nutritious and Sustainable Food Supplies* (Bonn: Center for Development Research (ZEF))

Behrendt H, Megevand C and Sander K 2013 Deforestation trends in the Congo Basin: wood-based biomass energy *Report* 77940 World Bank, Washington DC

Bilonda M K 2020 Burning of biomass in the Democratic Republic of Congo *Biomass Burning in Sub-Saharan Africa: Chemical Issues and Action Outreach* (Dordrecht: Springer) pp 57–70

Bishoge O K, Kombe G G and Mvile B N 2020 Renewable energy for sustainable development in sub-Saharan African countries: challenges and way forward *J. Renew. Sustain. Energy* **12** 052702

Burnley C 2011 Natural resources conflict in the Democratic Republic of the Congo: a question of governance *Sustain. Dev. L. Pol'y* **12** 7

Cabré M M and Sokona M Y 2016 Renewable energy investment in Africa and nationally determined contributions (NDCs) *GEGI Working Paper* Global Economic Governance Initiative, Boston, MA

Casati P, Moner-Girona M, Khaleel S I, Szabo S and Nhamo G 2023 Clean energy access as an enabler for social development: a multidimensional analysis for Sub-Saharan Africa *Energy Sustain. Dev.* **72** 114–26

Corbino A 2021 Virunga National Park: a possible governance model for Green Deal implementation in the Democratic Republic of Congo *J. Urban Regen. Renew.* **15** 6–14

Dalder J, Oluleye G, Cannone C, Yeganyan R, Tan N and Howells M 2024 Modelling policy pathways to maximise renewable energy growth and investment in the Democratic Republic of the Congo using OSeMOSYS (open source energy modelling system) *Energies* **17** 342

De Angelis P, Tuninetti M, Bergamasco L, Calianno L, Asinari P, Laio F and Fasano M 2021 Data-driven appraisal of renewable energy potentials for sustainable freshwater production in Africa *Renew. Sustain. Energy Rev.* **149** 111414

De Wit M J, Guillocheau F and De Wit M C (ed) 2015 *Geology and Resource Potential of the Congo Basin* (Berlin: Springer)

Demissie G 2010 Geothermal resource indications of the geologic development and hydrothermal activities of DRC *Third Africa Rift Geothermal Conf. (Djibouti)* (Nairobi: African Geothermal Association) pp 22–5

Deng C, Song F and Chen Z 2020 Preliminary study on the exploitation plan of the mega hydropower base in the lower reaches of Congo River *Glob. Energy Interconnect.* **3** 12–22

Deogratias O A, Pacifique M S and Hilaire M V 2022 Current energy capacity of the DRC and the need for the development of geothermal energy *9th African Rift Geothermal Conf. (Djibouti)* (Nairobi: African Geothermal Association)

Dominguez C and Foster V 2011 The Central African Republic's infrastructure: a continental perspective *Policy Research Working Paper* World Bank, Washington DC

Dountio E G, Meukam P, Tchaptchet D L P, Ango L E O and Simo A 2016 Electricity generation technology options under the greenhouse gases mitigation scenario: case study of Cameroon *Energy Strat. Rev.* **13** 191–211

Egbende L, Helldén D, Mbunga B, Schedwin M, Kazenza B, Viberg N and Alfven T 2023 Interactions between health and the sustainable development goals: the case of the Democratic Republic of Congo *Sustainability* **15** 1259

Enongene K E and Fobissie K 2016 The potential of REDD+ in supporting the transition to a green economy in the Congo Basin *Int. Forest. Rev.* **18** 29–43

Enow-Arrey F 2020 Renewable energy deployment policy-instruments for Cameroon: implications on energy security, climate change mitigation and sustainable development *Bull. Korea Photovolt. Soc.* **6** 56–68

Falchetta G, Adeleke A, Awais M, Byers E, Copinschi P, Duby S and Hafner M 2022 A renewable energy-centred research agenda for planning and financing Nexus development objectives in rural sub-Saharan Africa *Energy Strat. Rev.* **43** 100922

Feumoe A N S, Mouzong M P and Ngatchou E H 2022 3-D geophysical inversion-modeling and intrusion estimation using gravity data of convergence zone between Pan-African-belt and Congo Craton, Centre-South of Cameroon: its geothermal implications on limit between the two geotectonic units *Geophysica* **57** 23–44

Flora F M I, Donatien N, Donatien N, Tchinda R and Hamandjoda O 2019 Impact of sustainable electricity for Cameroonian population through energy efficiency and renewable energies *J. Power Energy Eng.* **7** 11

Fobissie K, Chia E, Enongene K and Oeba V O 2019 Agriculture, forestry and other land uses in nationally determined contributions: the outlook for Africa *Int. For. Rev.* **21** 1–11

Gamaidandi D 2021 Water-energy-food nexus research: assessment of household indicators in DRC *Master's Thesis* PAU Institute of Water and Energy Sciences, Tlemcen

Gond V, Dubiez E, Boulogne M, Gigaud M, Peroches A, Pennec A and Peltier R 2016 Forest cover and carbon stock change dynamics in the Democratic Republic of Congo: case of the wood-fuel supply basin of Kinshasa *Bois Forets Trop.* **327** 19–28

Heimpel E, Ahrends A, Dexter K G, Hall J S, Mamboueni J, Medjibe V P and Harris D J 2024 Floristic and structural distinctness of monodominant *Gilbertiodendron dewevrei* forest in the western Congo Basin *Plant Ecol. Evol.* **157** 55–74

Ijoma G N, Mutungwazi A, Mannie T, Nurmahomed W, Matambo T S and Hildebrandt D 2022 Addressing the water-energy nexus: a focus on the barriers and potentials of harnessing wastewater treatment processes for biogas production in Sub Saharan Africa *Heliyon* **8** e09385

Jagger P, Bailis R, Dermawan A, Kittner N and McCord R 2019 SDG 7: affordable and clean energy—how access to affordable and clean energy affects forests and forest-based livelihoods *Sustainable Development Goals: Their Impacts on Forests and People* (Cambridge: Cambridge University Press) pp 206–36

Kalonda O K, Tshiwis I N, Atalatala B M and Umba-Di-Mbudi C N Z 2023 Evolution of sedimentation in a headrace canal for hydroelectric production case of the Shongo Basin of the Inga Complex from February 2020 to May 2021 (Kongo Central Province/DR Congo) *J. Geosci. Environ. Prot.* **11** 404–26

Kenfack J, Bossou O V and Tchaptchet E 2017 How can we promote renewable energy and energy efficiency in Central Africa? A Cameroon case study *Renew. Sustain. Energy Rev.* **75** 1217–24

Kevin N H, Feng L J, Wilfried K N G, Grichen M R F and Romaric M D 2017 Urban transport in the Congo: case of the city of Brazzaville, problems, and prospects *Am J. Eng. Res.* **6** 59–66

Kidmo D K, Deli K and Bogno B 2021 Status of renewable energy in Cameroon *Renew. Energy Environ. Sustain.* **6** 2

Kohnert D 2024 *SSRN* 10.2139/ssrn.4730179 17 Feb. 2024 The impact of the industrialized nation's CO2 emissions on climate change in Sub-Saharan Africa: case studies from South Africa, Nigeria and the DR Congo

Kusakana K 2016 A review of energy in the Democratic Republic of Congo *Conf.: ICDRE (Copenhagen, Denmark)*

Labordena M, Patt A, Bazilian M, Howells M and Lilliestam J 2017 Impact of political and economic barriers for concentrating solar power in Sub-Saharan Africa *Energy Policy* **102** 52–72

Malabo Montpellier Panel 2018 Mechanized: transforming Africa's agriculture value chains *Report* Malabo Montpellier Panel

Maniatis D 2011 Methodologies to ensure aboveground biomass in the Congo Basin forest in a UNFCCC REDD+ context *Doctoral Dissertation* University of Oxford

Manjong N B, Oyewo A S and Breyer C 2021 Setting the pace for a sustainable energy transition in central Africa: the case of Cameroon *IEEE Access* **9** 145435–58

Marijnen E and Schouten P 2019 Electrifying the green peace? Electrification, conservation and conflict in Eastern Congo *Confl. Secur. Dev* **19** 15–34

Masamba J B, Balomba P M and Savy C K 2023 Analysis of the satisfaction of household users of the improved cooking stove and liquefied petroleum gas in Kinshasa, Congo *Eur. J. Soc. Sci. Stud.* **9**

Maupin A 2017 *Energy Challenges in Southern Africa: Balancing Renewable Energy Source Options in the Democratic Republic of Congo* (Johannesburg: South African Institute of International Affairs)

Mayala T S, Ngavouka M D N, Douma D H, Hammerton J M, Ross A B, Brown A E and Lovett J C 2022 Characterisation of Congolese aquatic biomass and their potential as a source of bioenergy *Biomass* **2** 1–13

Mboumboue E and Njomo D 2016 Potential contribution of renewables to the improvement of living conditions of poor rural households in developing countries: Cameroon's case study *Renew. Sustain. Energy Rev.* **61** 266–79

Migolet P, Goïta K, Pambo A F K and Mambimba A N 2022 Estimation of the total dry aboveground biomass in the tropical forests of Congo Basin using optical, LiDAR, and radar data *GISci. Remote Sens.* **59** 431–60

Milburn R 2014 The roots to peace in the Democratic Republic of Congo: conservation as a platform for green development *Int. Aff.* **90** 871–87

Monyei C G, Akpeji K O, Oladeji O, Babatunde O M, Aholu O C, Adegoke D and Imafidon J O 2022 Regional cooperation for mitigating energy poverty in sub-Saharan Africa: a context-based approach through the tripartite lenses of access, sufficiency, and mobility *Renew. Sustain. Energy Rev.* **159** 112209

Muanza P M 2015 Presentation of geothermal potential and the status of exploration in Democratic Republic of Congo *Proc. World Geothermal Congr. (Melbourne, Australia, 19–25 April)*

Muh E, Amara S and Tabet F 2018 Sustainable energy policies in Cameroon: a holistic overview *Renew. Sustain. Energy Rev.* **82** 3420–9

Mukandala P S, Mahinda C K and de Goma O V 2020 Geothermal development in the Democratic Republic of the Congo—a country update *Geoconvention (Calgary, 15–17 May)*

Munyangeyo A 2021 Assessment and proposals for improved transportation safety on Congo River *Master's Thesis* Høgskolen på Vestlandet, Bergen

Nagabhatla N, Cassidy-Neumiller M, Francine N N and Maatta N 2021 Water, conflicts and migration and the role of regional diplomacy: Lake Chad, Congo Basin, and the Mbororo pastoralist *Environ. Sci. Policy* **122** 35–48

Nfah E M and Ngundam J M 2009 Feasibility of pico-hydro and photovoltaic hybrid power systems for remote villages in Cameroon *Renew. Energy* **34** 1445–50

Ngbolua K N, Ndanga B A, Gbatea K A, Djolu D R, Ndaba M M, Masengo A C and Mpiana P T 2018 Environmental impact of wood-energy consumption by households in Democratic Republic of the Congo: a case study of Gbadolite City, Nord-Ubangi *Int. J. Energy Sustain. Dev.* **3** 64–71

Niyibizi A 2015 SWOT analysis for renewable energy in Africa: challenges and prospects *Renew. Energy L. Pol'y Rev.* **6** 276

Njoh A J, Etta S, Essia U, Ngyah-Etchutambe I, Enomah L E, Tabrey H T and Tarke M O 2019 Implications of institutional frameworks for renewable energy policy administration: case study of the Esaghem, Cameroon community PV solar electrification project *Energy Policy* **128** 17–24

Nsafon B E K, Butu H M, Owolabi A B, Roh J W, Suh D and Huh J S 2020 Integrating multi-criteria analysis with PDCA cycle for sustainable energy planning in Africa: application to hybrid mini-grid system in Cameroon *Sustain. Energy Technol. Assess.* **37** 100628

Nsah K T 2022 The ecopolitics of water pollution and disorderly urbanization in Congo-Basin plays *Orb. Litt.* **77** 314–32

Ntirumenyerwa Mihigo B P and Cliquet A 2020 Payment for ecosystem services in the Congo Basin: filling the gap between law and sustainability for an optimal preservation of ecosystem services *Sustainability and Law: General and Specific Aspects* (Cham: Springer) pp 667–86

Odhipio D A, Mukandala P S, Kawa G N, Kasay G M and Mambo V S 2020 Identification of thermal springs in Eastern DRC, case study of Katanga, Kivu and Ituri Provinces *Proc. 8th African Rift Geothermal Conf.* (Nairobi: African Geothermal Association) pp 1–12

Olanrewaju B T, Olubusoye O E, Adenikinju A and Akintande O J 2019 A panel data analysis of renewable energy consumption in Africa *Renew. Energy* **140** 668–79

Omole D O and Ndambuki J M 2014 Sustainable living in Africa: case of water, sanitation, air pollution and energy *Sustainability* **6** 5187–202

Onyeji-Nwogu I 2017 Harnessing and integrating Africa's renewable energy resources *Renewable Energy Integration* (New York: Academic) pp 27–38

Ouedraogo N S 2019 Opportunities, barriers and issues with renewable energy development in Africa: a comprehensible review *Curr. Sustain./Renew. Energy Rep.* **6** 52–60

Ozor N, Nyambane A, Onuoha C M, Makokha M O and M'mboyi F 2020 Nationally determined contributions (NDCs) implementation index, monitoring and tracking tools for selected countries in Africa *Report* The African Technology Policy Studies Network (ATPS) and Pan African Climate Justice (PACJA) https://atpsnet.org/wp-content/uploads/2020/07/NDC-Implementation-Index-Report.pdf

Piergallini L J 2019 Three essays on Africa: Part 1: Climate change, urbanization, and government responses in sub-Saharan Africa. Part 2: Cinderella country: charcoal vs clean energy in the DR Congo. Part 3: Tangled! Patronage politics and provincial elites in the DR Congo *Doctoral Dissertation* The Claremont Graduate University, Claremont, CA

Reed E and Miranda M 2007 *Assessment of the Mining Sector and Infrastructure Development in the Congo Basin Region* (Washington DC: World Wildlife Fund, Macroeconomics for Sustainable Development Program Office) p 27

Schoneveld G and Zoomers A 2015 Natural resource privatisation in Sub-Saharan Africa and the challenges for inclusive green growth *Int. Dev. Plan. Rev.* **37** 95–118

Schure J 2012 Woodfuel and producers' livelihoods in the Congo Basin *Forest-people Interfaces* (Wageningen: Wageningen Academic) pp 87–104

Schure J 2014 Woodfuel for urban markets in the Congo Basin: a livelihood perspective *Doctoral Dissertation* Wageningen University and Research, Wageningen

Schure J, Levang P and Wiersum K F 2014 Producing woodfuel for urban centers in the Democratic Republic of Congo: a path out of poverty for rural households? *World Dev.* **64** S80–90

Schure J, Marien J M, De Wasseige C, Drigo R, Salbitano F, Dirou S and Nkoua M 2012 Contribution of woodfuel to meet the energy needs of the population of Central Africa: prospects of sustainable management of available resources *The Forests of the Congo Basin—State of the Forest 2010* (Luxembourg: Publications Office of the European Union) pp 109–22

Showers K B 2009 Congo River's Grand Inga hydroelectricity scheme: linking environmental history, policy and impact *Water History* **1** 31–58

Sogena F X L, Lutete T M, Kimilita P D and Lobo N 2018 Transport and business improvement in the province of South-Ubangi (Democratic Republic of the Congo) *Int. J. Innov. Appl. Stud.* **24** 551–8

Tamba J G 2017 Energy consumption, economic growth, and CO_2 emissions: evidence from Cameroon *Energy Sources* B **12** 779–85

Tangka J 2021 Sustainable energy: case study of Cameroon *Energy and Environmental Security in Developing Countries* (Cham: Springer) pp 587–607

Tegegne Y T, Lindner M, Fobissie K and Kanninen M 2016 Evolution of drivers of deforestation and forest degradation in the Congo Basin forests: exploring possible policy options to address forest loss *Land Use Policy* **51** 312–24

Tongele T N 2021 Human ways of life and environmental sustainability: Congo Basin case study *J. Civ. Eng. Arch.* **15** 547–59

Tyukavina A, Hansen M C, Potapov P, Parker D, Okpa C, Stehman S V and Turubanova S 2018 Congo Basin forest loss dominated by increasing smallholder clearing *Sci. Adv.* **4** eaat2993

Ulimwengu J M, Funes J, Headey D D and You L 2009 Paving the way for development: the impact of road infrastructure on agricultural production and household wealth in the Democratic Republic of Congo *Agricultural and Applied Economics Association (AAEA) Conf. Annual Meeting (Milwaukee, WI, 26–28 July)*

Vallaeys V, Lambrechts J, Delandmeter P, Pätsch J, Spitzy A, Hanert E and Deleersnijder E 2021 Understanding the circulation in the deep, micro-tidal and strongly stratified Congo River estuary *Ocean Modell.* **167** 101890

Varela Pérez P, Greiner B E and von Cossel M 2022 Socio-economic and environmental implications of bioenergy crop cultivation on marginal African drylands and key principles for a sustainable development *Earth* **3** 652–82

Warner J, Jomantas S, Jones E, Ansari M S and De Vries L 2019 The fantasy of the Grand Inga hydroelectric project on the river Congo *Water* **11** 407

Warnock R L 2013 Creating sustainable economic and ecological growth in the Congo Basin: bushmeat consumption and biodiversity protection *Masters Thesis* Western Michigan University, Kalamazoo, MI

Wiese L, Wollenberg E, Alcántara-Shivapatham V, Richards M, Shelton S, Hönle S E and Chenu C 2021 Countries' commitments to soil organic carbon in nationally determined contributions *Clim. Policy* **21** 1005–19

Woomer P L, Zozo R, Lewis S and Roobroeck D 2023 *Technology Promotion and Scaling in Support of Commodity Value Chain Development in Africa* (London: IntechOpen)

World Health Organization 2016 Burning opportunity: clean household energy for health, sustainable development, and wellbeing of women and children *Report* World Health Organization, Geneva

Yimen N, Monkam L, Tcheukam-Toko D, Musa B, Abang R, Fombe L F and Dagbasi M 2022 Optimal design and sensitivity analysis of distributed biomass-based hybrid renewable energy systems for rural electrification: case study of different photovoltaic/wind/battery-integrated options in Babadam, Northern Cameroon *IET Renew. Power Gener.* **16** 2939–56

Yufenyuy M, Pirgalıoğlu S and Yenigün O 2023 The asymmetric effect of biomass energy use on environmental quality: empirical evidence from the Congo Basin *Environ. Dev. Sustain.* 1–34

Zahar A 2023 Ambition in nationally determined contributions: the case of hydropower *Climate Technology and Law in the Anthropocene* ed A Zahar and L Reins (Bristol: Bristol University Press)

Chapter 8

Autonomous agent power contracting

Antonio Bertuccio, Jacqueline Stagner and Rupp Carriveau

This study investigates energy negotiation using Genius 10.4, an automated negotiation platform written in Java that utilizes artificial intelligence to help achieve less biased, more ethical, and probable solutions. In this context, Genius is used to negotiate hourly between one or more energy suppliers and a central grid, based on challenges related to wind energy, such as wind condition and battery storage reserves. Each representative party acting on behalf of an energy supplier exhibits distinct weightings and evaluation criteria, providing the constraints necessary for a timely agreement. This work illustrates the relative accessibility of using artificial intelligence to conduct simple energy contracting when facilitated by the Genius platform

8.1 Introduction

In the ever-growing area of machine learning and artificial intelligence, negotiation is a promising application. In autonomous negotiations, individual parties can interact with one another to produce mutually beneficial agreements. These negotiations achieve agreements though rounds of offers and counteroffers based on rules and procedures set in the negotiation process. Automated parties can execute different approaches of bargaining theory, such as game-theoretic, axiomatic, and heuristic strategies, for a given scenario.

Game-theoretic methods employ mathematical models and analysis of strategic interactions among rational agents to analyse or predict the outcome of various situations. This variety of strategies has its application in many fields, such as economics, system science, and computer science [1, 2].

Axiomatic methods rely on a set of basic assumptions or principles to derive logical consequences or theorems that describe the properties or behavior of a system. Axiomatic methods can provide clarity, generality, objectivity, and self-containedness to a subject. In logic, the axiomatic approach is a strategy for generating a system in accordance with stated principles through logical deduction (such as elimination) from basic premises, presented as axioms. Axioms are formed

doi:10.1088/978-0-7503-6179-8ch8

from a few primordial concepts that are developed and defined arbitrarily. These axioms could be considered the source of truth for a subject [3].

Heuristic methods utilize general rules, experience, intuition, or common sense to find satisfactory solutions to complex problems that are difficult to solve by exact or optimal methods, e.g. trial and error. This methodology is beneficial to an agent when tasked with generating new ideas to solve a challenging problem [4, 5].

The work discussed in this chapter uses the automated negotiating platform Genius, version 10.4. The framework of this negotiation platform will be discussed and will introduce how automated negotiation can be used to achieve agreement on the purchase price of electricity from a wind farm to a central grid. In this scenario of a central grid purchasing electricity from a wind farm, the purpose of the negotiations is to agree upon a fair, un-biased price [6, 7].

Several variables, such as market demand, manufacturing costs, legislative frameworks, and the reliability of power sources, affect energy pricing. Historically, conventional energy sources such as coal, natural gas, nuclear, and hydro have offered a consistent and dependable power supply, which helps to keep energy prices stable over time.

Energy prices are based on a combination of both market-based mechanisms and regulatory frameworks. There are two key aspects to take into consideration when discussing pricing for an energy source: levelized costs of energy (LCOE) and power purchase agreements (PPAs).

The LCOE is a metric used to assess the cost-effectiveness of different energy generation technologies. It represents the average cost per unit of electricity generated over the lifetime of a power plant, considering all costs involved [8]. The LCOE aids in the comparison of the economic viability of different energy sources [9]. Furthermore, there are various factors that influence the LCOE, such as capital costs, operational costs, fuel costs, discount rates, and capacity factors [10]. PPAs are long-term contracts between energy producers and buyers (usually utilities or large corporations) that outline the terms for the sale of electricity [11]. For the PPA to be successful, both parties in wind energy contracting must secure a mutually beneficial PPA, where the PPA is fair and un-biased to both parties in question: a central gird and a selected number of wind farms. In the work presented here, the energy producer and consumer engage in a negotiation (with no pre-determined pricing) in relation to the energy producer's current wind condition [12]. From there, the energy producer will physically deliver the energy to the energy consumer.

However, over time a PPA may become very one sided, due to market uncertainty, as there is a possibility of changes in energy pricing [13]. Technological advancements, as technology becomes more sophisticated in wind energy contracting, energy production, and storage, can alter the cost dynamics and efficiency of energy generation, potentially making the original PPA terms less favorable for one party [14]. Regulatory changes, due to shifts in government policies [15], operational risks, due to equipment failures, and supply chain disruptions [16] can disrupt the cost of producing and supplying energy. Therefore, it is highly favorable to consider autonomous negotiating in the future, as artificial intelligence could be a great benefit to combating the ideology of one-sided negotiating in PPAs.

Although automated negotiation systems are in their infancy of negotiating regarding energy markets, they have proven to be efficient in testing. They enhance negotiation management via reducing transaction times and facilitating real-time price adjustments, enabling quicker responses to supply and demand changes, leading to stability in pricing [17]. Furthermore, automated negotiation systems possess the ability to process complex data and scenarios and communicate them with higher efficiency in comparison with humans. Thus, this specific variation of machinery possesses complex problem management systems in place for negotiating [18].

For intermittent and fluctuating sources of electricity, such as wind, using static pricing may be an obstacle to wind farms entering the market. The cost of providing electricity can vary significantly when the wind is blowing and when it is not blowing and back-up sources of electricity, such as batteries, are required to fulfill contract obligations. In this scenario, more frequent negotiations would better enable the pricing to reflect the actual costs of providing the electricity. However, more frequent manual negotiation between humans is time-consuming and costly. In the scenario discussed here, automated negation software is used to consider the current wind conditions and customer demand and agree upon a price hourly.

When comparing automated negotiations to human negotiations, human negotiations typically last for longer time intervals. Also, human negotiators may possess biases, which influence negotiation. Automated negotiators are diverse in the way they can participate in the negotiation space. They can act as mediators/auditors and take into consideration different issues that multiple parties may manufacture in a presented situation. However, by adjusting the preferences of the negotiators, they can also be biased. Both types of negotiating are useful and give the user the ability to specify what approach the negotiators take. In this specific project, bias is intentionally avoided, since the negotiators' jobs are to find an agreed pricing that is fair to both the wind farm and the central grid. In this chapter, a proof-of-concept negotiation is presented. It demonstrates the potential for such negotiations to be used in the energy market.

8.2 The negotiation scenario

Figure 8.1 shows the scenario created to demonstrate the use of automated negotiation for setting energy prices. In this scenario, the wind farm has a capacity of 200 MW. The current wind conditions are low, i.e. it is not windy. The wind farm will open the bidding by asking for $100/MWh to provide electricity to the central grid; however, it is wiling to accept an offer as low as $30/MWh. The central grid has a demand of 200 MW for electricity. It will open the bidding by offering to purchase electricity for $20/MWh, but will pay up to $100/MWh.

8.3 Negotiating in Genius

To create the scenario shown in figure 8.1, it is necessary to become familiar with the terminology used in the automated negotiation software used, Genius 10.4. Genius is a Java-based negotiation platform [19]. The platform consists of domains and sub-domains, respectively. A domain is the environment in which a negotiation occurs, while the sub-domain are the parties that are in negotiation with each other. For a

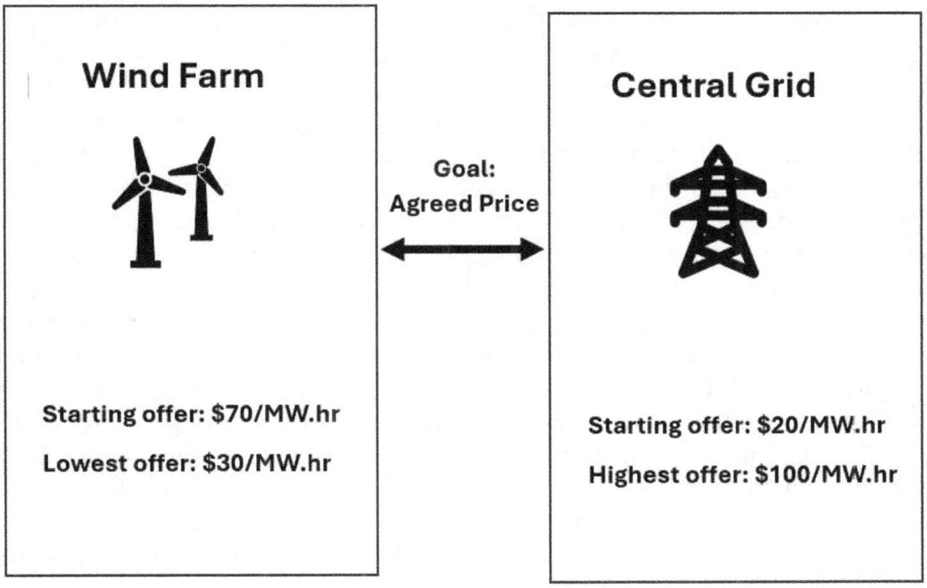

Figure 8.1. Scenario for automated negotiation.

negotiation to proceed, the sub-domains need to interact with each other. This is done through issues. Like the characteristics of the parties that are provided in figure 8.1, issues are set for each sub-domain. Issues can either be discrete or integer values. Discrete values are words that used to express comparative values, for example, low, medium, and high. Integer values are categorized as numerical minimum and maximum values, for instance, min: 20, max: 100.

With the issues established, preferences can be assigned to each issue. At the start of a simulation, the preferences are set to create the first offer. This initial preference is determined by the agent selecting the optimal value out of a range of values for a given issue, discrete or integer. This selection process is aided using evaluation numbers. For discrete values, an evaluation number is assigned to every word, (the discrete range of values), the highest number is chosen for the initial preference, there is no limit for how high or low an evaluation number can be for discrete values. However, for integer values, the evaluation numbers that represent the minimum and maximum integer values must not exceed a combined limit of 1.

In addition to considering the evaluation number of issues, a weighting is set for each issue. A weighting can be characterized as how much the negotiating agent of a sub-domain 'cares' about a given issue. If the weighting is decreased, then the agent 'cares' less about a given issue, and vice versa. This categorization is the one that affects a given negotiation over time, as in some circumstances, an agent (or negotiating party) may concede or hold on to their position faster or slower depending on the weighting.

In order to create the negotiation space, Genius uses XML, a hardware and software agnostic data storage platform. XML files are used to store the characteristics of the domain and its sub-domains. The user manually enters the information for the domain, such as issues, (discrete and integer), evaluation numbers, and weightings. This is accomplished through the use of IDE software, such as Visual Studio Code.

At this point, the agents have established connections with one another in the negotiation space, and have their assigned preferences. However, they are not yet equipped to negotiate with each other, until a style of bargaining is assigned to them. Therefore, a personality agent is assigned to each agent before the negotiation commences. The personality agent is what provides the agents the ability to be fully autonomous, giving them the strategies and protocols necessary to negotiate. There are two types of personality agents programmed in Genius: BoulwareNegotiationParty and ConcederNegotiationParty. These alternative agents contain a timing constraint, as the user can limit both the timing and the number of sequences (rounds) per negotiation. The difference between these two agents is the strategy employed in a negotiation. Boulware's ideology of mild stubbornness tends to influence its behavior to perform longer negotiations (more rounds), as it hesitates to concede with other agents, whereas Conceder will concede.

When a resolution is established, a graphical analysis of the interaction between agents is provided. This graphical representation plots the utility versus the rounds. The utility for each agent is expressed by the individual agents' preferences throughout the negotiation and ranges from 0 to 1, with a higher value of utility being preferred over a lower value [20]. The utility of an agent is expressed by individual agents' preferences throughout the negotiation; the farther all the agents' ideologies are from one another, the farther their utilities are, and vice versa. As agents interact with each other in each round, their utilities revise. Utility can also be expressed by the positive result of a negotiation—the closer the utilities of each agent are to one another, the quicker the negotiation will end. In a round, each agent contributes an offer; once each agent expresses their offer, all agents in the negotiation have a choice to either concede or continue negotiating. The chronological list of rounds over a consecutive period is labeled as the negotiation path. At the resolution of the negotiation path some key metrics are provided: the verdict, time of the negotiation (in seconds), number of rounds, and the number of agents involved in the negotiation.

8.4 Creating the negotiation scenario in Genius

In Genius, the domain (i.e. negotiation of price for 1 h of electricity) was divided into two parties: the wind farm and the central grid. There are a total of two sub-domains (agents) that will act as mediators in the negotiation. From there, the agents will collectively agree on an agreed price.

With the issues established, preferences can be assigned to each issue. At the start of a simulation, the preferences are set to create the first offer. This initial preference is determined by the agent selecting the optimal value out of a range of values for a given issue, discrete or integer. This selection process is aided using evaluation numbers. For discrete values, an evaluation number is assigned to every word (the discrete range of values), the highest number is chosen for the initial preference, and there is no limit for how high or low an evaluation number can be for discrete values.

However, for integer values, the evaluation numbers that represent the minimum and maximum integer values must add to 1. Qualitative issues, such as wind condition, are represented by discrete values: LOW, MEDIUM, and HIGH. Quantitative issues, such as agreed pricing, are represented by integer values. The integer values range between minimum and maximum values [21]. In this scenario, the wind farm has a LOW wind condition. The agreed pricing presents a range of $30/MWh to $100/MWh that the wind farm and central grid must converge upon, as Genius 10.4 does not allow one to manufacture a negotiation with multiple Min: and Max: values.

In addition to considering the evaluation number of issues, a weighting is set for each issue. A weighting can be characterized as how much the negotiating agent of a sub-domain 'cares' about a given issue. If the weighting is decreased, then the agent 'cares' less about a given issue, and vice versa. This categorization is the one that affects a given negotiation over time, as in some circumstances, an agent (or negotiating party) may concede or hold on to their position faster or slower depending on the weighting. The weighting for price was the same for both agents. Figure 8.2 displays the Genius negotiation scenario used for this study.

The goal of this exercise is to negotiate a fair, agreed pricing between a wind farm and a central grid. This is accomplished by setting the mediators of the negotiation

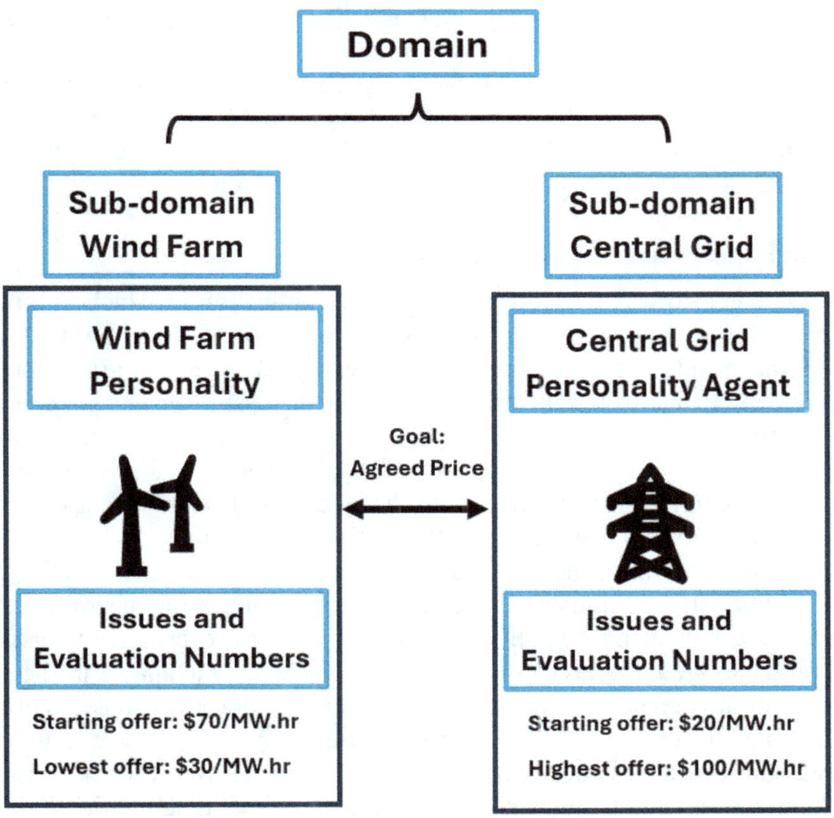

Figure 8.2. Set-up of the negotiation scenario in Genius.

Table 8.1. Agents' preferences for negotiation.

Agent	Issue name	Value type	Value range, (#) = evaluation number	Weighting number
Wind farm	Wind condition	Discrete	Low (6) Medium (2) High (2)	0.5
	Agreed price	Integer	Min: 30 (0.10) Max: 100 (0.90)	0.5
Central grid	Wind condition	Discrete	Low (3) Medium (4) High (3)	0.5
	Agreed price	Integer	Min: 30 (0.90) Max: 100 (0.10)	0.5

(the agents) to possess the same weighting, effectively giving them no bias towards a conclusion.

Finally, the agent personality for both agents must be set. For this scenario, Boulware was used for both. Table 8.1 displays the agents' preferences for the negotiation scenario. When this scenario is run in Genius, 50 rounds of negotiating occur in 0.125 s and produce an agreed price of $52/MWh.

8.5 Conclusions and future outlook

This simple scenario was developed to show an application in which autonomous negotiation can be used to quickly agree upon the price of electricity that a wind farm, or other producers of intermittent energy, sells to a central grid. Since wind conditions can change from one hour to the next, the cost of producing energy can also change from one hour to the next. Thus, by knowing the current conditions and state of electricity storage, a producer can set the acceptable limits for providing electricity to a central grid and negotiate the price hourly. As well, a central grid could perform this same negotiation with multiple producers.

Complex negotiations in the energy sector can be achieved through the implementation of additional constraints/issues, such as battery optimization of storage capacity in combination with wind conditions. As wind increases, energy contracts can be fulfilled, and storage batteries can be charged. The capacity available in the storage batteries becomes a relevant factor in setting negotiation prices for subsequent rounds of negotiations. In future work, Genius can be used to negotiate prices hourly, using additional issues such as battery capacitance. Furthermore, future works possess the option to implement additional wind farms, with would manufacture the opportunity to give arise to multi-wind farm negotiations for a given region.

Furthermore, automated negotiating systems bring numerous advantages to the wind energy sector beyond just boosting efficiency and eliminating bias. Firstly, they enhance market efficiency by cutting transaction costs and streamlining the bidding process, which leads to better resource allocation and more precise pricing in energy markets [22]. Secondly, automated systems allow for dynamic price adjustments based on real-time supply and demand changes, which is particularly important in renewable energy markets where generation can be unpredictable [23]. This adaptability helps manage the LCOE effectively, ensuring profitability despite fluctuating production levels. These systems excel at handling complex, long-term agreements such as PPAs and grid access contracts, simplifying the negotiation process and securing optimal terms for all parties involved [24]. Additionally, automated negotiations promote collaboration among various stakeholders, including energy producers, grid operators, and government regulators, by facilitating smoother and more efficient communication, aligning their interests for joint ventures or collaborative projects. Together, these benefits lead to more robust, responsive, and profitable operations in the wind energy sector [25].

Energy pricing is influenced by various factors, including market demand, manufacturing costs, and legislative frameworks. Traditional sources such as coal, natural gas, nuclear, and hydropower offer stability, but face unique challenges. LCOEs help compare technologies, while power purchase agreements PPAs ensure fair contracts between producers and buyers. Understanding these dynamics is crucial for a sustainable future. Utilizing artificial intelligence in PPAs could aid in fair, two-sided negotiations in many fields; this study considered wind energy contracting.

The evolving field of machine learning and artificial intelligence holds significant promise for the future of negotiation, enabling parties to achieve mutually beneficial agreements. The ability to adjust parameters and consider fluctuating constraints showcases autonomous negotiation's versatility and efficiency compared to traditional human negotiations.

References

[1] Davis M D and Brams S J 2024 Game theory *Encyclopedia Britannica* https://britannica.com/science/game-theory

[2] Roth A E 1985 *Game-Theoretic Models of Bargaining* (Cambridge: Cambridge University Press)

[3] Encyclopaedia Britannica 1998 Axiomatic method *Encyclopedia Britannica* https://britannica.com/science/axiomaticmethod

[4] Korobkin R B and Guthrie C 2004 Heuristics and biases at the bargaining table *Soc. Sci. Res. Netw.* **87** 795–808

[5] Mor B D 2022 The heuristic use of game theory: insights for conflict resolution *World Polit. Sci.* **3** 3

[6] Lin R, Kraus S, Baarslag T, Tykhonov D, Hindriks K and Jonker C M 2012 Genius: an integrated environment for supporting the design of generic automated negotiators *Comput. Intell.* **30** 48–70

[7] Yang L, Peng R, Li G and Lee C 2020 Operations management of wind farms integrating multiple impacts of wind conditions and resource constraints *Energy Convers. Manage.* **205** 112162

[8] Kanellakopoulou M 2022 Levelised cost of energy (LCOE)—an overview *Pexapark* https://pexapark.com/blog/lcoe/

[9] Curating Team 2023 How LCOE is helping to strengthen the PPA market *FrankNez* https://franknez.com/how-lcoe-is-helping-to-strengthen-the-ppa-market/

[10] Liu D, Liu X, Guo K, Ji Q and Chang Y 2023 Spillover effects among electricity prices, traditional energy prices and carbon market under climate risk *Int. J. Environ. Res. Public Health* **20** 1116

[11] Curating Team 2023 How LCOE is helping to strengthen the PPA market *FrankNez* https://franknez.com/how-lcoe-is-helping-to-strengthen-the-ppa-market/

[12] Novergy Solar 2024 Exploring the 8 different types of power purchase agreements (PPAs) *Novergy Solar* https //novergysolar.com/types-power-purchase-agreements-ppa-and-which-one-best-for-you/

[13] McGuireWoods 2023 5 key strategies for negotiating renewable energy project EPC contracts *McGuireWoods* https://mcguirewoods.com/client-resources/alerts/2019/3/5-key-strategies-for-negotiating-renewable-energy-project-epc-contracts/

[14] Agarwal K 2023 Energy contract management: a comprehensive guide *SpotDraft* https://spotdraft.com/blog/energy-contract-management

[15] US Energy Information Administration 2020 Trends and expectations surrounding the outlook for energy market *Report* US Department of Energy, Washington, DC https://eia.gov/outlooks/aeo/pdf/trends_and_expectations_2020.pdf

[16] Pierpont B 2020 A market mechanism for long-term energy contracts *WRI/RFF Market Design for the Clean Energy Transition Workshop (16–17 December)* https://media.rff.org/documents/pierpont-wri-rff-workshop-12-2020.pdf

[17] Sullivan Y and Fosso Wamba S 2024 Artificial intelligence and adaptive response to market changes: a strategy to enhance firm performance and innovation *J. Bus. Res.* **174** 114500

[18] Kellogg R and Reguant M 2021 *Energy and Environmental Arkets, Industrial Organization, and Regulation* **vol 5** (Amsterdam: Elsevier) pp 615–742

[19] Baarslag T n.d. Genius—a negotiation simulator https://automatednegotiation.gitlab.io/genius/

[20] Baarslag T, Pasman W, Hindriks K and Tykhonov D 2019 Using the Genius framework for running autonomous negotiating parties *TUDelft* https://ii.tudelft.nl/genius/sites/default/files/userguide.pdf

[21] Bruck M, Sandborn P and Goudarzi N 2018 A levelized cost of energy (LCOE) model for wind farms that include power purchase agreements (PPAs) *Renew. Energy* **122** 131139

[22] Gerding E H and Ketter W 2014 A hybrid approach to automated negotiation for energy trading: combining mechanism design with multi-agent learning *Proc. of the 13th Int. Conf. on Autonomous Agents and Multi-Agent Systems (AAMAS)* pp 1201–8

[23] Liu H and Zhang Y 2020 Decentralized negotiation strategy in renewable energy markets *IEEE Trans. Ind. Inf.* **16** 6662–70

[24] Ketter W, Peters M and Avrachenkov K 2013 Automated negotiation in smart grids *Proc. of the 12th Int. Conf. on Autonomous Agents and Multi-Agent Systems (AAMAS)* pp 981–8

[25] Parkes D C and Wellman M P 2015 Economic reasoning and automated negotiation *Proc. of the 24th Int. Joint Conf. on Artificial Intelligence (IJCAI)* pp 1355–61

IOP Publishing

Renewable Energy Systems
The way forward
David S-K Ting and Jacqueline A Stagner

Chapter 9

A prerequisite to computational fluid dynamics of airplane condensation trails

Devin Roland and Ahmad Vasel-Be-Hagh

Jet condensation trails (contrails) are a reality of air transportation that influences environmental health by altering the net heat flux of the Earth's surface. If aviation turns toward hydrogen-powered engines to address this industry's CO_2 challenges, the contrail characteristics will change from that of the current contrails emitted by conventional, kerosene-powered aircraft. Such a transition presents a concern that the net warming effect on the planet from the alteration of the afore-mentioned heat flux will negate the beneficial impact from the reduction of emitted CO_2 gas. That scenario would prove the implementation of hydrogen-powered aircraft to be fruitless from a global warming perspective. To investigate the validity of this concern, computational fluid dynamics (CFD) simulations could potentially be utilized to simulate the creation, evolution, and persistence of future contrails under hydrogen engine conditions. The goal would be to accurately predict contrail characteristics before the physical version could be observed. In order to better understand how such a simulation could be created, the literature pertaining to modeling contrails with CFD has been reviewed. This chapter shares this review on the available CFD software, physical models (viscous modeling, microphysics, and particle tracking), and conditions that could be used to describe contrail development, evolution, and persistence. Validation data and experiments discussed in the literature will also be reviewed. The goal of this chapter is to provide the prerequisite knowledge one needs to have before modeling contrails.

9.1 Introduction

Aviation's CO_2 emissions account for around 2.6% of CO_2 generation globally, which constitutes 3.5% of the global human-related radiative forcing [1]. Radiative forcing, defined as the change in the net radiation, is a common way of quantifying climate impact. Net radiation is the difference between the incoming and outgoing

doi:10.1088/978-0-7503-6179-8ch9

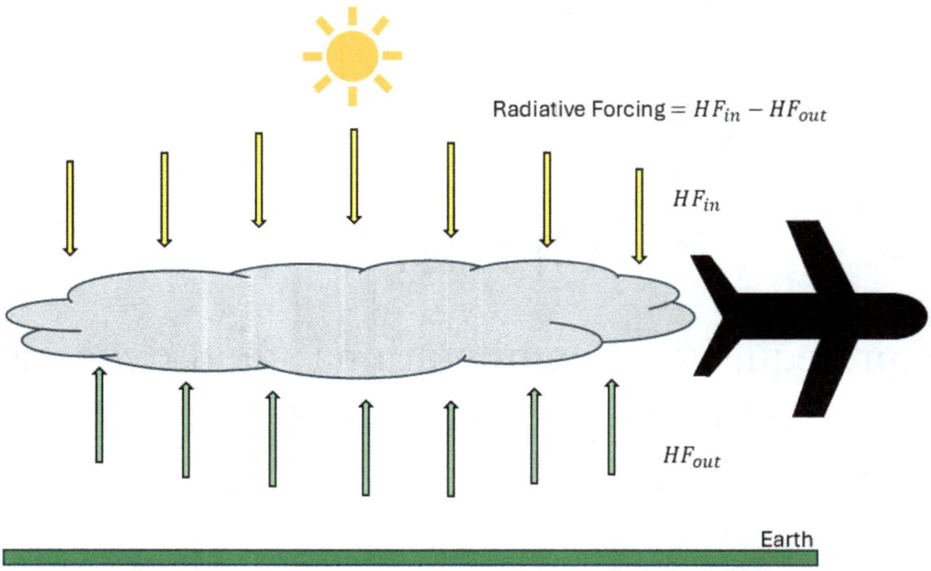

Figure 9.1. Concept of radiative forcing. Radiative forcing is the difference between incoming heat flux (HF_{in}) from the Sun and outgoing heat flux (HF_{out}) reflected back from the Earth.

radiation at the planet's surface. Figure 9.1 presents a graphic showcasing the radiative forcing concept. The aviation industry's contribution to radiative forcing is especially concerning because aviation is a sector that is expanding quickly (4.5%–4.8% per year) [1]. This expansion and emission magnitude put a spotlight on alternative fuels as a potential solution [1]. However, when shifting from carbon-based to alternative fuels, one must also consider the potential parallel changes in aviation's non-CO_2 pollution.

Radiative forcing due to aviation's non-CO_2 pollution, particularly condensation trails (contrails), is also a significant source of climate change [2] (see figure 9.2). An approximate value for the net radiative forcing effect of clouds originating from contrails is 37.5 mW m^{-2}—although this value has great uncertainty associated with it [3]. The climate effect of clouds originating from contrail sources (in the year 2005) has been gauged to be among 10 to 80 mW m^{-2}, which encompasses the previously stated value [4]. A different study gave a slightly different range of 30 to 90 mW m^{-2} [2]. This radiative forcing is deemed to outweigh the sole effect of CO_2 by 3 to 1 [2].

Research on contrails has regained attention due to the potential change in its environmental impact with the aviation industry shifting from conventional kerosene to alternative fuels, mainly hydrogen [5]. As mentioned, the shift from kerosene is to reduce the CO_2 emissions the aviation industry generates. However, it is important to ensure that the contrails of these alternative fuels will not introduce additional radiative forcing to an extent that would offset the climate benefit of CO_2 reduction. Since the characteristics of contrails produced by airplanes consuming alternative fuels are largely unknown, using CFD for predictive analysis would be necessary.

Figure 9.2. Visual example of a contrail.

Studying contrails through field measurements and laboratory-scale experiments is both challenging and expensive. This highlights the importance of using CFD to address the many questions about how changes in fuel impact contrails. Alterations in fuel lead to changes in the aircraft's propulsion system and aerodynamics, thus altering the thermodynamics of the exhaust gas and the aircraft's wake, all of which would impact the contrail formation, evolution, persistence, and eventually, its effect on radiative forcing and near ground climate. Radiative forcing and effects on near ground climate are measured via various parameters, such as climate sensitivity (the change in global temperature of the Earth in reaction to risen CO_2 in the atmosphere [6]). The physics of contrails is complex, spanning a broad range of scales from micro-scale ice nucleation to mesoscale climate effects. Simulating the processes underlying contrails, from formation to conversion into persisting cirrus clouds with measurable climate effects, requires modeling multiple phases, condensation and freezing, micro-scale particle motion, entrainment, turbulence, and compressibility effects.

Developing a simulation that encompasses a viscous model, a multiphase model, a particle tracking model, and a model to simulate compressibility effects in a compatible way that converges to reliable results is challenging. One has many options for modeling each of these underlying processes. Therefore, it would be very beneficial to know what models have been used in the literature so far to simulate contrails. It is also important to understand what contrail characteristics the simulations in the literature have analysed. By reviewing the literature, this chapter outlines the various models that approximate the desired physical phenomena. The chapter presents the software packages that enable utilizing these models. The chapter also discusses the dimensions and shape of computational space chosen by various studies, boundary and initial conditions, and the available data to validate contrail simulations. The goal of this chapter is to provide a preliminary prerequisite for beginners to understand how CFD simulations in the literature have been set up and what kind of questions surrounding the contrail problem have been answered through these simulations.

9.2 Software tools

Once the scope of a simulation is defined, i.e. the contrail formation, its evolution as it interacts with the aircraft's aerodynamics wake, its persistence, its climate effect, or maybe all of these, and the appropriate sub-models are chosen to represent various physical processes underlying each stage of a contrail life, the question may arise as to how to implement and incorporate these models into the simulation. What platforms are most suitable for handling the implementation of said models and their interaction and merger with one another? A literature review has revealed a variety of simulation software tools (i.e. codes or programs) utilized to simulate different aspects of contrails. Each of these tools has unique characteristics and behavior alongside similarities with its kin. This section discusses some of these various tools.

9.2.1 EULAG

The EULAG (Eulerian-semi-Lagrangian [7]) model/solver has been one of the most popular tools for simulating contrails. EULAG is a software developed by the German Aerospace Center (DLR) [3] that handles the dynamic (or turbulent) processes of the bulk fluid [8]. The EULAG model is anelastic, meaning that the overall elastic response regarding linear stress and strain characteristics is accompanied by a time delay in its return to equilibrium [9]. The model is also non-hydrostatic, which signifies the use of the incompressible form of the Navier–Stokes equations [10]. This method leads to more accurate solutions as opposed to a hydrostatic model, and when hydrostatic balance of forces is not achieved for small-scale simulations, this method is necessary [10]. The EULAG is often combined with a model called Lagrangian Cirrus Module [8] or Lagrangian Cloud Module (LCM) [11] that enables the detailed study of the microphysics involved in contrail formation and evolution [12]. All of the literature reviewed in this work that use the EULAG software for contrail study (not orographic cirrus cloud study [13]) also implement the LCM extension to capture the microphysical processes [3, 12, 14–19]. The LCM extension will be discussed in greater detail in section 9.3.2.

The EULAG software is based on Large-Eddy Simulations (LES) [8] and has been utilized in a variety of works in both two-dimensional (2D) and three-dimensional (3D) configurations. It has been applied to further comprehend contrail evolution and reaction to environmental and aircraft factors by studying the geometric depth and adiabatic heating crystal loss of young and old contrails [19] and to better understand optical and microphysical influence on orographic cirrus clouds due to heterogeneous freezing [13]. It has also been used to study homogeneous, nucleation-based contrail-cirrus development [14], contrail-to-cirrus transition [20], and even as a source of validation data for other models [3] (among other applications). When coupled to the LCM, the package also allows for the implementation of Lagrangian particle tracking (LPT) in a similar fashion to other software [8]. More on LPT will be discussed later in this work.

9.2.2 CEDRE

Another popular software tool used for modeling contrails is CEDRE (short for Calcul d'coulements Diphasiques Ractifs pour l'Énergétique [21]), a code developed by Office National D'Etudes et de Recherches Aerospatiales (ONERA) in Chatillon, France [22]. CEDRE is a software that can handle multiple types of physics due to its system of coupled solvers [23]. Its hosted solvers address a variety of physical phenomenon and aspects, including the gas flow, radiation, stochastic gaseous particles, dispersed condensed phase, and solid conduction [24]. Each of these has at least one solver associated with it. The gas flow is treated by CHARME (simulates reactive flows in 3D [25] in Reynolds average Navier–Stokes (RANS) or LES formulations [23]), stochastic gaseous particles by PEUL (the Probabilistic Eulerian Lagrangian model: Eulerian discretization to solve dissipation rate, turbulent kinetic energy, momentum, mass, and energy and stochastic Monte Carlo to solve the probability density function for thermochemical makeup [23]), and solid conduction by ACACIA (solid boundary conductive heat transfer [23]) [24]. The radiation and dispersed condensed phase have two solvers each that can be chosen [24]. For radiation, the user can pick REA (radiative transfer for solid fuel rocket engines [23]) including DOM (discrete ordinate method [23]) or ASTRE (a radiative transfer solver [23]) paired with Monte Carlo [24]. The dispersed condensate phase has the choice of SPARTE (a Lagrangian dispersed phase solver [23]) if Lagrangian treatment is desired or SPIREE (an Eulerian dispersed phase solver for two-phase flow [23]) if Eulerian treatment is preferred [24]. The CEDRE software was used in various works such as to study the creation of contrails in an airplane's near field (including flows from both the bypass and engine core) [26], to study contrails newly created just behind an aircraft (again with flows from the bypass and engine core included) with dilution of the contrail occurring because of an intertwining of the contrail with the vorticies or jet [27], and to study the growth of ice particles [28]. In addition to these, the CEDRE fluid solver, CHARME, has been used to simulate an aircraft wake through RANS and LES methods [29] and to observe the effect of complex geometry (wing/nacelle) on contrail creation [30]. If more information is desired with regards to these solvers, it can be discovered in [23].

9.2.3 Ansys

Ansys, developed by Ansys Inc. [31], is a widely used commercial tool that includes several CFD tools, such as Fluent and ICEM, for studying contrails, or at least some of their smaller scale characteristics, such as mechanisms leading to ice crystal formation. This tool combines different models, simulation capabilities, and post-processing tools into a user-friendly package. The software is operated through a graphical user interface (GUI), unlike text-based software such as EULAG, which uses command line interface (CLI).

Ansys Fluent, featuring a large swathe of learning resources and documentation provided by ANSYS Inc, has been extensively validated in the literature for a broad range of applications. The software hosts models simulating viscous effects (both RANS and LES), multiphase flow, condensation/evaporation, freezing/melting, and different modes of heat transfer. Thus Ansys Fluent is a viable contender for exploring various aspects of contrails; however, in the context of contrails, the software has not been used as often as EULAG. This is most likely because simulating contrails is computationally heavy, and GUI-based CFD tools such as Fluent, with comparably lower computational efficiencies than their CLI counterparts, are not ideal candidates for such a complex, large-scale phenomenon.

Ansys (Fluent) was used to study contrail detection in the visible light spectrum [32] and to study flow both in near wake and running over an aircraft [33]. Ansys (ICEM) has also been used to create the geometry and the computational mesh for contrail simulations conducted via other CFD tools, such as STAR-CCM+ [2]. Ansys has also been employed to generate the mesh for simulations conducted by CEDRE [26] and for Ansys (Fluent) itself [33]. For more information on Ansys Fluent, the reader may study this tool's documentation [31].

9.2.4 STAR-CCM+

STAR-CCM+, a commercial software developed by Simcenter, has been broadly compared in the literature with Ansys Fluent as a reliable host for CFD simulations. Like Ansys, STAR-CCM+ has not been used as often as EULAG and CEDRE to investigate the characteristics of contrails. However, our review indicates that STAR-CCM+ can technically simulate the formation and evolution of contrails as it offers a wide variety of turbulence and multiphase models suitable for modeling aerosol particles, the condensation of vapor on them, the freezing of the condensed water, and the interaction of the ice crystals with the turbulent airflow in the background. The tool offers eddy-viscosity models (e.g. RANS), Reynolds stress transport models, and transient models (e.g. LES, Detached-Eddy Simulation) to model turbulence. It offers seven different multiphase models, of which the Fluid Film model can simulate the condensation of the atmospheric moisture and water content existing in the exhaust gases on the aerosol particles. It also offers the Lagrangian Multiphase model for simulating the aerosol particles themselves. This tool also provides a discrete element method that allows users to investigate the effects of particle shape and particle–particle contact. This software has been utilized in the literature to simulate the beginnings of near field contrail creation [2]. For more information on Star-CCM+, the reader can visit the STAR-CCM+ website by Siemens [34].

9.2.5 FLUDILES

FLUDILES is an LES code based on the Navier–Stokes equations in their 3D and compressible form [35]. The Navier–Stokes equations are reserved for the gas phase and the vapor phase (scalar field) is treated with a transport equation [35]. The discrete phase (or dispersed phase) is resigned to an LPT method [35] (see section 9.3.3 for a discussion on LPT). FLUDILES has been used to simulate the expanding plume left behind an aircraft [11] and to handle the gaseous phase when studying microphysical transformations [35]. It should be noted that FLUDILES is often combined with various models (ice growth, aerosols, micro-physics) to give the solver greater capability required to model contrails [35]. Examples of these additional codes include ULP-ONE [35] and LCM [11] (see section 9.3.2).

9.3 Models

This section reviews the models that have been used to numerically handle the treatment of the physical processes that underlie contrails. These models include viscous, microphysical, and particle tracking. The viscous model will handle the turbulent fluid motion, the microphysical model will treat the processes of evaporation/condensation, sublimation/deposition, and freezing/melting, and the particle tracking model will manage the motion of the discrete phase. See table 9.1 for an organized list of the mentioned models.

9.3.1 Modeling flow turbulence

Contrails form in the aerodynamic wake of airplanes, a very turbulent environment. Thus, turbulence dominates the dynamics that lead to the formation and evolution of the contrails. Hence, a proper turbulence model must be incorporated into the simulations to provide an approximate solution (even if that solution is very accurate) for velocity, pressure, and density fluctuations and their variations with time and space. Two commonly used turbulence models for simulating contrails are LES and the family of RANS techniques, such as k-ϵ and k-ω. Some software tools introduced in section 9.2, such as CEDRE, offer both LES and RANS options [23],

Table 9.1. List of models and the associated physical processes that they describe.

Model Type	→	**Purpose (Describes the...)**
Viscous	→	Turbulent fluid motion
Microphysical	→	evaporation, sublimation, melting
		⇕ ⇕ ⇕
	→	condensation, deposition, freezing
Particle Tracking	→	Motion of the discrete phase

Table 9.2. Comparison of RANS versus LES. Through a review of the literature, RANS has not been observed to simulate contrails outside of the near field and young age. However, LES was utilized in the literature to treat contrails from near to far field and from young to old age.

#	Factor	RANS	LES
1	Computational expense	Less	More
2	Domain of study behind aircraft	Near field	Near to far field
3	Age of contrails	Young	Young to old

while others only support one of these models, such as EULAG which is only LES-based [8].

A question may arise as to whether LES or RANS is preferred for simulating contrails. In general, LES is a more advanced model that provides a more accurate estimation of turbulence; however, it demands more computational power, which is an important matter given the large extent of the contrail simulations. So, the answer to this question depends on the application, as modeling certain aspects of contrails might not need utilizing LES. For instance, while the literature found LES more applicable than an RANS scheme at times due to RANS producing too much energy dissipation, it applied RANS to an analysis residing in the wake just behind the wings [28]. A comparison of RANS versus LES can be found in table 9.2.

9.3.1.1 RANS

The literature appears to trust applying RANS models to simulations focused on the early stages of contrail formation, i.e. infantile contrails. This section reviews a few examples of such applications. For example, a work focused on the study of the growth of ice particles employed RANS to model the turbulence effect [28]. This steady-state study focused on the wake behind the wing. The literature has also employed RANS to investigate the creation of contrails in an airplane's near field (including flows from both the bypass and engine core) [26]. A follow up investigation also used RANS to understand how an intertwining of the contrails newly created just behind an aircraft with the vorticies or jet causes the dilution of contrails [27]. Another study incorporated unsteady RANS to consider contrail creation in the near wake alongside the effect of turbofan geometry [2].

9.3.1.2 LES

The literature reveals a trend regarding the variance of application between RANS and LES methods. LES appears in a wider variety of works than RANS. Also, as was mentioned above, works that have implemented RANS focused their attention on the near field of the wake (in other words, young contrails), while LES appears in studies that present a more holistic view, with attention given to young and aged contrails (both close to and far from the aircraft).

LES was used to observe the evolution of contrails from the latter portion of the dispersion phase onward for 30 min (~1000 s to ~2800 s since leaving the aircraft) [36]. Note that the dispersion phase, and later, would not be considered near field. A focus on this scale of contrail age is also seen where a research effort followed the creation and evolution of contrails via simulation from several seconds to around a half hour [37]. Again, note that this range of times would include more than just the near field. The authors of this study also focused on contrail evolution mingling with the complex turbulence associated with the wake [37]. The literature includes several other studies employing LES to evaluate the evolution of contrails within the vortex and dissipation regions of their life cycle [38–40]. It is worth emphasizing that LES is considered highly reliable, to the extent that it has been used to validate a former contrail model in the literature [3].

9.3.2 Microphysical models

Microphysical models handle small-scale, water-particulate processes such as evaporation, condensation, sublimation, deposition, melting, and freezing. Essentially, they describe the evolution of the water/water-encrusted particulates in the flow (e.g. growth or shrinkage of ice crystals). The sections below will dive into two separate coupled (microphysical model to software) models that treat these processes, as seen in the literature. It is also possible that the microphysical model can be a user-defined function [2, 26, 27] or internal to the simulation software, as is the case with Ansys Fluent (see above) (figure 9.3).

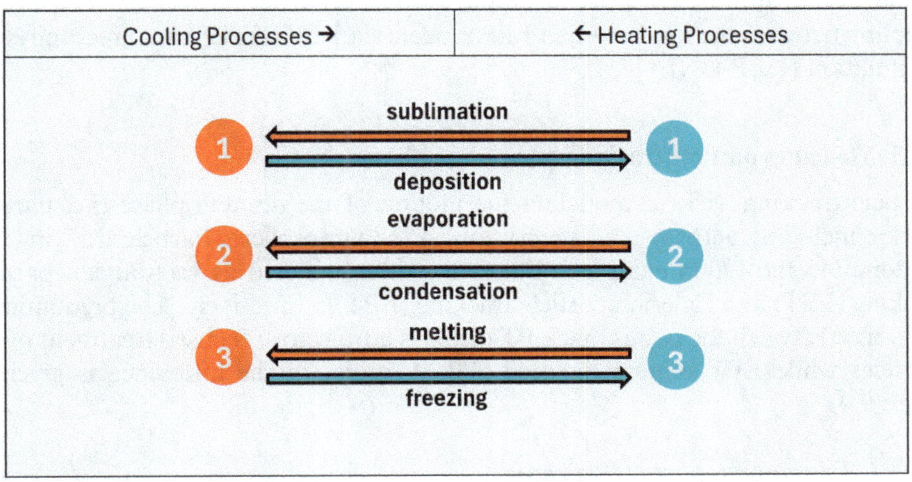

Figure 9.3. Range of processes treated by microphysical model. Left to right: warm particles experiencing cooling processes. Right to left: cool particles experiencing warming processes.

9.3.2.1 Lagrangian cloud module (LCM)

One of the most commonly used microphysical models in the literature is the LCM [8, 11]. This module is an attachment to the EULAG software package, previously discussed in section 9.2 [8]. The LCM is used in a variety of capacities when studying contrail microphysics, including the expulsion of latent heat and deposition, heterogeneous nucleation, homogeneous freezing, particle growth effects due to radiation, aggregation, sedimentation, and explicit microphysics [41]. The literature has employed this model to investigate cirrus evolution [42]. Reviewing the contribution of this model to understanding cloud microphysics puts the capabilities of this model into perspective. For example, LCM provided an understanding of the differences in habit as the ice crystals grew and the influence on particle motion due to turbulent dispersion [42].

9.3.2.2 ULP-ONE

The ULP-ONE, a microphysical model founded on a trajectory box model, accounts for both microphysics and the interaction of aerosols [35]. The 'box' is a domain containing the exhaust that moves along a trajectory dictated by the output of the CFD software (FLUDILES in this case, see section 9.2) [35]. Properties related to the aerosol interplay vary with the box location (temperature, for example) [35]. The aerosol particles are considered hydrophilic and take on water and sulfuric acid from the box domain [35]. Once a threshold is reached (10% water-sulfuric acid by mass), the particles are considered 'activated' [35]. At that point, the particles may absorb more water and potentially freeze if the conditions are correct [35]. The absorption process is driven by a transport equation that depends on the saturation vapor pressure and the water vapor content [35]. The classical theory is implemented to work out the heterogeneous ice nucleation rate [35]. Furthermore, both ambient air entrainment and particle formation and coagulation are taken into account (in certain scenarios) [35]. The ULP-ONE has not been used nearly as often as the LCM. Still, it was used in an effort to study microphysics detailed by the coupling between the implemented LES model (via FLUDILES) and microphysical formulation (ULP-ONE) [35].

9.3.3 Modeling particle tracking

'Particle tracking' (PT) is modeling the motion of the discrete phase (i.e. particle phase), including aerosols and ice crystals in the atmosphere. Particle tracking can be done via the Lagrangian and Eulerian methods, noted as Lagrangian particle tracking (LPT) and Eulerian particle tracking (EPT), respectively. The predominant difference between the two is that LPT displays a trajectory-focused treatment of the particles while EPT is point-focused [43]. A visual of the difference is given in figure 9.4.

9.3.3.1 Lagrangian particle tracking

LPT is a tracking method that individually tracks each discrete particle or parcel of particles in the flow field and observes its motion through the continuous phase as it

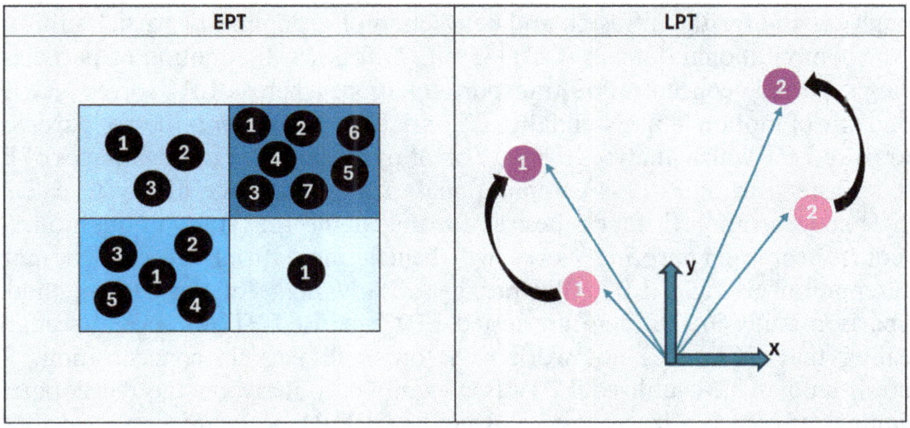

Figure 9.4. Comparison of EPT versus LPT. EPT: darker shade of cell indicates higher particle concentration. LPT: darker shading indicates a forward time step.

changes with time [44]. LPT operates based on Newton's second law (versus EPT, which is governed by a transport equation) and calculates the behavior of individual particles (as opposed to the mean particle behavior) [45]. It should be emphasized that the LPT method does not necessarily track every individual particle but rather groupings of particles termed 'simulation ice particles' (SIP) (or computational particles [2]) if the number of particles would result in high computational expense for individual calculations [8]. Some works will call these groupings of particles 'parcels' instead [44]. Each SIP will house anywhere between 10 to 10^5 physical ice crystals within it (with the most common value landing around 10^4) [8]. Two factors realize the SIP–real ice crystal number correlation: the simulation's spatial grid size and the conditions dominating while the crystals are being formed [8].

LPT is a more computationally expensive method than an alternative, such as EPT, but that also provides a wealth of information if the computational expense can be justified [45]. LPT is advantageous due to the explicit representation of particle motion that is provided by the method (in contrast to the concentration-based transport equation method of EPT) [45]. However, it also has disadvantages, such as increased computational expense and simulation run times [45]. In addition, some engineering applications may be more suited to particle concentration information rather than particle motion [45]. In that case, a method such as EPT may be more applicable (more details in the discussion below) [45]. LPT is the literature's dominant method for treating and tracking the discrete phase because particle motion significantly influences contrail development, evolution, and persistence; thus, obtaining a detailed description of particle motion is essential in predicting such characteristics.

9.3.3.2 Eulerian particle tracking

EPT, a less commonly utilized method for treating the discrete phase than LPT, differs from LPT in that it does not explicitly solve for the motion of each particle (or SIP [8]) like LPT does [45]. Instead, EPT solves for the particle concentration through solving for the diffusion and convection of groupings of particles throughout the computational domain (CD) [45]. EPT 'follows' the motion of particles by solving a particle concentration transport equation, whereas LPT solves Newton's second law of motion [45]. Essentially, EPT analyses the *concentration* of particles in contrast to LPT which analyses the *position* of particles [45]. The advantages of EPT include faster run times (less computational expense) and the ability to calculate particle concentration (if that is desired for the engineering study in question) [45]. Of course, some engineering works will benefit more from the explicit motion representation given by LPT, illustrating a disadvantage of the EPT method. A comparison study in the literature tested EPT against LPT for several scenarios, revealing that EPT does not work well for small particle concentrations [45]. Contrail studies have utilized EPT. For example, a study on the developmental beginnings of contrails in near wake using the CEDRE software employed EPT to track particles [26].

9.4 Design of computational domain

This section guides setting up the CD for contrail CFD simulations. This discussion includes the location of the CD within the atmosphere, the dimensions of the CD, the boundary conditions (BC), and the initial conditions (IC) necessary to accurately describe the settings for the 'real estate' that the models of the simulation will operate within.

9.4.1 Atmospheric location

An important consideration for creating the CD is where the domain will reside in the atmosphere. This is crucial due to the dependency of contrail formation on atmospheric conditions that vary with elevation and atmospheric layer. The atmosphere is composed of five distinct layers: the troposphere, the stratosphere, the mesosphere, the thermosphere, and the exosphere (which is the transition between the void of space and the thermosphere) [46]. The transitioning layers between these sections are coined 'pauses' [47]; there are three main pauses termed the tropopause, stratopause, and mesopause, as well as an outermost transition boundary between the thermosphere and exosphere called the thermopause or exobase [46]. These regions outline the general structure of the Earth's atmosphere.

With this layout in mind, the question of where the CD will reside is still unanswered. When studying the nature of each of the atmospheric spheres, it is known that weather systems and precipitation occur exclusively in the troposphere [46]. In addition to this, most clouds form here as well [46]. A common region that hosts the cruising altitude for aircraft is the upper troposphere and lower stratosphere (UTLS) [40], defined as the region ± 5 km around the tropopause. With this in mind, the choice of the UTLS as the home of the CD would make a great deal of sense. The literature reflects this reasoning with the chosen locations of applied CDs. One study used a section of the UTLS when studying contrail evolution left in the path of aircraft duos [18]. Similarly, other works modeled the atmospheric conditions to be likened to the UTLS [38, 39]. In another work, the initial conditions for the atmosphere were set to resemble that of an aircraft cruising at 35 000 ft (10.668 kilometers) [27], which falls into the range for the UTLS [46] again showing the applicability of the UTLS for the location of the CD. The UTLS is stated to be the danger zone for pollution caused by aviation [36], which is intuitive as emitted pollutants would be present at cruising altitudes. This location (UTLS), when combined with the other parameters such as CD dimensions (quantity and magnitude), boundary conditions, and initial conditions, describes the computational domain that will host the desired contrail simulation.

9.4.2 Dimensions and shape

When approaching a CFD problem, it is very important to select a suitable shape (e.g. rectangular prismatic, cylindrical-axisymmetric, i.e. a rectangular plane revolved around a center line), size (magnitude of the boundaries), and the number of dimensions (2D versus 3D) of the computational domain. Each of these parameters will influence factors such as the computational expense associated with running the simulation or the applicability of the simulation to represent the physical phenomena it is attempting to replicate accurately.

Although the literature has employed both 2D [2, 12, 14, 20, 48] and 3D domains [2, 3, 26–28, 32, 36, 38–40, 49, 50] to investigate different characteristics of contrails in various stages of their lives, the 3D set-ups have been used more often. In addition, some studies have used a hybrid 2D/3D approach [15–18, 41, 51]. With this in mind, it would be useful to understand when a 2D type could be more appropriate than its counterpart. The applications of a 2D versus a 3D CD are in alignment in certain cases while being different in others. Seeing that we live in a 3D world, the usage of a computational domain defined in 2D presents the opportunity to orient the domain in a variety of ways. Examples could include planar parallel to the flight path or planar perpendicular to the flight path; however, the latter was seen much more often in the literature [14, 16–18] with the former seen scarcely [2].

A 2D domain type comes with strong and weak points. One advantage of the 2D methodology is that excluding the third dimension could permit faster simulation times and the applicability of less powerful computing. Thus, they allow for studying contrails across more extensive domains. However, the LES viscous model is incompatible with 2D domains since turbulence is a 3D phenomenon [52]. Therefore, using a 3D formulation would be more useful if the LES model is the desired approach to turbulence modeling. In spite of this, the 2D works referenced above still implemented this treatment of the viscous processes by assigning a very thin thickness to the domain to address the numerical incompatibility, although the physical inaccuracy would still remain. The literature has implemented LES techniques in such 2D domains to simulate the microphysics of contrails occurring in the vortex phase [12, 48] and to investigate contrail-to-cirrus transformation under various parameters and conditions [14, 20]. Implementing an RANS technique would make more sense if a 2D study is desired. However, if turbulent processes are to be meaningfully simulated, a 3D CD would be more prudent.

Whether 2D or 3D, the dimensions of the CD's boundaries and meshing (cell size) vary across the literature, depending on the models used and the scope of the studies, i.e. the contrail characteristics of interest. One 2D case used a CD that was perpendicular to the airplane's motion and measured 3 km (vertically) by up to 80 km (horizontally), aiming to observe the development of homogeneously nucleated cirrus from both natural and contrail sources [16]. The resolution of the mesh was $\Delta x = \Delta z = 10$ m in this and another similar study [16, 51]. Another 2D work looked to simulate contrail transformation to cirrus [14]. The first case measured 1 km by 5.7 km (z and x, respectively), and the second case measured 2 km in the z-direction, while the domain's x-dimension varied from 17 to 34 km between different simulations [14]. The first mesh sizing was $\Delta x = \Delta z = 5$ m, and the second mesh sizing was $\Delta x = 15$ m and $\Delta z = 10$ m [14]. The provided shear dictated the variance in the second x-dimension [14].

A 3D case examining the beginning of contrail development in the vortex phase used a CD that measured 384 m by 400 m by 600 m (the x, y, and z directions, respectfully) [41], significantly smaller than the 2D domains due to the increased computational costs of adding the third dimension to the domain. These quantities corresponded to the transverse, direction of flight, and vertical (in that order) [41]. Considering the mesh resolution, the cell sizing was $\Delta x = 1$ m, $\Delta y = 2$ m, and $\Delta z = 1$ m, indicating a higher resolution mesh compared to 2D simulations [41].

Another work, which used CEDRE to study the developmental beginnings of contrails in the near wake, used a 3D CD measuring 4.5b by 2b by 2b (x, y, and z, respectively), where b was the 34 m wingspan of the aircraft [26]. It should be noted that the CD was set up to treat the simulation of only half the aircraft [26]. The airplane was split from tip to tail by the x–z plane [26]. Therefore, the full aircraft would correspond to the CD being doubled in the y-direction if the simulation were changed to treat the entire airplane [26]. The meshing was not regular, in contrast to the previously mentioned study, and was completed using ANSYS ICEM [26]. However, four areas of fine resolution meshing were the vortex sheet ($\Delta = 0.07$ m), vortex ($\Delta = 0.07$ m), engine jet ($\Delta = 0.2$ m), and the area between vortex and jet ($\Delta = 0.2$ m) [26].

Another 3D study investigated the microphysical treatment by combining FLUDILES (CFD software) and ULP-ONE (microphysical model) [35]. In that work, the CD was defined to be rectangular prismatic (like the above) and had the dimensions $L_y = 6R$ (direction of flight) and $L_x = L_z = 42R$ (square, cross-sectional plane to direction of flight) [35]. Here, R represents the jet radius and has a value of 0.5 m [35]. This study conducted the meshing in a slightly different manner [35]. Uniformity is observed in the mesh when $x > -11.1R$ and $z < 11.1R$ ($\Delta x = \Delta y = \Delta z = 0.15R$) [35]. Upon leaving that zone, the mesh expands by a monotonic tanh law (hyperbolic tangent) [35]. Another study worth mentioning was completed using an axisymmetric format [2]. The study observed how the exhaust of a CFM56-3 engine influenced near wake contrail creation [2]. The domain included the engine (as opposed to the entire aircraft) and was defined initially as a rectangular plane measuring 5D_b by 76D_b (exhaust direction by transverse direction) with one 76D_b side lying along the axis of the engine (half the engine is included from a side view) [2]. Then, that plane was rotated by 90° to create the CD (average flow taken as axisymmetric) [2]. It should be noted that D_b is the diameter of the bypass flow leaving the engine and is equal to 1.586 m [2].

9.4.3 Boundary condition types

This section provides a guide for setting up the boundary conditions for simulating contrails, including what and where they need to be. For axisymmetric-cylindrical, the computational domain can be chosen so that the revolution line corresponds with the engine's center line, with a boundary condition of 'symmetry' imposed on the center line [2]. Commercial airplanes mostly use turbofans. Thus, the inlet boundaries generally include three sections: the engine's core, the engine's bypass, and the free-stream charging the flow around the engine [26]. Literature has assigned two types of conditions to these boundaries, i.e. constant pressure [2] or constant velocity [26]. The engine walls are no-slip in general [2]. The domain's outlet is a pressure far field [2] utilized to model a free-stream condition at infinity via specifying a free-stream Mach number and static conditions. The side boundaries are set as either symmetry or slip [26].

9.4.4 Initial conditions and boundary condition values

This section discusses the initialization (initial conditions) needed to run a simulation. One must allocate initial values to humidity, temperature, velocity, volume fraction, and pressure, among any other dependent or independent variables of interest, to communicate to the computer where to start calculations.

Initializing humidity is important due to its effect on condensation and nucleation as described by the Schmidt–Appleman criterion (SAC) [3]. The literature displays a variety of methods for initializing the humidity of the simulation.

In a 2D domain, one way to initialize humidity is to divide the planar domain vertically in half by a tall, horizontal band that stretches the entire width (x-direction) [16]. Depending on the scope of the simulation, one would initially assign subsaturation or supersaturation conditions with respect to ice to this band. Various studies have initialized the humidity with different values, including 30%, 60%, and 120% [2, 16, 26]. One needs to define the initial humidity by setting the initial mass fraction of vapor phase water.

Temperature is another parameter influencing SAC; thus, its initialization is crucial [3]. Assigning realistic temperatures that represent the atmospheric conditions is critical to simulation accuracy as the temperature is a key factor for nucleation [3]. Atmospheric temperature varies with elevation; however, this variance is comprised of a series of increasing and decreasing temperature gradients as altitude increases [46]. Therefore, the choice of cruising altitude is a determining factor in the operational and initial temperature present during cruising flight. A reasonable choice for the background cruising temperature could be somewhere within 217–220 K range, representing a cruising altitude of 35 000 ft (10.668 km) within the upper troposphere conditions [16, 17, 27, 28, 37, 40].

Of course, this temperature will increase after the injection of the hot exhaust gas. The conservation of energy equation accounts for this temperature increase. This would require defining a thermal boundary condition at the inlet boundaries, e.g. a constant temperature boundary condition. These temperatures would be supplied depending on the engine conditions. For instance, for commercial turbofans, the literature suggests applying 547–580 K and 242–253 K to the engine core and bypass flows, respectively [2, 26].

For pressure, one needs to initialize the model using values simulative of the upper troposphere cruising conditions. Various studies have adopted slightly different values; however, they all range from 220 hPa to 250 hPa [2, 27, 28, 37, 50].

Particulates present in the wake of an aircraft have a monumental influence on the development of a contrail by providing nucleation sites that kick start the condensation and then freezing processes [3]. The characterization of the wake particulates includes factors such as particle size and particle number, both requiring initializations. According to the literature [2, 26], we recommend a concentration of 10^{11} particles m^{-3} and a beginning diameter of 20–40 nm [2], in line with available *in situ* measurements [26]. Obviously, the initial total number of particles would depend on the domain size. Note that to make such simulations computationally viable, one generally needs to group the actual physical particles into parcels called simulation particles [12]. The literature suggests using 10^7 physical particles per simulation particle ratio [2]—of course, one can choose a different ratio given their computational resources.

Velocity can be broken down into two different types if the aircraft is assumed to be at cruising conditions: aircraft cruising velocity and jet exhaust velocity. Values assigned depend on the engine type and flight stage. For instance, for a CFM56-3 engine, the cruising velocity, core's exhaust velocity, and bypass's exhaust velocity can be set at nearly 252, 480, and 311 m s^{-1}, respectively [2]. Note that these are suggested reasonable values for cruising, and one can slightly deviate from them. For instance, in a different study, the cruising speed of the aircraft was 237 m s^{-1} while the core and bypass flows of the engine were 475 m s^{-1} and 316 m s^{-1}, respectively [26].

9.5 Works of validation

Constructing a CFD simulation of such a complex phenomenon, i.e. the formation and evolution of an artificial cloud in the turbulent wake of an airplane, is a very complex task. The subject of contrail-specific CFD simulation lends itself to incorporating several models coexisting in harmony as one package. This process also involves choosing an appropriate simulation platform, the simulation domain, and initial/boundary conditions. This complexity introduces numerous areas where results could become skewed. Thus, validating such simulations is essential, that is, comparing the simulation results to the literature, particularly to experimental data collected *in situ*.

There are only a few studies sharing experimentally measured data that one could use to validate contrail simulations. One study measured the water vapor content of contrails via an AIMS-H_2O mass spectrometer and WARAN (WAter vapoR ANalyzer) tunable diode laser system [53]. Through two sets of data in the study, the relative humidity with respect to ice (RH_i) showed an increase from 89% to 95% and 90% to 97% for transition from clear sky to cirrus and contrail [53]. The atmospheric conditions present during this data collection could be fed into the simulation, and simulated results could be compared against the reported data. This would be a prudent method of validating the simulations' output. Analysing the measured data also revealed that, whether sub-saturated or supersaturated, RH_i moved toward saturation when transitioning from clear sky to contrail interior [53]. A new simulation could be validated via this as well by running all three cases and ensuring that the output RH_i values trend the same as the study.

Another experimental study measured particle concentration associated with contrails [54]. Soot and ice crystal concentrations were collected (via a cloud and aerosol spectrometer (CAS)) and discussed [54]. The study found that the highest soot particle concentration was 4.6×10^{15} kg^{-1}, and the highest ice crystal concentration was 3.7×10^{15} kg^{-1} [54]. Thus, to validate a CFD model, one could attempt to employ the model to predict the measured values by applying the recorded atmospheric conditions to the numeric study. In addition to these field measurements, very limited wind tunnel studies also contribute to the pool of experimental data available for validation purposes [28].

9.6 Concluding remarks

With the advent of hydrogen-powered aviation on the horizon, the characteristics of tomorrow's contrails have the potential to change dramatically. This is concerning due to the radiative forcing effects that contrails impose on the climate [2]. Since hydrogen power's main goal is to reduce the CO_2 emissions of the aviation industry, which in turn is associated with negative climate effects, it is essential to understand the scale of radiative forcing effects of the new contrails. If the climate effects from the radiative forcing end up outweighing the benefits of the reduction in CO_2 emissions, then this would prove largely counterproductive to the efforts of implementing hydrogen power in the first place. This then calls for developing tools for predicting contrail characteristics before the advent of hydrogen-powered aviation.

There is a possibility that the predictions mentioned above can be executed via the application of CFD simulation. In fact, since measuring hydrogen contrails *in situ* is impossible because such airplanes do not exist yet, and investigating them in laboratories is very difficult due to the need for relatively large chambers with climate control, CFD appears to be the most viable option for conducting such studies. CFD simulations have been reported in the literature to study conventional contrails for many years. Therefore, applying this methodology to hydrogen-generated contrails is worth researching and developing. The development of a CFD simulation comes with a number of considerations including the software used, the models implemented, the set-up of the CD, and the validation of the simulation results.

The preceding text presented some details necessary for consideration when creating a CFD contrail simulation by reviewing the literature in the field of CFD and contrails. By using this collection of information, it is hoped that future studies will be accelerated in their efforts to simulate contrail phenomenon with CFD methodology. In addition to the information provided, some guidelines could be helpful for the aspiring researcher with regard to possible paths forward and potential limitations when approaching the contrail problem. CFD simulations can easily become computationally heavy and result in lengthy calculation times. However, the desire for simulations to run faster must be balanced with the level of detail and range of application needed to provide meaningful results. For faster run times, RANS can be implemented for the viscous model; however, in the literature, RANS was only seen to be used in the near field (young contrails). With that in mind, LES may be a better choice, since it was implemented in the literature for both near field and far field simulations (young and old contrails). Note, this path will increase computation time. What software package to use then comes into question. Some software packages allow the choice of either viscous model, such as STAR-CCM+ or Ansys Fluent, but other software packages are more specific. If RANS is desired, an LES-based package such as EULAG or FLUDILES is not applicable. EULAG would be appropriate if the LCM package is of interest. For example, a researcher may want to use the LPT capability of the LCM which would lead to an applicable use of the EULAG software. However, if EPT is more relevant to the study in question, CEDRE would be more applicable due to the included SPIREE solver [24].

Particle tracking prompts considerations in its own right based on the information the researcher aspires to gain. EPT is more pertinent to studies that would benefit from particle concentration information or less computational expense [45]. However, a limitation of EPT is that it does not work well with low particle concentrations [45]. If having information about each particle's (or group of particle's) motion is a more desirable outcome, then LPT is the more fitting choice. However, this abundance of useful information comes at a cost; LPT pays that cost in computational expense [45]. When studying the literature, it is observed that the microphysical model choice is often related to the software used. For example, EULAG is seen with LCM and FLUDILES with LCM or ULP-ONE when studying the literary works. In contrast, Ansys Fluent has its own built-in models for the microphysics, and other software packages may gravitate more toward user-defined functions. Once these considerations have taken place, the computational domain characteristics and conditions can be attended to. A good deal of these will be tailored to the scenario desired for simulation; however, it should be noted that the choice of a 2D or 3D CD will affect the viscous model choice since LES is only meaningful in 3D [52]. With these guidelines regarding paths and limitations and the information provided in the prior pages, it is hoped that the forthcoming researcher will garner a clearer sense of how to approach creating a CFD contrail simulation. See figure 9.5 for a visual concept of how software and models come together to form an overall simulation system.

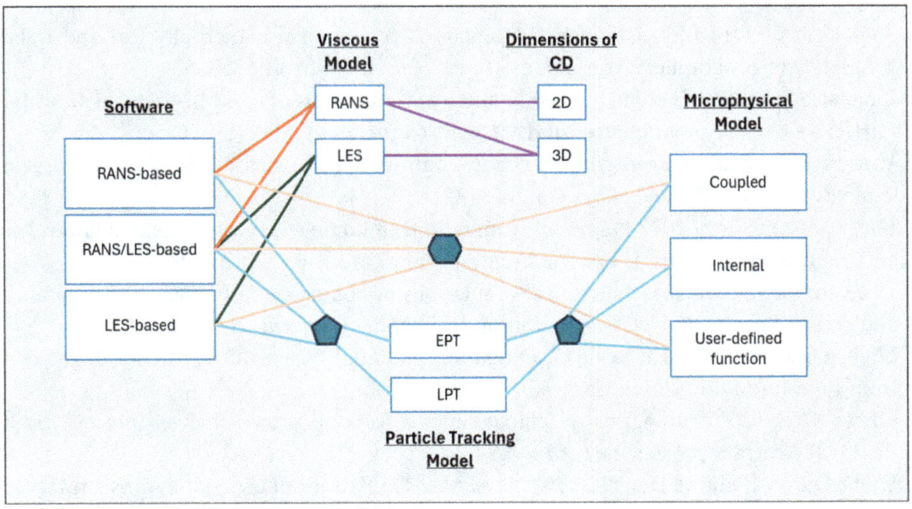

Figure 9.5. Categories for consideration. Lines indicate how models are connected to the software or to each other. Nodes (pentagons and hexagon) indicate any combination of model with software or model with model (depending on the scenario).

References

[1] Staples M D *et al* 2018 Aviation CO_2 emissions reductions from the use of alternative jet fuels *Energy Policy* **114** 342–54

[2] Cantin S *et al* 2022 Eulerian–Lagrangian CFD-microphysics modeling of a near-field contrail from a realistic turbofan *Int. J. Eng. Res.* **23** 661–77

[3] Cabrera E and Melo de Sousa J M 2024 Addressing confidence in modeling of contrail formation from e-fuels in aviation using large eddy simulation parametrization *Energies* **17** 1442

[4] Burkhardt U and Kärcher B 2011 Global radiative forcing from contrail cirrus *Nat. Clim. Change* **1** 54–8

[5] Singh D K, Sanyal S and Wuebbles D J 2024 Understanding the role of contrails and contrail cirrus in climate change: a global perspective *Atmos. Chem. Phys.* **24** 9219–62

[6] Knutti R, Rugenstein M A A and Hegerl G C 2017 Beyond equilibrium climate sensitivity *Nat. Geosci.* **10** 727–36

[7] Grabowski W W and Smolarkiewicz P K 2002 A multiscale anelastic model for meteorological research *Mon. Weather Rev.* **130** 939–56

[8] Sölch I and Kärcher B 2010 A large-eddy model for cirrus clouds with explicit aerosol and ice microphysics and Lagrangian ice particle tracking *Q. J. R. Meteorol. Soc.* **136** 2074–93

[9] Benoit W 2005 Mechanical properties: anelasticity *Encyclopedia of Condensed Matter Physics* ed F Bassani, G L Liedl and P Wyder (Oxford: Elsevier) pp 271–80 https://www.sciencedirect.com/science/article/pii/B012369401900574X

[10] Marshall J *et al* 1997 Hydrostatic, quasi-hydrostatic, and nonhydrostatic ocean modeling *J. Geophys. Res. Oceans* **102** 5733–52

[11] Bier A, Unterstrasser S and Vancassel X 2022 Box model trajectory studies of contrail formation using a particle-based cloud microphysics scheme *Atmos. Chem. Phys.* **22** 823–45

[12] Unterstrasser S and Sölch I 2010 Study of contrail microphysics in the vortex phase with a Lagrangian particle tracking model *Atmos. Chem. Phys.* **10** 10003–15

[13] Joos H *et al* 2014 Influence of heterogeneous freezing on the microphysical and radiative properties of orographic cirrus clouds *Atmos. Chem. Phys.* **14** 6853–52
[14] Unterstrasser S and Gierens K 2010 Numerical simulations of contrail-to-cirrus transition—Part 1: an extensive parametric study *Atmos. Chem. Phys.* **10** 2017–36
[15] Unterstrasser S 2016 Properties of young contrails-a parametrisation based on large-eddy simulations *Atmos. Chem. Phys.* **16** 2059–82
[16] Unterstrasser S *et al* 2017 Numerical simulations of homogeneously nucleated natural cirrus and contrail-cirrus. Part 1: how different are they? *Meteorol. Z.* **26** 621–42
[17] Unterstrasser S *et al* 2017 Numerical simulations of homogeneously nucleated natural cirrus and contrail-cirrus. Part 2: interaction on local scale *Meteorol. Z.* **26** 643–61
[18] Unterstrasser S 2020 The contrail mitigation potential of aircraft formation flight derived from high-resolution simulations *Aerospace* **7** 170
[19] Grewe V *et al* 2017 Mitigating the climate impact from aviation: achievements and results of the DLR WeCare project *Aerospace* **4** 34
[20] Unterstrasser S and Gierens K 2010 Numerical simulations of contrail-to-cirrus transition—Part 2: impact of initial ice crystal number, radiation, stratification, secondary nucleation and layer depth *Atmos. Chem. Phys.* **10** 2037–51
[21] ONERA 2024 Simulation pour l'nergtique et la propulsion *ONERA* https://cedre.onera.fr/fr (Accessed: 7 May 2024)
[22] McAlister K W, Lambert O and Petot D 1984 Application of the ONERA model of dynamic stall *Report* No. NASA-A-9824 National Aeronautics and Space Administration Moffett Field CA Ames Research Center https://ntrs.nasa.gov/citations/19850004554
[23] Refloch A *et al* 2011 CEDRE software *Aerosp. Lab* **2** AL2–AL11 https://hal.science/hal-01182463
[24] Scherrer D *et al* 2011 Recent CEDRE applications *Aerosp. Lab* **2** 1–28 https://hal.science/hal-01182477
[25] Dupoirieux F and Bertier N 2011 The models of turbulent combustion in the CHARME solver of CEDRE *Aerosp. Lab* **2** AL2 https://hal.science/hal-01181233/
[26] Khou J C *et al* 2015 Spatial simulation of contrail formation in near-field of commercial aircraft *J. Aircr.* **52** 1927–38
[27] Khou J C *et al* 2017 CFD simulation of contrail formation in the near field of a commercial aircraft: effect of fuel sulfur content *Meteorol. Z.* **26** 585–96
[28] Guignery F *et al* 2012 Contrail microphysics in the near wake of a realistic wing through RANS simulations *Aerosp. Sci. Technol.* **23** 399–408
[29] Bouhafid Y, Bonne N and Jacquin L 2024 Combined Reynolds-averaged Navier–Stokes/large-eddy simulations for an aircraft wake until dissipation regime *Aerosp. Sci. Technol.* **154** 109512
[30] Guignery F *et al* 2009 Numerical simulation of contrail formation using a simplified wing-injector configuration *1st AIAA Atmospheric and Space Environments Conference (San Antonio, TX, June)* p 3868
[31] Ansys Inc 2024 Ansys Fluent—fluid simulation software *ANSYS* https://www.ansys.com/products/fluids/ansys-fluent (Accessed: 10 March 2024)
[32] Sui G *et al* 2023 Contrail polarization scattering characteristics simulation based on jet flow field *IEEE Photon. Technol. Lett.* **36** 705–8
[33] Lobanova M A and Tsirkunov Y M 2012 Numerical simulation of a jet-vortex wake behind a cruise aircraft *Proceedings of the 6th European Congress on Computational Methods in Applied Sciences and Engineering (Vienna, Austria, 10–14 September)*

[34] Siemens 2024 Simcenter STAR-CCM+ CFD software https://plm.sw.siemens.com/en-US/simcenter/fluids-thermal-simulation/star-ccm/ (Accessed: 10 March 2024)

[35] Vancassel X, Mirabel P and Garnier F 2014 Numerical simulation of aerosols in an aircraft wake using a 3D LES solver and a detailed microphysical model *Int. J. Sustain. Aviat.* **1** 139–59

[36] Chlond A 1998 Large-eddy simulation of contrails *J. Atmos. Sci.* **55** 796–819

[37] Lewellen D C and Lewellen W S 2001 The effects of aircraft wake dynamics on contrail development *J. Atmos. Sci.* **58** 390–406

[38] Picot J *et al* 2014 Large-eddy simulations of contrails in a turbulent atmosphere *Atmos. Chem. Phys. Discuss.* **14** 29499–546

[39] Picot J *et al* 2015 Large-eddy simulation of contrail evolution in the vortex phase and its interaction with atmospheric turbulence *Atmos. Chem. Phys.* **15** 7369–89

[40] Shirgaonkar A and Lele S 2005 Simulations of aircraft wake-exhaust mixing using LES: contrail formation in the near-field *35th AIAA Fluid Dynamics Conference and Exhibition (Toronto, ON, 6–9 June)* p 4909

[41] Unterstrasser S 2014 Large-eddy simulation study of contrail microphysics and geometry during the vortex phase and consequences on contrail-to-cirrus transition *J. Geophys. Res. Atmos.* **119** 7537–55

[42] Sölch I and Kärcher B 2011 Process-oriented large-eddy simulations of a midlatitude cirrus cloud system based on observations *Q. J. R. Meteorol. Soc.* **137** 374–93

[43] Smolarkiewicz P K and Margolin L G 1997 On forward-in-time differencing for fluids: an Eulerian/semi-Lagrangian non-hydrostatic model for stratified flows *Atmos. Ocean* **35** 127–52

[44] Garnier F *et al* 2014 Effect of compressibility on contrail ice particle growth in an engine jet *Int. J. Turbo Jet-Engines* **31** 131–40

[45] Saidi M S *et al* 2014 Comparison between Lagrangian and Eulerian approaches in predicting motion of micron-sized particles in laminar flows *Atmos. Environ.* **89** 199–206

[46] Sandford D J 2008 Dynamics of the stratosphere, mesosphere and thermosphere *PhD Thesis* University of Bath https://researchportal.bath.ac.uk/en/studentTheses/dynamics-of-the-stratospheremesosphere-and-thermosphere

[47] Mohanakumar K 2008 *Stratosphere Troposphere Interactions: An Introduction* (Dordrecht: Springer Science)

[48] Unterstrasser S, Gierens K and Spichtinger P 2008 The evolution of contrail microphysics in the vortex phase *Meteorol. Z.* **17** 145–56

[49] Lewellen D C 2020 A large-eddy simulation study of contrail ice number formation *J. Atmos. Sci.* **77** 2585–604

[50] Paoli R *et al* 2017 Three-dimensional large-eddy simulations of the early phase of contrail-to-cirrus transition: effects of atmospheric turbulence and radiative transfer *Meteorol. Z.* **26** 597–620

[51] Unterstrasser S and Görsch N 2014 Aircraft-type dependency of contrail evolution *J. Geophys. Res. Atmos.* **119** 14–15

[52] Zhiyin Y 2015 Large-eddy simulation: past, present and the future *Chin. J. Aeronaut.* **28** 11–24

[53] Kaufmann S *et al* 2014 In situ measurements of ice saturation in young contrails *Geophys. Res. Lett.* **41** 702–9

[54] Kleine J *et al* 2018 *In situ* observations of ice particle losses in a young persistent contrail *Geophys. Res. Lett.* **45** 13–553

IOP Publishing

Renewable Energy Systems
The way forward
David S-K Ting and Jacqueline A Stagner

Chapter 10

Sustainability transitions: comparing urban sophistication and rural simplicity

Ariva Sugandi Permana and Chantamon Potipituk

This chapter explores sustainability in urban and rural areas amidst rapid urbanization. Urban areas face challenges such as resource management and waste reduction but also offer innovative solutions such as renewable energy and green buildings. Rural areas, with simpler lifestyles and a closer connection to nature, adopt practices such as sustainable agriculture but face resource access challenges. Bridging the gap between urban and rural sustainability is fundamental, requiring knowledge sharing, collaboration, policy coordination, and community engagement. The transition towards sustainability requires a comprehensive approach that fosters innovation and community empowerment for a resilient and sustainable world.

10.1 Introduction

In the modern era, the concept of sustainability has emerged as a crucial consideration in both urban and rural settings. We need to provide a brief overview of the importance of sustainability in these diverse environments, examining the unique challenges and opportunities they present. We seek to explore the different approaches to sustainability transitions and their impact on communities and the environment by comparing urban sophistication and rural simplicity,

Sustainability is a pressing global issue, with implications for both developed and developing regions. Urban areas, characterized by high population density and complex infrastructure, face distinct sustainability challenges such as resource management, waste disposal, and energy consumption (World Economic Forum 2018). On the other hand, rural settings, known for their natural landscapes and agricultural activities, grapple with issues such as land use, biodiversity conservation, and community resilience (Williams *et al* 2021). Understanding the importance

doi:10.1088/978-0-7503-6179-8ch10

of sustainability in these contrasting contexts is essential for fostering a comprehensive approach to environmental stewardship.

In the context of sustainability, we refer to 'sustainability transitions: comparing urban sophistication and rural simplicity' as the study or analysis of changes towards sustainable practices in both urban and rural settings. The term 'urban sophistication' signifies the complex and advanced systems in place in urban or city environments. This includes advanced waste management systems, energy-efficient buildings, public transportation systems, and infrastructure designed to minimize environmental impact, social life, and economic activities of the citizens. On the other hand, 'rural simplicity' pertains to the simpler, less complex systems often found in rural or countryside areas. This includes traditional farming practices, small-scale renewable energy projects, community-led conservation efforts, and the lifestyle of the rural inhabitants. Therefore, the comparison between the two could involve analysing the strengths and weaknesses of each approach, understanding how sustainable practices can be implemented in different contexts, and exploring how lessons learned in one setting can be applied in the other. It underscores the diverse strategies needed to achieve sustainability through collaboration while exploring the compatibility across different types of communities. Figure 10.1 summarizes this explanation.

We adopt a straightforward approach to defining urban and rural areas, based primarily on the predominant occupations of the residents. Therefore, in our view, urban areas are regions where the majority of the population is engaged in service-oriented jobs. Conversely, rural areas are regions where the majority of the residents are involved in agricultural activities.

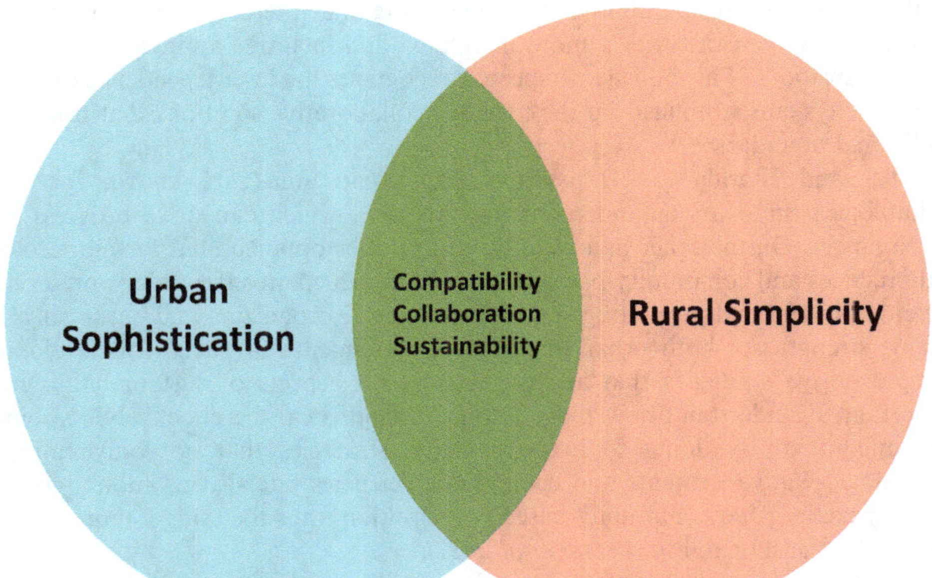

Figure 10.1. Transitions: urban sophistication–rural simplicity.

The significance of sustainability in urban and rural settings cannot be overstated. In urban areas, the need for sustainable practices is underscored by the sheer scale of environmental impact and resource consumption (Gorman-Murray and Lane 2012). As centers of economic activity and innovation, cities play a pivotal role in driving sustainable development and shaping global trends. Conversely, rural areas, often overlooked in discussions of sustainability, are vital for preserving ecosystems, supporting agricultural sustainability, and maintaining cultural heritage. Recognizing the interconnectedness of urban and rural sustainability is fundamental for achieving holistic and equitable progress.

The juxtaposition of urban sophistication and rural simplicity offers valuable insights into the diverse manifestations of sustainability. Urban centers, with their technological advancements and intricate networks, exemplify the potential for sustainable urbanization and smart infrastructure. Meanwhile, rural communities, with their close ties to nature and traditional practices, epitomize the harmonious coexistence of humans and the environment. Through examining these contrasting paradigms, we can uncover valuable lessons and best practices for promoting sustainability in both settings.

We seek to address the evolving nature of sustainability transitions in response to contemporary challenges. As societies grapple with issues such as climate change, resource depletion, and social inequality, the imperative for sustainable solutions becomes increasingly urgent. Understanding how urban and rural contexts adapt to these challenges and innovate towards sustainability is crucial for informing policy, technology, and community-based interventions. We aim to shed light on the evolving landscape of sustainability transitions by delving into these dynamics.

This chapter catalyzes reimagining the relationship between urban and rural areas in the context of sustainability. Acknowledging the interconnectedness of these environments, we can foster a more integrated and inclusive approach to sustainability transitions. This holistic perspective recognizes the interdependence of urban and rural systems, emphasizing the need for collaborative solutions that transcend traditional boundaries.

Adler and Florida (2021) asserted that urban areas are known for their technological innovations and economies of scale, which can drive progress and development. On the other hand, rural settings offer opportunities for agroecological practices and community-based conservation, which are essential for preserving the environment and promoting sustainable living. Recognizing and harnessing the unique strengths of both urban and rural environments is crucial for developing comprehensive strategies that address the diverse aspects of sustainability. It is important to understand how these distinct strengths can be effectively combined and integrated. By doing so, we can create strategies that not only embrace technological advancements and economic efficiencies but also promote agroecological practices and community-based conservation, thus fostering a more holistic approach to sustainability (Peano *et al* 2021).

In sustainability transitions, particularly within urban–rural dynamics, it is crucial to emphasize the significance of inclusivity and equity. Acknowledging the varied needs and aspirations of both urban and rural populations is vital for ensuring that sustainability initiatives are socially equitable and culturally attuned. We can empower communities to collectively develop sustainability strategies that resonate with their unique circumstances and values by embracing a participatory approach that is tailored to specific locations, we can also empower communities to collectively develop sustainability strategies that resonate with their unique circumstances and values. It is essential to recognize the importance of inclusivity and equity in sustainability transitions, especially within the context of urban–rural dynamics. Eizenberg and Jabareen (2017) also argued that understanding the diverse needs and aspirations of urban and rural populations is indispensable for ensuring that sustainability efforts are socially just and culturally sensitive.

This introduction lays the groundwork for a comprehensive exploration of sustainability transitions in urban sophistication and rural simplicity. Through the examination of the importance of sustainability in both settings, recognizing their unique challenges and opportunities, and advocating for an integrated and inclusive approach, we aim to contribute to a nuanced understanding of sustainability and inspire actionable pathways toward a more sustainable future.

Our goal is to stimulate conversation, thoughtful analysis, and joint efforts towards shifts in sustainability that acknowledge the variety of both urban and rural environments. By nurturing a profound comprehension of the links between the complexity of city life and the straightforwardness of rural living, we can foster a more comprehensive and robust strategy for sustainability. This approach surpasses traditional binary thinking and appreciates the wealth of varied landscapes and communities.

As we embark on this exploration of sustainability transitions, it is essential to acknowledge the dynamic and evolving nature of sustainability challenges and opportunities in urban and rural settings. The global context of climate change, urbanization, and technological advancement continuously shapes the sustainability landscape, presenting new complexities and possibilities for transformative change (Ruth and Coelho 2015). Keeping attuned to these dynamics, we can adapt our strategies and interventions to effectively address emerging sustainability imperatives.

The journey towards sustainability in urban sophistication and rural simplicity is characterized by its multi-dimensionality, requiring a nuanced understanding of environmental, social, economic, and cultural factors. Through this chapter, we aim to lay the groundwork for a comprehensive examination of sustainability transitions, inviting readers to engage with the complexities, contradictions, and potential synergies inherent in urban and rural sustainability paradigms.

10.2 Understanding sustainability

10.2.1 Definition and principles of sustainability

Sustainability refers to the capacity to meet the needs of the present without compromising the ability of future generations to meet their own needs. It embodies a holistic approach that seeks to balance environmental, social, and economic considerations (Spijkers 2018). The principles of sustainability encompass intergenerational equity, ecological integrity, social justice, and economic resilience

(Khan and Stinchcombe 2023). Sustainability aims to foster systems that are regenerative, inclusive, and adaptable, promoting harmony between human activities and natural ecosystems.

10.2.2 The role of sustainability in urban and rural development

In urban development, sustainability is essential in addressing the complex challenges posed by rapid urbanization, resource consumption, and environmental degradation (Rana 2011). Sustainable urban development encompasses strategies for efficient land use, integrated transportation systems, green infrastructure, and equitable access to resources and services. Promoting compact living, walkable neighborhoods, renewable energy deployment, and sustainable water management, urban sustainability initiatives encourage the enhancement of quality of life, reduction of ecological footprint, and fostering resilient, vibrant urban communities.

Likewise, in rural development, sustainability is fundamental for preserving natural ecosystems, supporting agricultural livelihoods, and enhancing community well-being (Scoones 1998). Sustainable rural development encompasses approaches that promote agroecological practices, biodiversity conservation, and equitable access to resources. Fostering sustainable land management, diversified agricultural systems, and community-based conservation initiatives, rural sustainability efforts mean strengthening local economies, preserving cultural heritage, and ensuring the long-term viability of rural communities.

Sustainability in rural development emphasizes the importance of social cohesion, traditional knowledge, and participatory decision-making processes. Rural sustainability initiatives contribute to the resilience and vitality of rural landscapes and communities by empowering rural communities to engage in sustainable resource management, value-added agricultural activities, and diversified livelihood opportunities (Fretes-Cibils *et al* 2008).

Ziervogel *et al* (2016) asserted that sustainability principles provide a comprehensive framework for addressing the interconnected challenges of urban and rural development, guiding efforts to create inclusive, regenerative, and resilient communities. Sustainable urban and rural development can move towards a more equitable, harmonious, and prosperous future, where human well-being is intricately linked with the health of the environment and the vitality of local communities. The principles of sustainability serve as a compass for urban and rural development, guiding decision-making processes, policy formulation, and community engagement toward long-term viability and well-being. With the integration of sustainability principles into urban and rural development strategies, stakeholders can work towards creating environments that are conducive to human flourishing while safeguarding the natural resources and ecosystems on which life depends.

The role of sustainability in urban and rural development extends beyond environmental considerations to encompass social and economic dimensions. Sustainable urban and rural development seeks to address social equity, cultural diversity, and economic prosperity, recognizing the interconnectedness of environmental, social, and economic well-being. In urban development, sustainability principles guide the design of infrastructure, buildings, and public spaces to minimize environmental impact, enhance energy efficiency, and promote healthy, livable environments. Sustainable urban development also emphasizes the importance of social inclusion, affordable housing, and accessible public transportation, aiming to create cities that are equitable, vibrant, and adaptable to changing needs and challenges (Wheeler 2013).

Bunch *et al* (2011) asserted that sustainable natural resource management and community development strategies ensure the long-term health of ecosystems and the well-being of rural residents. Sustainable rural development also focuses on preserving cultural heritage, supporting traditional knowledge, and enhancing local economies through sustainable agriculture, eco-tourism, and value-added enterprises, contributing to the vitality and resilience of rural communities. The principles of sustainability provide a robust foundation for guiding urban and rural development towards inclusive, regenerative, and resilient outcomes. Urban and rural development can strive towards creating environments that promote human well-being, preserve natural ecosystems, and foster vibrant, cohesive communities, ultimately contributing to a more sustainable and harmonious future for all (Nitivattananon *et al* 2010).

The role of sustainability in urban and rural development extends to fostering innovation, knowledge sharing, and capacity building. Sustainable urban and rural development initiatives often serve as platforms for experimentation, learning, and collaboration, driving the adoption of innovative technologies and practices. Knowledge exchange, skill development, and community empowerment contribute to building adaptive, knowledge-based societies capable of responding to evolving challenges.

Sustainability guides urban and rural development towards resilience, preparedness, and adaptation in the face of environmental and social uncertainties. Sustainable urban and rural development strategies prioritize risk reduction, climate resilience, and disaster preparedness, recognizing the importance of building communities that are capable of withstanding and recovering from shocks and stresses. This initiative strives to minimize waste generation (reduce), promote recycling, and reuse, and therefore, advocate the 3Rs. It also optimizes resource use through circular economy approaches. Embracing circular economics, urban and rural development can contribute to reducing environmental impact, conserving resources, and creating economic opportunities through the efficient use and reutilization of materials. Work towards creating environments that are conducive to human well-being, while safeguarding the natural resources and ecosystems on which life depends can be accomplished through sustainable urban and rural development. This can contribute to building vibrant, equitable, and regenerative communities that thrive in harmony with the environment.

10.3 Challenges and opportunities in the complexities of urban development

10.3.1 Urbanization and urban living

Urbanization, the rapid growth of cities, has dramatically changed the world around us. Cities are now home to large populations and complex infrastructure, showcasing human ingenuity and progress. However, these urban areas also bring significant challenges when it comes to sustainability, which has become crucial for our long-term well-being and success (figure 10.2).

Cities, with their towering skyscrapers, bustling streets, and vibrant cultures, are the epitome of urban sophistication. They are centers of economic activity, technological innovation, and cultural diversity. Yet, this urban sophistication is not without its drawbacks. The very characteristics that make cities dynamic and exciting also contribute to their environmental footprint. High energy consumption, waste generation, air and water pollution, and the urban heat island effect are just a few of the sustainability challenges that cities face.

Yusop and Permana (2016) argue that cities offer opportunities for more efficient use of resources. High population density in cities can lead to economies of scale in public services such as transportation and waste management. However, the concentration of people and activities can lead to overconsumption and waste. The challenge lies in managing these trade-offs to enhance the sustainability of urban living. Urbanization also has significant environmental implications. The expansion of cities often encroaches on natural habitats, leading to a loss of biodiversity. The high demand for resources in cities can strain ecosystems both locally and globally. We may agree that cities are major contributors to climate change, with most greenhouse gas emissions originating from urban areas. Addressing these environmental implications is crucial for the sustainability of our planet.

Figure 10.2. Traffic congestion: a sign of urbanization. (This Bangkok downtown traffic image has been obtained by the author(s) from the Wikimedia website where it was made available by TomaszSwatowski under a CC BY 4.0 licence. It is included within this article on that basis. It is attributed to TomaszSwatowski.)

Despite these challenges, there are numerous examples of sustainable urban practices around the world. Cities are experimenting with innovative solutions to enhance sustainability (Buuse *et al* 2022). For instance, many cities are investing in renewable energy, green buildings, and public transportation to reduce their environmental impact. Others are promoting circular economy practices to reduce waste and make better use of resources. These case studies provide valuable lessons on how cities can navigate the path toward sustainability.

10.3.2 Urban sophistication: challenges and opportunities

Urban sophistication, a term often used to describe the advanced state of development in cities, is characterized by complex social structures, diverse economies, and intricate infrastructures. This level of development is a testament to human ingenuity and progress, reflecting our ability to create complex systems and structures that facilitate modern life. However, this sophistication comes with its own set of challenges, many of which are environmental. Cities, with their dense populations and high levels of activity, consume vast amounts of resources (Kennedy *et al* 2007). This consumption is not just limited to energy for powering buildings and vehicles but also includes water for various uses, materials for construction, and food for feeding the population. The high demand for these resources puts pressure on natural ecosystems, leading to issues such as deforestation, water scarcity, and soil degradation. Cities generate enormous amounts of waste (figure 10.3). This includes solid waste from households and businesses, wastewater from various activities, and air pollutants from vehicles and industries. Managing this waste is a significant challenge for cities. Inadequate waste management can lead to pollution of the air, water, and soil, affecting the health of the city's residents and the quality of the environment. Cities also contribute significantly to greenhouse gas emissions, primarily through the burning of fossil fuels for energy and transportation (Permana *et al* 2018). These emissions are a major driver of climate change, which poses a range of risks to cities, including rising sea levels, increased frequency and intensity of extreme weather events, and disruptions to urban services.

Urban areas are often plagued by issues such as air and water pollution, traffic congestion, and inadequate waste management (Permana *et al* 2015). Air pollution, primarily caused by vehicle emissions and industrial activities, can lead to a range of health problems, including respiratory diseases and heart conditions (Zakka *et al* 2017a). Water pollution, resulting from the discharge of untreated or inadequately treated wastewater, can contaminate water bodies and affect both human health and aquatic ecosystems. Traffic congestion, a common issue in many cities, not only leads to increased air pollution and greenhouse gas emissions but also affects the quality of life by causing stress and wasting time. Inadequate waste management, on the other hand, can lead to the accumulation of waste in the environment, contributing to pollution, disease transmission, and aesthetic issues.

Despite these challenges, cities also present unique opportunities for sustainability. They are hubs of innovation and creativity, with the potential to develop and

Figure 10.3. Waste: challenges and opportunities. (This image has been obtained by the author(s) from the Wikimedia website, where it is stated to have been released into the public domain. It is included within this article on that basis.)

implement sustainable solutions. Cities, with their concentration on knowledge, skills, and resources, are often at the forefront of technological and social innovations that can drive sustainability.

Cities can leverage their dense populations and economies of scale to promote public transportation, energy-efficient buildings, and waste recycling programs. Public transportation systems, by reducing the reliance on private vehicles, can decrease traffic congestion, air pollution, and greenhouse gas emissions (Permana *et al* 2018). Energy-efficient buildings, using technologies and practices such as insulation, efficient appliances, and renewable energy, can significantly reduce energy consumption and emissions. Waste recycling programs, by recovering and reusing materials, can reduce the demand for virgin resources and the amount of waste that ends up in landfills or the environment (Yusuf and Permana 2013, Towolioe *et al* 2016).

Cities can also foster sustainable behaviors through education and community engagement. Education is a powerful tool for promoting sustainability in cities. Through education, residents can gain a better understanding of the environmental challenges their city faces and learn about the actions they can take to mitigate these challenges. Schools, universities, and community centers can offer programs and workshops on topics such as recycling, energy conservation, and sustainable consumption (Towolioe *et al* 2016). These programs can equip residents with the knowledge and skills they need to make more sustainable choices in their daily lives.

For example, a program on recycling can teach residents about the importance of waste segregation, the types of materials that can be recycled, and the proper way to dispose of recyclable materials (Towolioe *et al* 2016). Similarly, a workshop on energy conservation can provide tips on how to reduce energy consumption at home, such as using energy-efficient appliances, turning off lights and electronics when not in use, and insulating homes to reduce heating and cooling needs.

Community engagement is another crucial aspect of promoting sustainability in cities. When residents are actively involved in decision-making processes, they are more likely to support and participate in sustainability initiatives. Cities can foster community engagement through participatory planning processes, public consultations, and volunteer programs. Participatory planning processes allow residents to have a say in the development of their city. They can provide input on issues such as the location of new parks, the design of public spaces, and the implementation of public transportation systems.

However, they also hold immense potential for driving sustainability. With their dense populations and economies of scale, cities can efficiently provide services such as public transportation, energy-efficient buildings, and waste recycling programs. These services not only reduce the environmental impact of cities but also improve the quality of life for their residents.

Public transportation systems, for instance, can significantly reduce the number of private vehicles on the road, leading to lower traffic congestion, air pollution, and greenhouse gas emissions (Permana *et al* 2018). Energy-efficient buildings, equipped with technologies such as insulation, efficient lighting, and heating and cooling systems, can drastically reduce energy consumption and CO_2 emissions. Waste recycling programs, meanwhile, can divert waste from landfills, reducing pollution and resource consumption while recovering valuable materials.

Cities are also hubs of innovation and creativity, making them ideal places for developing and implementing sustainable solutions. Researchers, entrepreneurs, and policymakers in cities are continually coming up with new ideas and technologies to address sustainability challenges. These innovations range from renewable energy technologies and green building designs to sustainable transportation systems and waste management solutions. Moreover, cities can foster sustainable behaviors among their residents through education and community engagement. Schools, community centers, and local media can raise awareness about sustainability issues and promote sustainable practices. Community engagement initiatives, such as volunteer programs and participatory planning processes, can empower residents to take part in sustainability efforts and make their city a better place to live (Zakka *et al* 2017b).

Urban sophistication, despite its challenges, presents unique opportunities for sustainability. By leveraging their resources, innovation potential, and community engagement, cities can lead the way toward a more sustainable future. However, achieving this requires concerted efforts from all stakeholders, including city governments, businesses, civil society, and residents. Through these collective efforts cities can transform their challenges into opportunities and make urban sophistication synonymous with sustainability.

10.3.3 Impact of urban living on sustainability

Urban living, with its dense population and concentrated infrastructure, has a profound impact on sustainability. The lifestyle choices and consumption patterns of urban residents, coupled with the environmental footprint of urban infrastructures, pose significant challenges to sustainability. However, urban living also has the potential to be more sustainable than rural living, given the right strategies and policies.

Cities, being the hub of economic activities and home to a large portion of the global population, account for over 70% of global energy consumption and CO_2 emissions (Grubler *et al* 2012). This high energy consumption and emission level are largely due to the construction and operation of buildings, transportation systems, and other urban infrastructures. These infrastructures, while necessary for the functioning of the city, consume vast amounts of energy and resources, contributing to environmental degradation. The construction sector, for instance, is a major consumer of energy and emitter of CO_2. Buildings, both residential and commercial, require energy for their construction, operation, and maintenance. The materials used in construction, such as concrete and steel, are energy-intensive to produce and transport. Moreover, once constructed, buildings require energy for heating, cooling, lighting, and other services. This energy is often derived from fossil fuels, leading to CO_2 emissions. Similarly, urban transportation systems, including roads, railways, airports, and ports, are significant contributors to energy consumption and CO_2 emissions. Vehicles, whether private cars or public buses, consume fossil fuels, leading to CO_2 emissions. Moreover, the construction of transportation infrastructure also requires energy and resources, further adding to the environmental footprint of cities.

Dye (2008) argued that urban living also has the potential to be more sustainable than rural living. This potential lies in the inherent characteristics of urban areas, such as high-density living and the availability of infrastructure and resources. High-density living, which is a characteristic feature of urban areas, can be more energy-efficient due to shared walls and shorter travel distances. Shared walls in apartment buildings reduce heat loss, leading to lower energy consumption for heating. Shorter travel distances in compact cities reduce the need for private car use, leading to lower energy consumption for transportation.

Cities, due to their scale and density, have the infrastructure and resources to provide services such as public transportation, waste management, and recycling programs more efficiently than rural areas. Public transportation systems, such as buses and metros, are more energy-efficient per passenger-kilometer than private cars. Waste management and recycling programs in cities can reduce the amount of waste going to landfills and recover valuable resources.

10.3.4 Environmental implications of urbanization

Zakka *et al* (2017b) asserted that urbanization, the process of population shift from rural areas to cities, has significant environmental implications. One of the most prominent impacts is the change in land use. As cities expand, agricultural lands, forests, and other natural habitats are often converted into urban areas to accommodate the growing population and their needs. This leads to habitat destruction and, consequently, a loss of biodiversity. The construction of buildings, roads, and other infrastructures often involves clearing these natural habitats, leading to a significant reduction in the number of plant and animal species in the area. This loss of biodiversity can disrupt ecosystems and lead to the extinction of species.

Urban areas also generate significant amounts of waste and pollution. The high concentration of people and activities in cities leads to the production of large quantities of solid waste, wastewater, and air pollutants. These wastes, if not properly managed, can contaminate the air, soil, and water, affecting the quality of these vital resources. Air pollution, for instance, can lead to respiratory diseases and other health problems among urban residents. Similarly, the improper disposal of solid waste and wastewater can pollute water bodies, affecting both human health and aquatic ecosystems.

Another environmental implication of urbanization is the creation of urban heat islands (Zhou and Chen 2018). Cities, with their concentration of buildings, roads, and other paved surfaces, tend to absorb more solar radiation during the day compared to rural areas. These surfaces then re-emit the heat at night, raising the temperature in urban areas. This phenomenon, known as the urban heat island effect, can make cities significantly warmer than surrounding rural areas. The urban heat island effect not only increases energy consumption for cooling but also exacerbates heat-related health issues. The higher temperatures in cities increase the demand for air conditioning, which in turn increases energy consumption and CO_2 emissions. Moreover, the elevated temperatures can exacerbate heat-related health issues such as heatstroke and cardiovascular problems, particularly among vulnerable populations such as the elderly. Therefore, mitigating the urban heat island effect through urban planning and design strategies, such as increasing green spaces and using reflective materials for buildings and roads, is crucial for enhancing the sustainability of urban areas.

10.3.5 Case studies of sustainable urban practices

Some have argued that cities are parasites of resources, for example, Hoselitz (2002) and Barles (2011), and we agree with their statement but many cities around the world are pioneering sustainable urban practices. These are some examples of urban best practices in Europe, America, and Asia.

10.3.5.1 Copenhagen, Denmark

The references for this case study are Fourtané (2020), Mathiesen *et al* (2019), Skibsted (2019), and Szumski (2022). Copenhagen, the capital city of Denmark, has set an ambitious goal to become carbon-neutral by 2025. This commitment is a testament to the city's dedication to sustainability and its proactive approach to combating climate change.

The city's strategy to achieve this goal is multi-faceted, involving significant investments in renewable energy, green buildings, cycling infrastructure, and an extensive district heating system.

(a) Renewable energy

Copenhagen has made substantial investments in renewable energy sources as part of its carbon neutrality goal. The city is harnessing the power of wind, solar, and bioenergy to reduce its reliance on fossil fuels. Offshore wind turbines in the Øresund Strait between Denmark and Sweden are a prominent feature of the city's renewable energy landscape. These efforts are not only reducing the city's carbon emissions but also setting a precedent for other cities around the world.

(b) Green buildings

Green buildings are another key component of Copenhagen's sustainability strategy. The city has implemented stringent building codes that require new constructions to be energy-efficient buildings. Many buildings in the city are equipped with green roofs, which provide insulation, absorb rainwater, and create habitats for wildlife. These measures are helping to reduce the city's energy consumption and enhance its urban biodiversity.

(c) Cycling infrastructure

Copenhagen is renowned for its cycling culture, and the city has invested heavily in cycling infrastructure to encourage this mode of transport. The city boasts over 400 kilometers of cycle paths, and more than half of all trips to work or school in Copenhagen are made by bicycle. The city is reducing traffic congestion, improving air quality, and promoting physical activity among its residents.

(d) District heating system

Copenhagen's district heating system is one of the largest and most successful in the world. The system uses waste heat from power plants to heat homes across the city. This not only makes efficient use of energy but also reduces the need for individual heating systems in each building.

10.3.5.2 Curitiba, Brazil

The references for this case study are Berzins (2020), Krosofsky (2021), Lozzy (2023), The Sustainable Travel (2019), and UN Department of Economic and Social Affairs (n.d.a). Curitiba, Brazil, is another example of a sustainable urban environment. The city has implemented a successful bus rapid transit (BRT) system that serves as a model for other cities. It has also developed numerous parks and green spaces, helping to preserve biodiversity and improve residents' quality of life.

(a) Curitiba's BRT System

One of the most significant innovations that Curitiba has implemented is its BRT system. This system was designed to provide quick, affordable, and efficient public transportation for the city's residents. The BRT system features roads with express lanes for buses and specially designed buses for quick boarding. The ticket prices are uniform for everyone, making the system accessible to all residents. This innovative transit system not only reduces traffic congestion but also lowers carbon emissions, contributing to the city's sustainability goals.

(b) Green spaces in Curitiba

Curitiba has also made significant strides in developing green spaces throughout the city. Since the 1970s, the city has planted 1.5 million trees and built 28 public parks. These green spaces not only enhance the city's aesthetic appeal but also provide habitats for various species, helping to preserve biodiversity. Moreover, they offer recreational spaces for residents, improving their quality of life. Curitiba has surrounded all the urban areas with fields of grass to combat flooding. This innovative approach to flood control has saved the city the cost and environmental expense of building dams.

(c) Other sustainability initiatives in Curitiba

Beyond its BRT system and green spaces, Curitiba has implemented several other sustainability initiatives. The city recycles around 70% of its garbage thanks to a program that allows residents to exchange their recycling for bus tokens, notebooks, and even food. This program not only protects the environment but also boosts education, increases food access, and facilitates transport for poor citizens.

10.3.5.3 Samut Sakhon, Thailand

The references for this case study are Ecological Alert and Recovery—Thailand (2019, 2020) and UN ESCAP (n.d.). In Bangkok, a city in a developing country, the Thai government is working closely with the UN-Habitat and the Regional Economic and Social Commission for Asia-Pacific (ESCAP) to implement the New Urban Agenda for achieving sustainable urbanization. They have piloted initiatives in the Nadee Subdistrict towards integrated solid waste management and protection of coastal areas. Nadee Subdistrict is in the province of Samut Sakhon, in the central part of Thailand. The subdistrict is known for its efforts towards sustainable waste management. The Thai government has also participated in the Smart Green ASEAN Cities project, which aims to support sustainable urbanization in the region, reduce the environmental footprint of urban areas, and improve the quality of life of residents.

10.4 A sustainable lifestyle through rural simplicity

10.4.1 Fostering sustainable living and harmony

Rural simplicity embodies a sustainable lifestyle that emphasizes harmony with nature, self-sufficiency, and a slower pace of living. In rural areas, individuals and communities often cultivate a deep connection with the natural environment, fostering a profound respect for the land and its resources. This way of life encourages sustainable practices such as organic farming, permaculture, and renewable energy usage, minimizing the ecological footprint and promoting environmental conservation. The simplicity of rural living is reflected in the resourcefulness and self-reliance of its inhabitants, who often grow their food, raise animals, and engage in traditional crafts and trades. This self-sufficiency not only reduces dependence on external sources but also promotes a sense of fulfillment and connection to the products of one's labor.

The rural lifestyle promotes a slower and more mindful approach to daily life, allowing individuals to savor the small joys and appreciate the beauty of the natural world (figure 10.4). Away from the hustle and bustle of urban centers, rural communities often prioritize human connection and a strong sense of community. This fosters a supportive network where individuals come together to share resources, skills, and experiences, strengthening social bonds, and promoting a sense of belonging. In terms of sustainability, rural simplicity offers a model for living in harmony with the Earth, responsibly utilizing resources, and minimizing waste. Practices such as composting, rainwater harvesting, and the use of natural building materials contribute to a more sustainable and eco-friendlier lifestyle. The emphasis on locally sourced goods and traditional craftsmanship supports the preservation of cultural heritage and reduces the environmental impact of mass production and global supply chains.

Figure 10.4. Rural life: simple activities, simple economy, mindful life. (This Culture of peace1 image has been obtained by the author(s) from the Wikimedia website where it was made available by Suriya Thonawanik under a CC BY 3.0 licence. It is included within this article on that basis. It is attributed to Suriya Thonawanik.)

The benefits of rural simplicity extend beyond environmental sustainability, encompassing physical and mental well-being. The abundance of natural spaces and the opportunity for outdoor activities promote a healthy and active lifestyle, while the slower pace allows for reduced stress and a greater sense of inner peace. Furthermore, a strong sense of community and connection to nature can contribute to improved mental health and overall life satisfaction.

10.4.2 Examination of rural living and its inherent sustainability

Rural living, often characterized by its simplicity and close connection to nature, inherently embodies many principles of sustainability. Unlike urban areas, rural communities typically have a lower population density, smaller ecological footprints, and a lifestyle that is more directly tied to the natural environment.

Agriculture, the primary occupation in many rural areas, provides a direct link between people and the land. Traditional farming practices often involve crop rotation, intercropping, and the use of natural fertilizers, which can enhance soil fertility and biodiversity while reducing the need for chemical inputs. Moreover, rural communities often rely on local resources for their needs, reducing the energy and emissions associated with transporting goods over long distances.

Rural homes, with their unique characteristics and inherent sustainability features, often stand in stark contrast to their urban counterparts. These homes, typically smaller in size and constructed using local, natural materials, have a significantly reduced environmental impact. Many rural homes incorporate sustainable design features, such as passive solar design or natural ventilation, which contribute to a reduction in energy consumption. The size of rural homes is also sustainable. Smaller homes require less energy for heating and cooling, simply because there is less space to heat or cool. This results in lower energy consumption and, consequently, lower greenhouse gas emissions.

Smaller homes require fewer materials to build, which reduces the environmental impact associated with the extraction, processing, and transportation of construction materials. The use of local, natural materials in the construction of rural homes further enhances their sustainability. Local materials, such as wood, stone, or clay, are often abundant and require less energy to extract and transport compared to manufactured materials such as steel or concrete. Moreover, these natural materials are typically renewable and have a lower environmental impact over their lifecycle. For instance, wood, when sourced sustainably, is a renewable resource that sequesters carbon, helping to mitigate climate change. Natural materials often have excellent thermal properties, which can enhance the energy efficiency of homes. For example, stone and clay have high thermal mass, meaning they can absorb heat during the day and release it at night, helping to regulate indoor temperatures and reduce the need for mechanical heating and cooling.

Many rural homes also incorporate sustainable design features that further reduce energy consumption. One such feature is passive solar design, which takes advantage of the sun's energy to heat homes in the winter and keep them cool in the summer. This is achieved through the right orientation of the home, the use of energy-efficient windows, and the incorporation of thermal mass. Passive solar design can significantly reduce the need for mechanical heating, leading to lower energy consumption and CO_2 emissions. Natural ventilation is another sustainable design feature commonly found in rural homes. This involves the strategic placement of windows, doors, and vents to encourage airflow, helping to cool the home and improve indoor air quality. Natural ventilation reduces the need for air conditioning, which is often energy-intensive and uses refrigerants that contribute to global warming.

Rural homes, with their smaller size, use of local and natural materials, and incorporation of sustainable design features, offer a more sustainable alternative to urban living. They embody a lifestyle that is in harmony with nature, characterized by lower energy consumption and a reduced environmental footprint. However, the sustainability of rural homes also depends on other factors, such as the lifestyle of the inhabitants and the sustainability of the local community and economy. While rural homes offer many sustainability advantages, achieving truly sustainable living requires a holistic approach that considers all aspects of sustainability.

10.4.3 Challenges of maintaining sustainability in rural areas

Despite the inherent advantages of rural living, such as a close connection to nature and lower population density, maintaining sustainability in rural areas is not without challenges. These challenges range from a lack of access to modern amenities and services, and economic hardships, to the impacts of climate change. One of the key challenges in rural areas is the lack of access to modern amenities and services. Infrastructure such as paved roads, electricity, and clean water, which are often taken for granted in urban areas, can be scarce in rural settings. The absence of these basic amenities makes it difficult for residents to maintain a comfortable and healthy lifestyle. For instance, the lack of paved roads can make transportation difficult, limiting access to markets, schools, and healthcare facilities. The absence of electricity can hinder various activities after dark, while the lack of clean water can pose serious health risks. Moreover, rural areas often face economic challenges. While agriculture, the primary occupation in many rural areas, is inherently sustainable, it is often not profitable. Factors such as fluctuating market prices, high input costs, and vulnerability to weather conditions can make agricultural activities economically unstable. This instability can lead to poverty, forcing rural residents into a cycle of economic hardship.

This economic hardship can, in turn, force rural residents to overexploit natural resources, undermining sustainability. For example, in the absence of alternative sources of income, rural residents may resort to activities such as over-farming, overgrazing, or deforestation. These activities can degrade the environment, reducing biodiversity, depleting soil fertility, and disrupting ecosystems.

Amid the hard challenges in rural living, rural communities can have the opportunities to enhance their economic stability through various strategies, as shown in table 10.1.

By implementing these strategies, rural communities are expected to enhance their economic stability, ensuring a sustainable and prosperous future. However, the

Table 10.1. Strategies to enhance economic stability in rural areas.

1	*Diversification of income sources*: Rural communities often rely heavily on agriculture for their income. While agriculture is a vital sector, it can be subject to fluctuations due to factors such as weather conditions and market prices. Diversifying income sources can help rural communities become more economically stable. This could involve promoting non-agricultural activities such as handicrafts, tourism, or small-scale manufacturing.
2	*Sustainable agricultural practices*: Implementing sustainable agricultural practices can increase productivity and reduce dependence on expensive inputs such as chemical fertilizers and pesticides. Techniques such as crop rotation, intercropping, and organic farming can improve soil health, increase crop yields, and reduce costs.
3	*Value addition and processing*: Rural communities can increase their income by adding value to agricultural products. This could involve processing agricultural produce into products such as jams, pickles, cheese, or wine. Value addition not only increases the income from agricultural products but also creates additional employment opportunities.
4	*Skill development and education*: Providing education and skill development opportunities can help rural communities access better-paying jobs. This could involve vocational training programs or adult education classes.
5	*Access to credit*: Access to affordable credit can help rural communities invest in income-generating activities. Microfinance institutions or cooperative banks can provide loans to farmers and small businesses at reasonable interest rates.
6	*Infrastructure development*: Developing infrastructure such as roads, electricity, and internet connectivity can attract businesses to rural areas, creating jobs and stimulating economic growth.
7	*Cooperative models*: Cooperative models, where farmers or artisans pool their resources and work together, can help rural communities achieve economies of scale and negotiate better prices for their products.
8	*Government support and policies*: Government policies can play a crucial role in enhancing rural economic stability. This could involve subsidies for sustainable farming practices, grants for rural businesses, or policies that promote rural development.

effectiveness of these strategies can vary depending on the specific context and needs of each community. Therefore, it is crucial to involve community members in decision-making processes and tailor strategies to the local context.

Climate change poses another significant challenge to rural sustainability. Changes in temperature and precipitation patterns can disrupt agricultural activities, which are often the backbone of rural economies. Increased temperatures can exacerbate water scarcity and increase the incidence of pests and diseases, reducing crop yields. Changes in precipitation patterns can lead to either droughts or floods, both of which can devastate agricultural production. These climatic changes threaten food security and rural livelihoods. With their livelihoods heavily dependent on agriculture, rural residents are particularly vulnerable to these changes. A single crop failure due to unfavorable weather conditions can push a farming family into poverty. Rural areas often lack the resources and infrastructure to adapt to these changes. Unlike urban areas, which may have the resources to invest in climate-resilient infrastructure or technologies, rural areas often lack these resources. This lack of adaptive capacity makes rural areas particularly vulnerable to the impacts of climate change.

While rural areas have inherent advantages that lend themselves to sustainability, they also face significant challenges. Addressing these challenges requires a multifaceted approach that not only improves access to modern amenities and services but also enhances economic stability and resilience to climate change. This could involve investments in rural infrastructure, the promotion of sustainable agricultural practices, and the implementation of climate adaptation measures.

10.4.4 Case studies of sustainable rural practices

Despite the difficulties faced, there are many instances of eco-friendly initiatives being implemented in rural communities. These range from sustainable agriculture techniques, renewable energy projects, and waste management systems to community-driven conservation efforts. These practices demonstrate the resilience and ingenuity of rural populations in addressing environmental concerns and promoting long-term sustainability. Here are some best practice examples in rural areas of developing countries.

10.4.4.1 Sikkim, India: organic farming

The references for this case study are Heindorf (2019) and Singh (2023). Sikkim, a state in India, has made a significant stride towards sustainable agriculture by becoming the world's first 100% organic state. This achievement is the result of a policy implemented by the Sikkim government that promotes organic farming and bans the use of synthetic pesticides and fertilizers. The policy, known as Sikkim's State Policy on Organic Farming, was implemented in 2003 to gradually convert all the state's agricultural land to organic farming. The policy implemented a phase-out of chemical fertilizers and pesticides and achieved a total ban on the sale and use of chemical pesticides in the state. The transition to organic farming has benefitted more than 66 000 farming families that practice organic and agroecological farming on more than 76 000 hectares of land. The policy has not only improved soil health and biodiversity but also increased farmers' incomes by tapping into the premium market for organic produce.

The implementation of Sikkim's policy was successful since it combines mandatory requirements, such as gradually banning chemical fertilizers and pesticides, with support and incentives, thus providing sustainable alternatives. More than 66 000 farming families benefitted from the policy. The policy reaches beyond organic production and proves truly transformational. For instance, Sikkim's tourism sector benefited greatly from the new organic image: between 2014 and 2017 the number of tourists increased by over 50%.

The Sikkim Organic Mission, designed in 2010, detailed all the measures necessary to achieve the target of becoming a fully organic state by 2015, a goal that was reached within just 12 years. This mission demonstrates the commitment of the Sikkim government to sustainable farming practices and the well-being of its farming community.

The organic farming policy of Sikkim serves as an inspiring example of how sustainable agricultural practices can not only improve environmental health but also enhance the economic well-being of farming communities. It underscores the potential of organic farming in promoting biodiversity, improving soil health, and providing economic benefits to farmers.

10.4.4.2 Costa Rica: renewable energy

The references for this case study are Wikipedia Contributors (n.d.), Young (2020), and Zúñiga (2020, 2021). Costa Rica, a small Central American nation known for its rich biodiversity and commitment to sustainability, has made significant strides in the field of renewable energy. The country has set an ambitious goal to become carbon-neutral by 2025, and renewable energy plays a key role in achieving this target. Costa Rica is blessed with abundant natural resources, which it has leveraged to generate electricity from renewable sources. The country's energy mix is diverse, with hydroelectric power being the most significant contributor. More than 67.5% of the country's energy comes from hydropower (figure 10.5).

In rural areas of Costa Rica, the installation of small-scale hydroelectric plants has been a common sight. These plants harness the power of flowing water to generate electricity. The use of this technology not only provides a sustainable source of energy but also has several other benefits. One of the key advantages of

Figure 10.5. Costa Rican renewable energy source. (This Represa de Cachi ICE Costa Rica CA has been obtained by the author(s) from the Wikimedia website, where it is stated to have been released into the public domain. It is included within this article on that basis.)

these small-scale hydroelectric plants is that they create jobs. The construction, operation, and maintenance of these plants require labor, providing employment opportunities for residents. This can help stimulate local economies and contribute to rural development.

The revenue generated from these plants can be reinvested into the community. This could be used for various development projects, such as improving local infrastructure or providing social services. This creates a virtuous cycle, where the benefits of renewable energy extend beyond just environmental sustainability. The success of Costa Rica's renewable energy sector is not just due to its natural resources. It is also a result of strong government support and favorable policies. The government has implemented various measures to promote renewable energy, such as providing financial incentives for renewable energy projects and implementing a regulatory framework that supports renewable energy development.

10.4.4.3 Bangladesh: solid waste management

The references for this case study are Ashikuzzaman (2020), UN Department of Economic and Social Affairs (n.d.b, 2021), and Waste Concern (n.d.a, n.d.b). Waste Concern, a non-profit organization based in Bangladesh, has been making significant strides in waste management, particularly in rural areas. Established in 1995, Waste Concern has been involved in the experiment, piloting, and disseminating appropriate technologies for the waste management, wastewater, and sanitation sectors. Over the last 29 years, Waste Concern has developed and disseminated innovative technologies suitable for Bangladesh and similar economies in the Asia Pacific region.

One of their key initiatives is turning organic waste into compost. This initiative not only reduces the amount of waste going to landfills but also provides a source of income for residents. The process involves collecting organic waste such as food scraps and yard waste, and then composting it to create nutrient-rich soil. The compost produced is sold to farmers, providing them with an affordable and sustainable alternative to chemical fertilizers. This initiative has been successful in reducing waste, improving soil health, and creating jobs in rural communities. By diverting organic waste from landfills, the initiative helps to reduce methane emissions, a potent greenhouse gas. The composting process also improves soil health by adding nutrients and improving soil structure, which can enhance crop yields and farm productivity. The initiative provides a source of income for residents. Residents can earn money by collecting organic waste and selling the compost produced. This not only provides a financial incentive for waste reduction but also creates jobs and stimulates local economies.

Waste Concern's initiative in Bangladesh serves as an inspiring example of how sustainable waste management practices can not only improve environmental health but also enhance the economic well-being of rural communities. It underscores the potential of composting in promoting soil health, reducing waste, and providing economic benefits to residents.

These examples demonstrate how rural communities, even in different parts of the world, can implement sustainable practices that benefit both the environment and the local economy. It is important to note that the success of these initiatives often depends on community involvement and supportive policies at the local and national levels. Therefore, fostering community engagement and advocating for supportive policies are crucial for promoting sustainability in rural areas.

10.5 Comparative analysis

10.5.1 Comparison of sustainability practices in urban and rural settings

Sustainability practices in urban and rural areas often differ due to the distinct characteristics and needs of each setting. Urban centers, with their higher population densities and increased resource demands, tend to focus on optimization and efficiency. Renewable energy projects, public transportation systems, and comprehensive recycling programs are commonly found in cities as ways to manage resources and reduce environmental impact. These large-scale, infrastructure-based approaches aim to serve the needs of the many who reside in urban areas.

In contrast, rural communities frequently rely on a more direct relationship with the natural environment. Sustainable agriculture techniques, such as crop rotation, no-till farming, and organic methods, are prevalent in rural areas as a means of cultivating the land while preserving its long-term productivity. Additionally, decentralized renewable energy projects, such as small-scale solar, wind, and micro-hydropower, are often more feasible and practical in rural settings where access to a centralized grid may be limited.

Another key difference lies in the community-driven nature of rural sustainability efforts. Rural populations tend to have a stronger sense of stewardship over their local ecosystems and natural resources, leading to community-based conservation initiatives to protect wildlife, forests, and other valuable habitats. These grassroots efforts can be more responsive to the specific needs and challenges of rural areas, fostering a sense of ownership and investment in sustainable practices.

While urban areas often benefit from greater access to resources, technologies, and infrastructure, rural communities face unique challenges in implementing sustainable practices. Limited access to specialized services, transportation, and educational opportunities can hinder the adoption and scaling of sustainable solutions. However, rural areas also have the advantage of smaller carbon footprints due to their lower population densities and less intensive industrial activities, presenting opportunities for targeted sustainability interventions. The comparison of both is shown in table 10.2.

Both urban and rural areas face unique sustainability challenges, but they also demonstrate innovative practices tailored to their local contexts. Integrating the strengths of both settings can lead to more comprehensive and effective sustainability solutions.

10.5.2 The lessons urban and rural areas can learn

Urban and rural communities can enhance their collaboration through various strategies. Recognizing interdependence, promoting regional collaboration, building

Table 10.2. Comparison of sustainability practices between urban and rural settings.

Urban	Higher population density requires more efficient resource management and waste reduction. Renewable energy sources such as solar and wind power are more viable in urban areas. Public transportation, biking, and walking are more practical for commuting. Recycling and composting programs are more easily implemented and accessible. Green buildings, urban farming, and green spaces help reduce the urban environmental footprint.
Rural	More reliance on renewable natural resources such as land, forests, and water. Sustainable agriculture techniques such as crop rotation, no-till farming, and organic methods. Smaller carbon footprint due to less dense populations and less industrial activity. Decentralized renewable energy projects such as solar, wind, and micro-hydropower. Community-based conservation efforts to protect local ecosystems and wildlife. Challenges with infrastructure and access to resources compared to urban areas.
Primary differences	Scale and density: Urban areas require more systematic sustainability approaches Resource reliance: Rural areas depend more directly on natural resources. Infrastructure: Urban areas have better access to sustainable technologies and services. Community engagement: Rural sustainability is often more community-driven.

common ground, fostering trust and commitment, understanding leadership roles, bridging the digital divide, and sharing knowledge and experiences are key. These are shown in table 10.3. These strategies can lead to more effective cooperation, mutual learning, and a more integrated approach to sustainability, benefiting both urban and rural areas.

10.6 Transitioning towards sustainability

Transitioning towards sustainability involves adopting practices that balance economic growth, social inclusion, and environmental protection. It requires collective efforts from urban and rural communities, governments, businesses, and individuals. Depending on the settings, strategies may include, but are not limited to, sustainable urban planning, smart cities and green spaces for urban settings, sustainable agriculture, natural resources management, community empowerment, and infrastructure development for rural areas.

Table 10.3. Strategies for urban–rural collaboration.

1	*Recognize interdependence*: Both rural and urban areas are interconnected and depend on each other for resources, commerce, and information exchange. Recognizing this interdependence can lead to more collaborative and holistic sustainability strategies.
2	*Promote regional collaboration*: Regional collaboration can improve social and economic opportunities for all people in a region by ensuring that rural and small-town interests and perspectives are included in decision-making processes.
3	*Build common ground*: Stakeholders can build common ground through connection or relationship to a place or geographic location. Shared fears can be addressed to acknowledge broad goals that satisfy all stakeholders.
4	*Foster trust and commitment*: Fostering existing social assets can build trust and create commitment in collaborative efforts.
5	*Understand leadership roles*: Understanding leadership roles throughout the collaborative process can find opportunities for innovation and incremental success.
6	*Bridge the digital divide*: Better connectivity can bridge the digital divide between urban and rural areas, leading to improved educational and economic opportunities.
7	*Share knowledge and experiences*: Sharing knowledge and experiences can help both urban and rural areas learn from each other's sustainability practices.

10.6.1 Strategies for enhancing sustainability

Given the distinct characteristics and contexts of urban and rural areas, it is essential to tailor sustainability strategies to the specific needs and opportunities of each setting. This approach will ensure the effectiveness and impact of sustainability initiatives in both environments.

10.6.1.1 Urban sustainability strategies

Urban areas, due to their dense population and infrastructure, have unique challenges and opportunities for sustainability. Some of the innovative practices are shown in table 10.4.

10.6.1.2 Rural sustainability strategies

Rural areas, with their unique landscapes and resources, have different sustainability practices. The key strategies are shown in table 10.5.

While urban and rural areas have distinct characteristics and challenges, there are several shared strategies with similar themes they can adopt to enhance sustainability. These strategies involve a combination of technological innovation, community involvement, and careful planning and management of resources, as shown in table 10.6.

Both urban and rural areas can enhance their sustainability through a combination of technological innovation, community involvement, and thoughtful planning and management of resources. By doing so, they can contribute to the broader goals of environmental sustainability, economic viability, and social equity.

Table 10.4. Key strategies for enhancing sustainability in urban areas.

Sustainable urban planning: This involves designing cities to balance environmental responsibility, social equity, and economic strength. It includes creating connections between people and enhancing connections to the natural environment, history, local culture, goods, and services.

Smart cities: The use of technology and data can help cities become more efficient and sustainable. This includes the use of the Internet of Things (IoT) for managing city resources and infrastructure.

Green spaces: Improving access to green spaces can enhance the livability of cities and contribute to the well-being of its residents.

Table 10.5. Key strategies for enhancing sustainability in rural areas.

Sustainable agriculture: This involves practices such as organic farming and permaculture design that aim to create self-sustaining ecosystems.

Natural resource management: This includes strategies for conserving water, managing waste, and protecting natural landscapes.

Community empowerment: Empowering rural communities, particularly women and youth, can enhance rural livelihoods. This includes promoting the effective participation of rural people in the management of their own social, economic, and environmental objectives.

Infrastructure development: Investing in rural infrastructure, such as roads, schools, and healthcare facilities, can improve the quality of life in rural areas.

10.6.2 Role of technology, policy, and community engagement in this transition

The roles of technology, policy, and community engagement in the transitions of urban and rural sustainability

10.6.2.1 Role of technology

In urban areas, technology is of paramount importance in enhancing sustainability. Smart cities use data and technology to create efficiencies, improve sustainability, and enhance the quality of life for people living and working in the city. This includes smart grids for electricity, efficient water systems, waste management, and urban farming. Cities are increasingly digitalizing their critical utilities and infrastructures to be more efficient, thus becoming smart and sustainable.

In rural areas, technology can revolutionize rural development and facilitate the accomplishment of sustainability goals. For instance, renewable energy technologies such as solar panels and wind turbines can provide reliable and sustainable power sources. Digital technology can also provide access to services and opportunities.

10.6.2.2 Role of policy

In urban areas, policies influence city planning, transportation, housing regulations, infrastructure development, and more. Effective urban policies address challenges

Table 10.6. Shared strategies in similar theme for urban and rural areas.

Themes	Urban areas	Rural areas
Technological innovation	Urban areas often lead to technological innovation. For example, smart cities use data and technology to create efficiencies, improve sustainability, and enhance the quality of life for people living and working in the city. This includes smart grids for electricity, efficient water systems, waste management, and urban farming.	Technology can also play a role in rural sustainability. For example, renewable energy technologies such as solar panels and wind turbines can provide reliable and sustainable power sources. Digital technology can also provide access to services and opportunities.
Community involvement	Engaging local communities in decision-making processes can lead to more sustainable outcomes. This can be facilitated through public consultations, participatory budgeting, and community-led initiatives.	Rural communities often have strong social networks and a close connection to the land. These can be leveraged to promote sustainable practices, such as community-managed forests, local conservation efforts, and cooperative business models.
Resource planning and management	Thoughtful urban planning can help to create more sustainable cities. This includes promoting mixed-use development to reduce the need for transportation, creating green spaces, and implementing zoning policies that protect natural and agricultural lands.	In rural areas, sustainable management of natural resources is crucial. This includes sustainable agricultural practices, watershed management, and conservation of biodiversity.

related to population growth, environmental sustainability, social equity, and economic vibrancy.

In rural areas, policies are indispensable in promoting and ensuring gender equality, empowering rural women through decent work and productive employment, not only contributing to inclusive and sustainable economic growth but also enhancing the effectiveness of poverty reduction and food security initiatives.

10.6.2.3 Role of community engagement

In urban areas, community engagement is a key driver that supports the acquisition of knowledge and requirements needed for innovation and creativity toward achieving an equitable community for social sustainability. It enables varied communities to express their distinct viewpoints, needs, and aspirations, enabling a balanced approach to planning.

In rural areas, community engagement serves as a powerful catalyst for sustainable development by promoting inclusivity, shared responsibility, and collaborative problem-solving. When communities actively participate in decision-making processes, they contribute unique insights, local knowledge, and a genuine understanding of their specific needs and challenges.

10.7 Conclusion and way forward

Sustainability is a crucial consideration in the development and growth of both urban and rural areas. As the world becomes increasingly urbanized, the challenges and opportunities of sustainable development in cities have come to the forefront. At the same time, rural communities continue to play a vital role in resource management and environmental preservation, presenting their unique sustainability considerations.

In the realm of urban development, the complexities of high population density, resource consumption, and infrastructure demands pose significant sustainability challenges. Urban areas must grapple with issues such as efficient resource management, waste reduction, and transportation systems that minimize environmental impact. However, these challenges also present opportunities for innovative solutions, such as the implementation of renewable energy, green buildings, and integrated public transit networks.

The sophistication of urban settings can also be both a blessing and a curse when it comes to sustainability. On one hand, the concentration of resources, technology, and expertise in cities can facilitate the development and adoption of sustainable practices. On the other hand, the fast-paced, consumerist nature of urban lifestyles can create barriers to the adoption of more sustainable behaviors.

In contrast, rural areas often display a closer connection to the natural environment and a simpler, more resource-efficient lifestyle. Sustainable agriculture, decentralized renewable energy projects, and community-driven conservation efforts are examples of sustainable practices that are more prevalent in rural settings. However, rural communities may face challenges in accessing specialized resources, infrastructure, and education necessary to scale up their sustainability initiatives.

Bridging the gap between urban and rural sustainability practices is imperative for a comprehensive and effective approach to sustainability. Strategies for enhancing sustainability in both settings include knowledge sharing, collaborative partnerships, policy coordination, educational programs, and community engagement.

The transition towards sustainability requires a multi-faceted approach that considers the complexities and interconnections of urban and rural development. By embracing innovative solutions, fostering cross-pollination of ideas, and empowering local communities, we can create a more resilient and sustainable world, one that balances the needs of both urban and rural populations.

As we look towards the future, the prospects for sustainability in both urban and rural living are promising, though not without their challenges. The global shift towards more sustainable practices and the recognition of the urgent need to address environmental degradation have set the stage for transformative changes in the way we live and develop our communities.

In the urban landscape, the future of sustainability is characterized by a focus on creating smart, livable cities. Advancements in renewable energy technologies, such as solar, wind, and geothermal power, will enable cities to reduce their reliance on fossil fuels and transition towards cleaner, more efficient energy sources. Innovative urban planning will prioritize compact, mixed-use developments, reducing the need for private vehicle usage and promoting walkable, transit-oriented communities.

The integration of digital technologies, such as the Internet of Things (IoT) and data analytics, will be indispensable in optimizing urban resource management, from water and waste systems to energy distribution and traffic management. This 'smart city' approach will empower urban dwellers to monitor and manage their environmental impact, fostering a culture of sustainable living.

In rural areas, the prospects for sustainability are equally promising, with a focus on revitalizing local food systems, renewable energy generation, and natural resource conservation. Advancements in precision agriculture, agroforestry, and regenerative farming practices will enable rural communities to produce food more efficiently and sustainably, reducing their environmental footprint while improving food security.

The decentralization of renewable energy, such as small-scale solar and wind projects, will provide rural areas with greater energy autonomy and resilience, reducing their reliance on centralized, fossil-fuel-based power grids. Furthermore, the preservation and restoration of natural habitats, coupled with community-driven conservation efforts, will safeguard the ecological integrity of rural landscapes, ensuring the continued provision of vital ecosystem services.

As urban and rural communities work together to share knowledge, collaborate on innovative projects, and align policies, the future of sustainability in both living environments holds great promise.

References

Adler P and Florida R 2021 The rise of urban tech: how innovations for cities come from cities *Reg. Stud.* **55** 1787–800

Ashikuzzaman M 2020 *Sustainable Solid Waste Management in Bangladesh: Issues and Challenges Sustainable Solid Waste Management in Bangladesh: Issues and Challenges* (Hershey, PA: IGI Global)

Barles S 2011 Are cities parasites or resource pools? *Books & Ideas* https://booksandideas.net/Are-Cities-Parasites-or-Resource.html (Accessed: 10 May 2024)

Berzins R 2020 Sustainability in Curitiba, Brazil *The Borgen Project* https://borgenproject.org/sustainability-in-curitiba/

Bunch M J, Morrison K E, Parkes M W and Venema H D 2011 Promoting health and well-being by managing for social–ecological resilience: the potential of integrating ecohealth and water resources management approaches *Ecol. Soc.* **16** 6

van de Buuse D, van Winden W and Schrama W 2022 Balancing exploration and exploitation in sustainable urban innovation: an ambidexterity perspective toward smart cities *Sustainable Smart City Transitions* (London: Routledge) pp 172–94

Dye C 2008 Health and urban living *Science* **319** 766–9

Ecological Alert and Recovery—Thailand 2019 *First Strategic Planning Workshop of the Sustainable Urban Development Committee for Nadee Sub-district* 31 July Strategy and Programme Management Division, UN-ESCAP

Ecological Alert and Recovery—Thailand 2020 Assessment on sustainable industrial waste management for a circular and inclusive economy: Nadee Pilot Project, Sakhon Province, Thailand *Report* UN ESCAP https://unescap.org/sites/default/d8files/Nadee_Research%20Outcomes%20Report.pdf

Eizenberg E and Jabareen Y 2017 Social sustainability: a new conceptual framework *Sustainability* **9** 68

Fourtané S 2020 Copenhagen: world's first carbon-neutral smart city by 2025 *Interesting Engineering* https://interestingengineering.com/science/copenhagen-worlds-first-carbon-neutral-smart-city-by-2025

2008 *Revisiting Ecuador's Economic and Social Agenda in Evolving Landscape* ed V Fretes–Cibils, M Giugale and E Somensatto (Washington, DC: The IBRD/World Bank)

2012 *Material Geographies of Household Sustainability* ed A Gorman-Murray and R Lane (London: Routledge)

Grubler A *et al* 2012 Urban energy systems *Global Energy Assessment* (Cambridge: Cambridge University Press)

Heindorf I 2019 Sikkim's state policy on organic farming and Sikkim Organic Mission, India *Panorama Solutions* https://panorama.solutions/en/solution/sikkims-state-policy-organic-farming-and-sikkim-organic-mission-India#:~:text=Sikkim%20is%20the%20first%20state,more%20than%2066%2C000%20farming%20families

Hoselitz B 2002 *Generative and Parasitic Cities* (London: Routledge) pp 5–23

Kennedy C, Cuddihy J and Engel-Yan J 2007 The changing metabolism of cities *J. Ind. Ecol.* **11** 43–59

Khan U and Stinchcombe M B 2023 Intergenerational equity and sustainability: a large population approach *Soc. Choice Welf*

Krosofsky A 2021 How Curitiba, Brazil became one of the most sustainable cities on Earth *Green Matters* https://greenmatters.com/p/curitiba-sustainable

Lozzy 2023 Visiting Curitiba: Brazil's most sustainable green city *Cuppa to Copa Travels* https://cuppatocopatravels.com/south-america/brazil/curitiba-brazil-travel-guide-2019/

Mathiesen B V, Auken I and Skibsted J M 2019 This is how Copenhagen plans to go carbon-neutral by 2025 *World Economic Forum* https://weforum.org/agenda/2019/05/the-copenhagen-effect-how-europe-can-become-heat-efficient/

Nitivattananon V, Yusuf M A, Permana A S and Lloyds I 2010 Experiences and lessons learned for future urban environmental management drawn from SEA-UEMA Project Final Partners Workshop, Southeast Asia Urban Environmental Management Applications Project *Report* The Canadian International Development Agency, and Asian Institute of Technology, Thailand

Peano C, Caron S, Mahfoudhi M, Zammel K, Zaidi H and Sottile F 2021 A participatory agrobiodiversity conservation approach in the oases: community actions for the promotion of sustainable development in fragile areas *Diversity* **13** 253

Permana A S, Sinniah G K, Utomo R P and Andisetyana Putri R 2018 Dual formal-informal transport modes towards quasi-seamless transit in developing city *Int. J. Built Environ. Sustain.* **5** 224–40

Permana A S, Towolioe S, Aziz N A and Ho C S 2015 Sustainable solid waste management practices and perceived cleanliness in a low-income city *Habitat Int.* **49** 197–205

Rana M M P 2011 Urbanization and sustainability: challenges and strategies for sustainable urban development in Bangladesh *Environ. Dev. Sustain* **13** 237–56

Ruth M and Coelho D 2015 Understanding and managing the complexity of urban systems under climate change *Integrating Climate Change Actions into Local Development* (BrightonLondon: Routledge) pp 317–36

Scoones I 1998 *Sustainable Rural Livelihoods: A Framework for Analysis* **vol 72** (Brighton: Institute of Development Studies) pp 1–22

Singh S 2023 Sikkim: India's first 100% organic state leading the way in sustainable farming *Indian Farm* https://www.indiafarm.org/agriculture/agroecology/sikkim-indias-first-organic-farming-state/

Skibsted J M 2019 This is how Copenhagen plans to go carbon-neutral by 2025 *Medium* https://medium.com/everything-thats-next/this-is-how-copenhagen-plans-to-go-carbon-neutral-by-2025-70849d2d67dc

Spijkers O 2018 Intergenerational equity and the sustainable development goals *Sustainability* **10** 3836

Szumski C 2022 Copenhagen's dream of being carbon neutral by 2025 goes up in smoke *Euractiv* https://euractiv.com/section/energy-environment/news/copenhagens-dream-of-being-carbon-neutral-by-2025-go-up-in-smoke/

The Sustainable Travel 2019 How Curitiba became Brazil's most sustainable city *The Sustainable Travel* https://thesustainabletravel.com/curitiba-became-brazils-sustainable-city/

Towolioe S, Permana A S, Aziz N A, Ho C S and Pampanga D G 2016 The rukun warga-based 3Rs and waste bank as sustainable solid waste management strategy *Plan. Malays* **IV** 181–96

UN Department of Economic and Social Affairs n.d.a *Sustainable Urban Planning (Curitiba City)* United Nations https://sdgs.un.org/partnerships/sustainable-urban-planning-curitiba-city

UN Department of Economic and Social Affairs n.d.b Waste concern: public/private partnership and community-based composting in Dhaka *United Nation Department of Economic and Social Affairs* https://sdgs.un.org/partnerships/waste-concern-publicprivate-partnership-and-community-based-composting-dhaka

UN Department of Economic and Social Affairs 2021 The Government of Bangladesh announces its solid waste management rules 2021 *United Nations Department of Economic and Social Affairs* https://sdgs.un.org/partnerships/government-bangladesh-announces-its-solid-waste-management-rules-2021

UN ESCAP n.d. Nadee Profile https://www.unescap.org/

Waste Concern n.d.a Assessment of solid waste management opportunities in Bangladesh *Waste Concern* https://wasteconcern.org/assessment-of-solid-waste-management-opportunities-in-bangladesh-2/

Waste Concern n.d.b Who are we? *Waste Concern* https://wasteconcern.org/

Wheeler S 2013 *Planning for Sustainability: Creating Livable, Equitable, and Ecological Communities* (London: Routledge)

Wikipedia Contributors n.d Renewable energy in Costa Rica *Wikipedia* https://en.wikipedia.org/wiki/Renewable_energy_in_Costa_Rica

Williams J M, Chu V, Lam W F and Law W W Y 2021 Rural sustainability: challenges and opportunities *Revitalising Rural Communities* Springer Briefs on Case Studies of Sustainable Development (Singapore: Springer)

World Economic Forum 2018 5 big challenges facing big cities of the future *World Economic Forum* https://weforum.org/agenda/2018/10/the-5-biggest-challenges-cities-will-face-in-the-future/

Young E 2020 10 facts about renewable energy in Costa Rica *The Borgen Project* https://borgenproject.org/10-facts-about-renewable-energy-in-costa-rica/

Yusop A M and Permana A S 2016 Spatial pattern, transportation and air quality nexus: the case of Iskandar Malaysia *Int. J. Built Environ. Sustain.* **3** 199–208

Yusuf M A and Permana A S 2013 Results and lessons learned from Mekong Region Waste Refinery—International Partnership: Towards Zero Waste, Zero Landfill and Reduce Greenhouse Gas Emissions *Research Book* Asian Institute of Technology—Energy and Environment Partnership Mekong Region, Thailand

Zakka S D, Permana A S, Ho C S, Baba A N and Agboola O P 2017a Implications of present land use plan on urban growth and environmental sustainability in a Sub-Saharan Africa City *Int. J. Built Environ. Sustain.* **4** 105–12

Zakka S D, Permana A S and Majid M R 2017b Urban spatial pattern and carbon emission interconnectivity in a Sub-Saharan City, Nigeria *Plan. Malays. J.* **15** 51–62

Zhou X and Chen H 2018 Impact of urbanization-related land use land cover changes and urban morphology changes on the urban heat island phenomenon *Sci. Total Environ.* **635** 1467–76

Ziervogel G, Cowen A and Ziniades J 2016 Moving from adaptive to transformative capacity: building foundations for inclusive, thriving, and regenerative urban settlements *Sustainability* **8** 955

Zúñiga A 2020 Costa Rica's electric grid is powered by 98% renewable energy for 6th straight year *The Tico Times* https://ticotimes.net/2020/12/18/costa-ricas-electric-grid-powered-by-98-renewable-energy-for-6th-straight-year

Zúñiga A 2021 Costa Rica's electric grid is powered by 98% renewable energy for 7th straight year *The Tico Times* https://ticotimes.net/2021/10/28/costa-ricas-electric-grid-powered-by-98-renewable-energy-for-7th-straight-year

www.ingramcontent.com/pod-product-compliance
Lightning Source LLC
LaVergne TN
LVHW081926300925
822345LV00005B/122